Colloid Chemistry and Its Applications

Colloid Chemistry and Its Applications

Edited by Helena Price

Larsen & Keller
www.larsen-keller.com

Colloid Chemistry and Its Applications
Edited by Helena Price
ISBN: 979-8-88836-063-7 (Hardback)

Larsen & Keller
Published by Larsen and Keller Education,
5 Penn Plaza,
19th Floor,
New York, NY 10001, USA

Cataloging-in-Publication Data

Colloid chemistry and its applications / edited by Helena Price.
p. cm.
Includes bibliographical references and index.
ISBN 979-8-88836-063-7
1. Colloids. 2. Surface chemistry. 3. Chemistry, Physical and theoretical. I. Price, Helena.
QD549 .C65 2023
541.345--dc23

For more information regarding Larsen and Keller Education and its products, please visit the publisher's website www.larsen-keller.com

Table of Contents

Preface

A colloid refers to a type of mixture in which one substance is suspended in another substance, consisting of microscopically scattered insoluble particles, such as aerosol, emulsion and foam. These mixtures can be translucent due to the Tyndall effect, in which light is scattered by the particles in the colloid. They can also have a slight color or can be opaque. A few major examples of colloids are mayonnaise, gelatin, muddy water, paper, plaster, whipped cream, milk, jelly, butter and tinted glass. Colloids are extensively used for domestic, commercial and medical purposes. They serve as thickening agents in industrial products including lotions, coatings, lubricants and toothpaste. This book elucidates the concepts and innovative models around prospective developments with respect to colloid chemistry. It consists of contributions made by international experts. This book aims to equip students and experts with the advanced topics and applications of this area of chemistry.

The information shared in this book is based on empirical researches made by veterans in this field of study. The elaborative information provided in this book will help the readers further their scope of knowledge leading to advancements in this field.

Finally, I would like to thank my fellow researchers who gave constructive feedback and my family members who supported me at every step of my research.

Editor

1

Colloidal Dispersions of Gelled Lipid Nanoparticles (GLN): Concept and Potential Applications

Mariana Carrancá Palomo, Victoria Martín Prieto and Plamen Kirilov *

Université de Lyon (UCBL), Biologie Tissulaire et Ingénierie Thérapeutique UMR 5305–Vecteurs Colloïdaux et Transport Tissulaire, Institut des Sciences Pharmaceutiques et Biologiques, 8 Avenue Rockefeller, 69373 Lyon Cedex 08, France; mariana.carranca@ibcp.fr (C.P.M.); victoriabcn91@gmail.com (M.P.V.)

* Correspondence: plamen.kirilov@univ-lyon1.fr

Abstract: The interest in using colloidal dispersions of gelled lipid nanoparticles (GLN) for different fields of application has increased in recent years, notably in cosmetic, dermatology, and/or pharmaceutics due to their capacity to immobilize compounds with poor water solubility. The pharmaceutical field desires to achieve lipophilic drug formulations which are able to conserve their stability, although it is well-known that emulsions and solid lipid nanoparticles (SLN) present a lack of stability over time, leading to system destabilization. Furthermore, stable colloidal dispersions of gelled oil particles do not affect the properties of the molecule to be delivered, and they result as an alternative for the previously appointed systems. This review is an attempt to present the reader with an overview of colloidal dispersions of GLN, their concept, formulation methods, as well as the techniques used for their characterization. Moreover, various application fields of organogel dispersions have been illustrated to demonstrate the potential application range of these recent materials.

Keywords: gelled lipid nanoparticles (GLN); colloid dispersions; organogels; low molecular-mass organic gelator (LMOG); phase transition parameters; UV protection; drug delivery

1. Introduction

Deficient bioavailability due to certain physicochemical properties of some active substances is still a challenge in pharmaceutical science. Thus, new formulation strategies are required to overcome this problem—for example for poorly water-soluble compounds [1].

In the last years, the study of formulations to deliver water-insoluble molecules has increased. Emulsions and solid lipid nanoparticles (SLN) are up to now the most studied systems to deliver this type of molecule, and are unstable and have limited loading capacity. Therefore, there is still a demand for readily stable hydrophobic carriers with a higher loading capacity. Colloidal dispersions of gelled lipid nanoparticles (GLN) present an attractive alternative to surpass these limitations, since they may act as hydrophobic reservoirs [2,3].

Organogels are semi-solid systems (soft materials) in which a three-dimensional network composed of cross-linked gelator fibers immobilizes an organic liquid [4]. Organogels are principally classified based on the type of organogelator used, but also based on the nature of the organic liquid and its intermolecular interactions [5,6]. These organogelators can be divided into polymeric organic gelators (POG) and low molecular-mass organic gelators (LMOG) [5]. POG are high molecular-mass molecules which are capable of jellifying organic solvents, forming physical crosslinking points. These crosslinking points can be due to the formation of hydrogen bonds (inducing the formation of α-helices and β-sheets) for polypeptide polymers or $\pi-\pi$ stacking interactions for p-conjugated polymers. LMOG are relatively small organic molecules (molecular weight < 2000 g/mol, usually even smaller than 500 g/mol) which are capable of forming a three-dimensional network around the organic liquid, thus gelling the organic liquid. LMOG have several advantages over POG: better biocompatibility, lower toxicity, easy to prepare, stronger stability, and high gelling capacities needing less quantities of gelator (<1 wt %) [5,6].

The LMOG are separated into different classes according to criteria based on their molecular structure. For example, they can be classified as alkane organic gelators, as n-octacosane (Figure 1a), organic gelators with one heteroatom as ethers (e.g., 60-crown-20-macrocycle; Figure 1b), organic gelators with two heteroatoms (e.g., *N*-octyl-D-gluconamide; Figure 1c), organic gelators with three heteroatoms (e.g., racemic 2-acryloylamide-dedecane-1-sulfonic acid or ADSA; Figure 1d), polymerizable organic gelators such as bis-urea gelators (Figure 1e), two-component organic gelators such as bis-(2-ethylhexyl) sodium sulfosuccinate or AOT (Figure 1f), and organometallic gelators such as palladium-CNC pincer bis(carbene) (Figure 1g) [5,6].

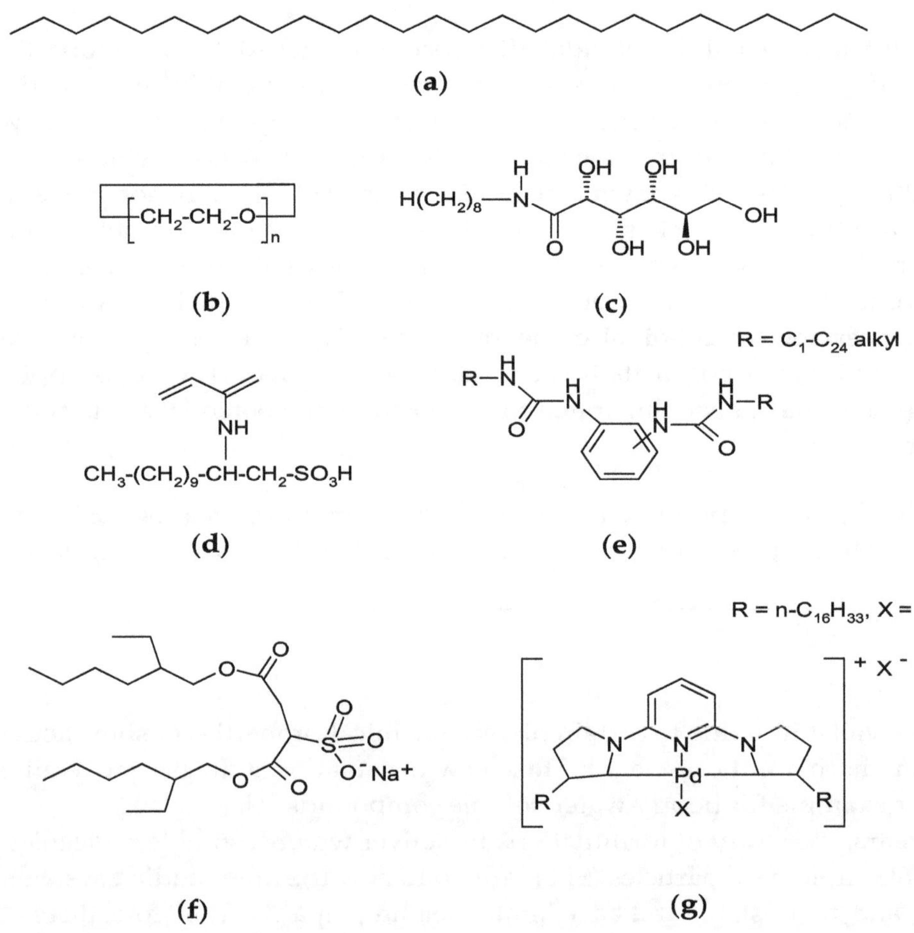

Figure 1. Examples of LMOG chemical structure: (**a**) n-octacosane; (**b**) 60-crown-20-macrocycle; (**c**) *N*-n-octyl-D-gluconamide; (**d**) 2-acryloylamide-dedecane-1-sulfonic acid (ADSA); (**e**) cyclic bis-urea gelator; (**f**) bis-(2-ethylhexyl) sodium sulfosuccinate (AOT); (**g**) palladium-CNC pincer bis(carbene).

Derived from this classification, several chemical compounds are used in cosmetics, such as 12-hydroxystearic acid (HSA) (Figure 2a), 1,3:2,4-di-*O*-benzylidene-D-sorbitol (DBS) (Figure 2b), sterols, lecithin, mono- and diglycerides, lecithin mixtures with sorbitan esters, fatty acids, fatty alcohols, waxes, and wax esters [5,6].

Plant metabolites—many of which are very structurally complicated—present attractive candidates as new LMOGs because they are available in renewable supply without extensive synthetic effort. Moreover, LMOG organogels can be classified into two groups depending on the kinetic properties of aggregates: "strong" organogel (solid fiber network) and "weak" organogel (fluid fiber network) [7–9].

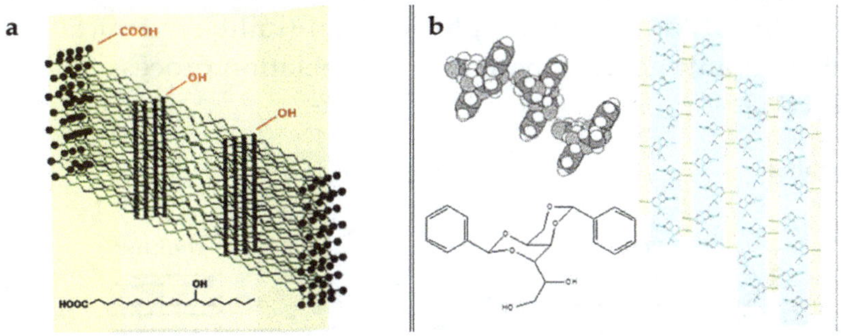

Figure 2. Chemical structure and representation of intermolecular interactions of: (a) HSA molecules; (b) DBS molecules.

In solution, LMOG self-assemble to form fibrous structures responsible for the gelation phenomenon [4]. Typically, the intermolecular forces that drive aggregation include hydrogen-bonding, $\pi-\pi$ stacking, dipole−dipole, and London dispersion forces; however, hydrogen-bonding is a very important component of the self-assembly driving force in a large fraction of reported organogelators [10]. These physical gels are usually thermoreversible (reversible gel–sol/sol–gel phase transition by heating and cooling, respectively), depending on the molecular structure of the gelling agent and the fluid which is rigidified. It is possible to form nanoscale superstructures such as nanofibres, nanoribbons, nanosheets, nanoparticles, etc., which are of interest for materials and nanostructures conception. The interconversion between the gel and solution/sol phases depends only on the disassembly and assembly of the constituent molecules [7,11].

LMOGs are mostly used, since a limited number of synthetic polymers (e.g., polypeptides and polyesters) with these organogelation properties have been reported [12–14].

2. Preparation of LMOG Organogel Self-Assembled Systems

The formulation of stable colloidal dispersions of GLN includes two successive steps. The preparation of the organogel from an organic liquid (vegetable or emollient oil) and an organic gelator as the first step. After organogel preparation, aqueous dispersions of GLN are obtained by their dispersion in a water phase containing a stabilizing and/or emulsifying agent(s) [12] (Figure 3).

Figure 3. Schematic representation of gelled lipid nanoparticle (GLN) colloidal dispersion preparation process.

2.1. Gelation Process

LMOG organogels are obtained by dissolving a small amount of the organogelator at high temperature to obtain an isotropic solution and then cooling this solution under its characteristic gelation transition temperature, known as Tgel. The mixture must be completely solubilized and homogenized before being cooled below its Tgel, which represents the temperature from which the organogel formation process started. The Tgel has proven to be dependent on the concentration of the organogelator, the properties of the organic liquid (or oil), such as its polarity or viscosity, and in some cases of the cooling process. Gelation occurs as the hot solutions cools, while the self-assembled networks are formed. Since LMOGs form a physical gel (unlike chemical gels), these organogels are thermally reversible organic gel-like materials. The gelation process in this type of organogel is represented in Figure 4 [12].

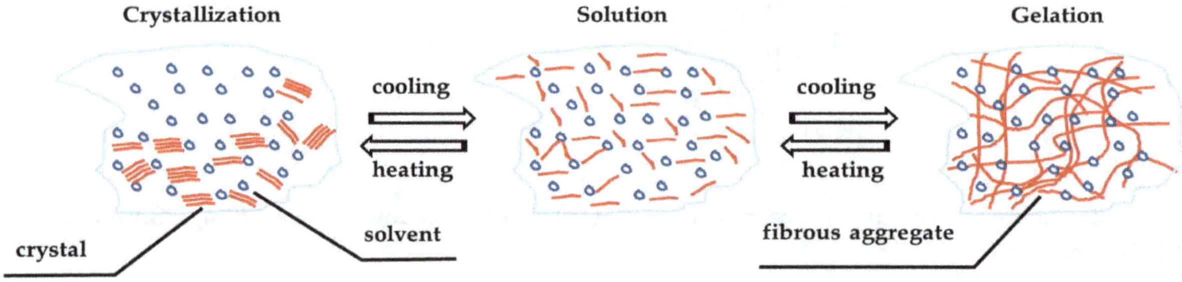

Figure 4. Organogel gelation–melting process representation.

It is important note that organogels formed by LMOGs tend to crystallize, leading to a more metastable organogel. It is possible to adhere a polymer as a gelation-driven compound with covalent bonds, converting it into a "polymer-based gelator" or supramolecular crosslinkable polymer organogelator that can form a more stable gel. For example, polysiloxanes, polyethers, and polycarbonates are useful polymers to confer a more stable characteristic to LMOGs [15]. Moreover, lecithin—a common organogelator usually used as stabilizing and dispersing agent to obtain stable organogel microemulsions—is expensive and difficult to obtain with high purity. When using lecithin of lesser purity, it can be used in the presence of synthetic polymers in the preparation of gel microemulsions in order to obtain the same stabilizing properties [16].

2.2. Phase Transition Parameters

The primary concept to consider in organogels is their phase transition parameters. These parameters are determined by rheology, as following the evolution of the elastic (G′) and viscous (G″) moduli versus temperature increase or decrease. The viscoelastic moduli permit the determination of the phase transition parameters of an organogel, such as gel–sol or sol–gel phase transition domains (PTDs), Tmelt (organogel melting temperature), Tgel, Tsol (organogel liquefaction temperature), as well as Tform (organogel formation temperature) [5,6].

Figure 5 shows the viscoelastic behavior of an LMOG organogel during its melting process (Figure 5a) and during its formation process (Figure 5b).

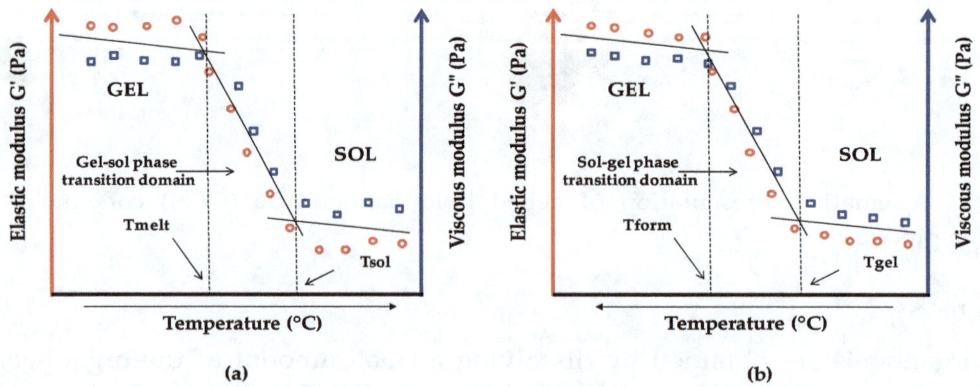

Figure 5. Schematic representation of phase transition parameters of an LMOG organogel: (**a**) organogel liquefaction process (temperature increase); (**b**) organogel formation process (temperature decrease).

The rheological behavior of an LMOG organogel versus temperature presents two phases: the gel phase and the sol phase, separated by a transitional phase, known as PTD. According to the considered process, there is a gel–sol PTD (organogel liquefaction) or sol–gel PTD (organogel formation). The Tmelt corresponds to an abrupt decrease of the viscoelastic moduli and characterizes the beginning of the organogel melting process. This process is progressive, and takes place over a temperature range determined by the gel–sol PTD, and from a certain temperature (known as Tsol—organogel liquefaction), the organogel is completely liquefied. Concerning the sol–gel process (organogel formation), the Tgel corresponds to the abrupt increase of the viscoelastic moduli. Tgel indicates the beginning of the gelation process, contrary to the Tform which indicates the organogel formation temperature, above which the material shows constant viscoelastic properties with the temperature decrease. The sol–gel PTD is closed between Tgel and Tform.

The phase transition parameters of organogels obtained from HSA and various oils (emollients and vegetables) are reported in Table 1. The data show that Tgel and Tmelt values increase with increasing HSA proportion, regardless of the nature of the oil. Concerning Tsol and Tform, these parameters also increase when the HSA wt % increases and show higher values with vegetable oils, which is related to their important dynamic viscosity. The PTD range variation depends on the considered process. In the case of sol–gel PTD, it is closer than those of the gel–sol one. Moreover, this difference decreases as the proportions of HSA are increased for both types of oils (emollients and vegetable oils).

Table 1. Phase transition parameters of organogels obtained from emollients (dicaprylyl ether, capric/caprylic triglycerides) or vegetable oils (soybean, safflower) and 3, 6, and 15 wt % of HSA, respectively. PDT: phase transition domain.

Emollients or vegetable oils	Tgel [1] (°C)	Tform [1] (°C)	Sol–Gel PTD (ΔT)	Tmelt [1] (°C)	Tsol [1] (°C)	Gel–Sol PTD (ΔT)
			Oil/HSA 3 wt %			
ether	49	45.5	3.5	48	63	15
triglycerides	47	44	3	48.5	61	12.5
soybean	51	47.5	3.5	53	67	14
safflower	54	50	4	53	68	15
			Oil/HSA 6 wt %			
ether	56	51.5	4.5	55	65	10
triglycerides	56	51	5	54	64	10
soybean	58	53.5	4.5	58.5	68	9.5
safflower	61	56	5	59	70	11
			Oil/HSA 15 wt %			
ether	63	56	7	62	71	9
triglycerides	63.5	55.5	8	61	70	9
soybean	66.5	60	6.5	66	71	5
safflower	65	56	9	63	72	9

[1] T ± 0.5 °C.

It is observed that when the proportion of HSA is increased, the sol–gel PTD increases and the gel–sol PTD decreases. The formation or destruction of the three-dimensional network of HSA fibers involves two very different processes. In the case of gelling, it is a highly cooperative process that is very energetically favorable. Moreover, during the melting process, there is a progressive breakdown of the network and the fluidization is complete only when all the HSA molecules are dissociated. These two processes are also very sensitive to the strength of the network junctions; that is, to their degrees of crystallization. The increase in HSA content increases the number of junctions in the network, which widens the sol–gel PTD because the network has a higher degree of interconnectivity. At low HSA wt %, a low density of junctions is observed, but these are strongly crystallized. Moreover, when the HSA content increases, the fiber network is more entangled, but the junction zones are more disordered and therefore break more rapidly during the gelling process.

2.3. *Organogel Nanoparticle Dispersion*

Tgel has an important part in the preparation of the GLN dispersions from a LMOG. As observed in Figure 6, the formation of these one depends on this parameter.

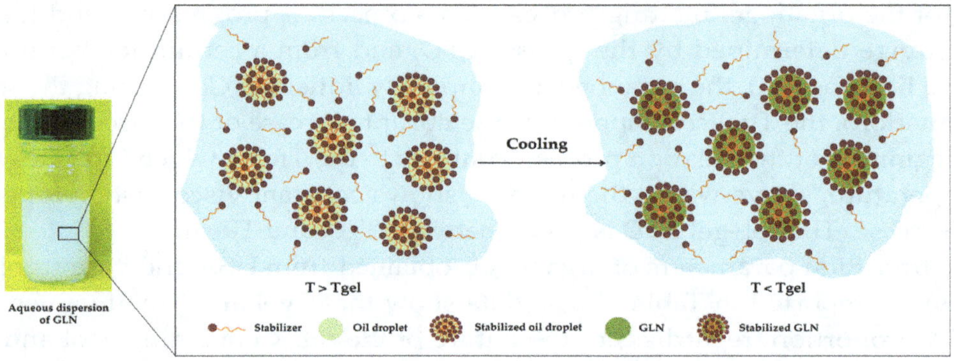

Figure 6. Schematic representation of GLN dispersion.

The mixture must be heated over the characteristic organogel Tgel, and then homogenized while still hot by ultrasound probe or using mechanical homogenization in the presence of emulsifying, thickening, or stabilizing agents. For example, polyethylenenimine (PEI) (Figure 7a) [17], acetylated glycol stearate (AGS) (Figure 7b), and polyvinyl alcohol (PVA) (Figure 7c) are used as stabilizing agents, and sodium hyaluronate (SH) as a thickening one (Figure 7d) [4]. Another common surfactant used as a stabilizing agent for these preparations is cetyltrimethylammonium bromide (CTAB) (Figure 7e) [12]. Even though both homogenization methods have been reported, sonication with ultrasound probe has proven to produce a more stable aqueous dispersion obtained from vegetable oils and HSA as organogelator [4], as well as for organogel dispersions with poly(3-hexylthiophene) (P3HT) (Figure 7f) [18].

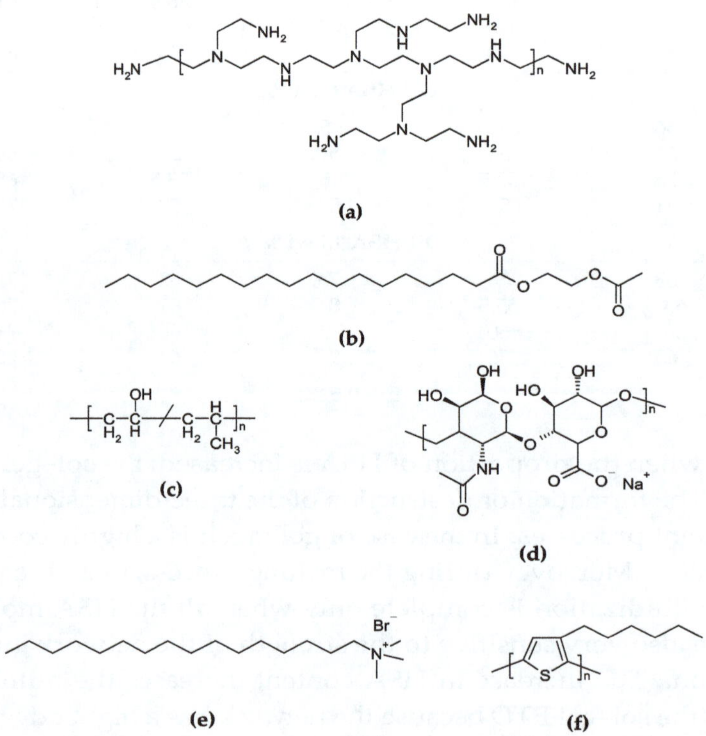

Figure 7. Chemical structure of common stabilizing agents used to prepare colloidal dispersions of GLN: **(a)** PEI; **(b)** AGS; **(c)** PVA; **(d)** SH; **(e)** CTAB; **(f)** P3HT.

3. GLN Physicochemical Characterization

The organogel dispersions of GLN possess characteristic physicochemical properties. The organogel components and preparation process influence the physicochemical properties of its dispersion, such as particle size and distribution, and ζ-potential, morphology, and physical stability [4]. As for the preparation of the organogel dispersions, it is of great importance to determine GLN phase transition parameters based on rheological measurements and to be compared to the "parent" organogel [12]. It is important to perform a proper characterization of GLN properties in order to assess their physical stability or aging and potential applicability.

3.1. Influence of Ingredient Amount and Nature on the Physical Stability of GLN Aqueous Dispersion

The physical stability of GLN dispersions is of great importance for various applications, and it is essential to control particle destructive/unstable processes and monitor their stability. Parameters such as type and concentration of surfactant and oil as well as viscosity of dispersed phase play an important role in the physical stability of the dispersion, and have an impact on particle size [12]. For example, a dispersion of vegetable oil with a greater viscosity (60 cP) presents a better stability compared to a dispersion of mineral oil (6 cP). The viscosity must be enough to maintain the emulsifying and stabilizing agent on the gelled particle surface, allowing the system to have an adequate physical stability [4]. This stability is also maintained in the GLN aqueous dispersion, due to the high electrostatic repulsion between the particles due to the stabilizing agent anchored in the particle's surface. [12].

While the concentration the organogelator has no influence on the particle size or on the physical stability, it has an effect on the viscosity of dispersions at high oil concentration. As described with other colloidal dispersions, particle size can be influenced by the formulation process parameters, such as shear speed during emulsification [12]. Dispersion can also be influenced by the pH, since the mean hydrodynamic diameter of a monodisperse population characterized by a high positive ζ-potential decreases with pH [2]. Some other parameters, such as the colloid size and surface area-to-volume ratio must be considered, and could have a strong influence for their potential uses.

3.2. Particle Mean Size, Distribution, and ζ-Potential

The light scattering method measures the time-dependent fluctuations of scattered light. Consequently, dynamic light scattering (DLS) provides information regarding the movement of the GLN in the colloidal dispersion, and uses Brownian motion to measure the mean hydrodynamic diameter and the size distribution of the particles. While smaller particles move faster, larger particles move slower, and the intensity of the fluctuations allows the particle size to be determined. Moreover, this method also permits the determination of the ζ-potential from the dynamic electrophoretic mobility. DSL studies show that by emulsifying bulk P3HT organogels into water-containing surfactant, two populations of particles are formed and there is an increase in size as a function of the P3HT concentration [18]. As for organogel dispersions obtain with HSA as gelator, the ζ-potential, polydispersity index (PDI), and particle size proved the influence of the concentration of the gelator on the stability of the dispersion [4]. Inclusive organogel dispersions made with dicarprylyl ether oil, HSA, and CTAB produced a stable colloidal system for longer than 45 days with GLN showing a mean diameter of around 260 nm, narrow PDI (0.01–0.1), and ζ-potential values varying between +25 and +50 mV [12].

3.3. Rheological Measurements

The rheological properties of organogel systems are of great importance with regards to preparing an aqueous dispersion of GLN and consequently to handling their properties and improve their potential applications. As described previously, gel–sol phase transition parameters of organogels are measured by following the variation of the elastic modulus (G′) and viscous modulus (G″) at increasing temperatures. The gelation process occurs as indicated by an abrupt increase of both moduli, while Tmelt is the exactly the opposite. When dispersed, the organogel showed similar rheological results, allowing the particle gelled state to be confirmed by demonstrating the existence of the Tmelt, Tsol, and gel–sol PTD [12].

Additionally, regarding the physicochemical properties of the organogels, colloidal dispersions of GLN possess some specific characteristics, giving them a broad spectrum of applications. In Figure 8a are reported the Tmelt, Tsol, and gel–sol PTD or phase transition parameters of gelled particles obtained from a vegetable (soybean, castor) or emollient (dicaprylyl ether, capric/caprylic triglycerides) oil, HSA as organic gelator, and PVA as stabilizing agent [12]. As previously noted, they are determined by rheology, measuring the evolution of the elastic (G′) and viscous (G″) moduli versus temperature. Tmelt is defined as the temperature below which gelled particles start to melt, while Tsol is the temperature from which the particles are totally transformed into oil droplets. The melting process is progressive, and takes place over a temperature gap, known as gel–sol PTD. Typically, Tmelt transition temperature increases with the concentration of organogelator and the viscosity of the oil used (Figure 8b) [4].

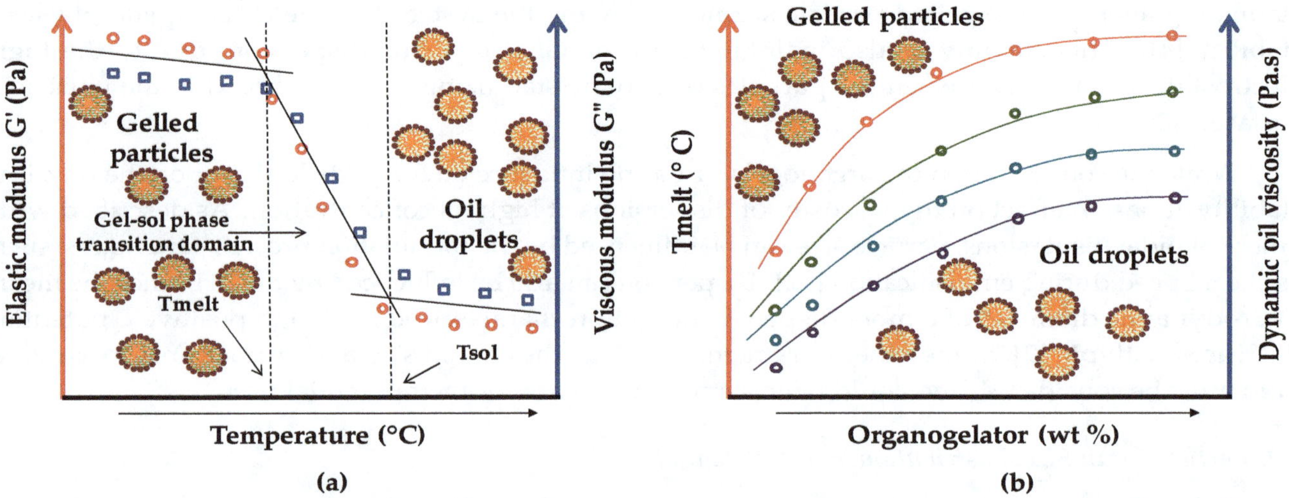

Figure 8. Gel–sol phase transition parameter evolution of GLN dispersion: (**a**) Tmelt, Tsol, and gel–sol PTD value determination according to the temperature increase; (**b**) Tmelt value evolution according the organogelator wt % and the dynamic oil viscosity increase.

Moreover, the Tmelt is easily influenced by the type organogelator and its concentration. POG gels have a lower Tmelt and higher gel strength compared to organogels obtained using LMOGs [5,6].

The formulation of the dispersed organogel had an impact on their rheological properties. An increase of viscosity was observed in the ratio gelator/solvents, as it was observed for lecithin/isopropyl myristate microemulsions [19]. In the case of HSA, the same phenomenon has been observed. Measuring the flow behavior of aqueous dispersions of GLN, there was an increase of the organogelator concentration from 0 wt % to 15 wt %, leading to an increase in shear viscosity from two to six orders of magnitude [20]. The characteristic rheological properties of organogel dispersions are strongly influenced by the interactions between the GLN. The gelled particle dispersions showed a non-Newtonian behavior when the shear rate increased, leading to a decrease of the system dynamic viscosity values [4]. In contrast, in the case of a polymer-based organogel dispersion, the same effect was observed when diluting the dispersion from 40 vol % to 20 vol %, but the system showed a constant viscosity when diluted at 10 vol % because the interactions between the particles became negligible [18].

3.4. GLN Morphology

To evaluate the morphology and structure of the aqueous GLN dispersions, transmission electron microscopy (TEM) and scanning electron microscopy (SEM) were used, respectively (Figure 9).

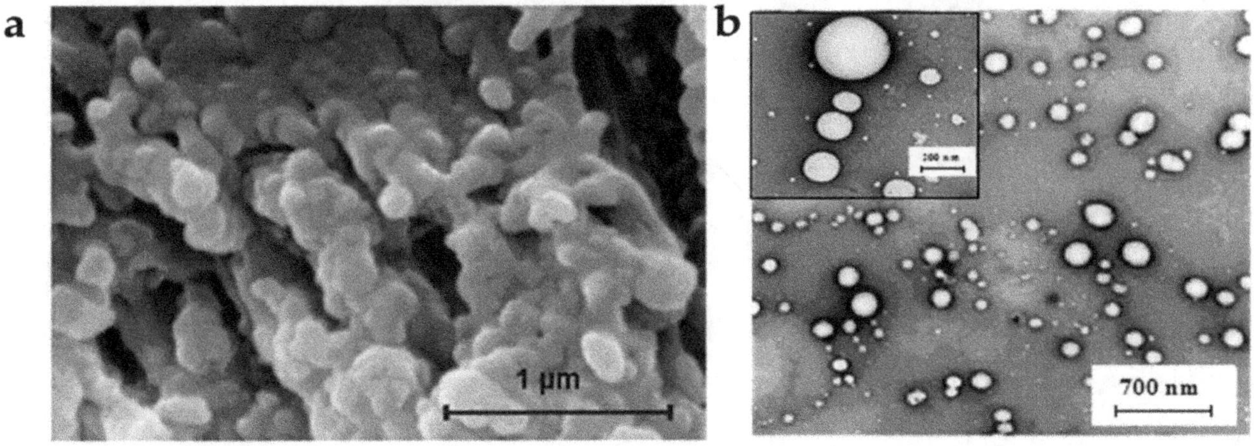

Figure 9. Electron microscopy pictures of GLN dispersion [4]: (**a**) SEM micrograph; (**b**) TEM micrograph.

These methods allow the data obtained from the DLS analysis to be completed. The DLS measurements provide insight into the size distribution of the gel particles within the dispersion, but they do not reveal information about their internal structure. In some applications, it is important that the particles retain porosity and a network structure original to the bulk organogel [18], while other GLN show spherical morphology without complementary information concerning their surface state [4]. These methods permit the accumulation of surfactant molecules at the surface of the gelled particles, confirming the ζ-potential values and indicating the stability of the dispersion [12].

4. GLN Potential Applications

Organogel dispersions of GLN are a suitable formulation for the preparation of new systems which could be used for cosmetic applications or as smart nanoparticles for drug delivery [4,12]. According to the data base, one of the first patent applications where the term organogel appeared was deposited organic solvent [21]. However, it was not until 1996 that the organogel dispersion concept appeared in European application patents. Notably, in the dermo-cosmetic field, companies like L'Oréal have patented dispersions of gel particles in a stabilized aqueous phase. [22].

Besides the cosmetic industry, organogel dispersions have been used as drug delivery systems and have proved their ability to enhance the permeation and improve drug release in dermal and transdermal formulations (Figure 9). It has been shown that organogel dispersions are effective in the improvement of sun protection factor (SPF) value, increasing the ability to absorb UV rays. This could be related to the optical properties of the particles and diffusing capacity provided by the organogelator. There is a synergistic improvement of two factors: the photo-protective ability and the photo-stability of the UVB filter [4]. Moreover, it has been used for topical treatments of early stages of skin cancers. In addition, these systems could be used for intravenous applications by adapting the formulation for this aim [3].

4.1. Hydrophobic Reservoirs for Sun Protection

Gelled sun protection nanoparticles (GSPN) in water could be employed as a nanocarrier for dermo-cosmetic applications or as hydrophobic reservoirs for photoprotection (Figure 10) [4–6,17,23].

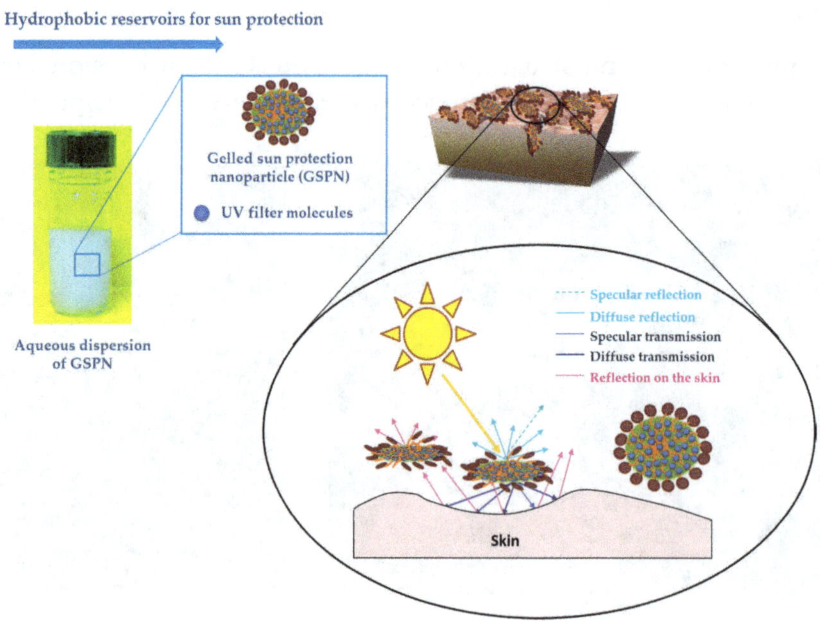

Figure 10. Representation of gelled sun protection nanoparticles (GSPN) and their photoprotection ability mechanism.

The preparation process and physicochemical properties of these GSPN aqueous dispersions have been described previously by Chaouat et al. [24]. Gelled particle dispersions of sunscreen organic mixture–1-(4-methoxyphenyl)-3-(4-*tert*-butylphenyl)propane-1,3-dione or avobenzone (UVA solar filter) (Figure 11a) dissolved in 2-ethylhexyl 2-cyano-3,3-diphenyl-2-propenoate or octocrylene (UVB solar filter, viscous liquid) (Figure 11b) and HSA as organic gelator were obtained by emulsification at high temperature and cooling at room temperature, using PVA as a stabilizing and dispersing agent. The results confirmed that the GSPN size was in agreement with the necessity of keeping the particles on the skin surface, avoiding their penetration through the *stratum corneum*. TEM observation of the dispersion showed spherical particles with a mean diameter of 600 nm—a size in accordance with the DLS measurements. The UV spectroscopy observations showed more important absorption of the gelled particles compared to that of corresponding emulsion droplets. A rheological investigation allowed the particle sol-gel phase transition parameters to be determined, confirming their gelled state. A comparative aging study of the emulsion and corresponding dispersion showed greatly enhanced physical stability after gelation. This last point was also proved by ζ-potential measurements.

According to Kirilov et al., aqueous dispersions of GSPN were also obtained with two organic liquids [4]. Vaseline and almond oils were used as organic media for the GSPN preparation. Gelled nanoparticles containing a model sunscreen molecule with mean size of 450 nm, ζ-potential value above −30 mV, and PDI of 0.18 were obtained by the ultrasound probe homogenization method. A comparative study of their dispersion aging showed a greatly enhanced GSPN stability after gelation. According to the UVB absorption evaluation, GSPN improved the photoprotective ability and the photostability of the immobilized UVB blocker as 2-ethylhexyl-p-dimethylaminobenzoate or EHMAB (Figure 11c). In addition, GSPN showed a high water resistance (~83%), even after 40 min of immersion.

Actually, studies concerning the immobilization of benzophenone-3 or BP-3—a commonly used UVA and UVB absorbed molecule—as GSPN are being carried out (Figure 11d).

The obtained results demonstrate the interest in GSPN and their aqueous dispersion as a concept for the preparation of original sun protection formulations with improved diffusion properties and waterproof ability.

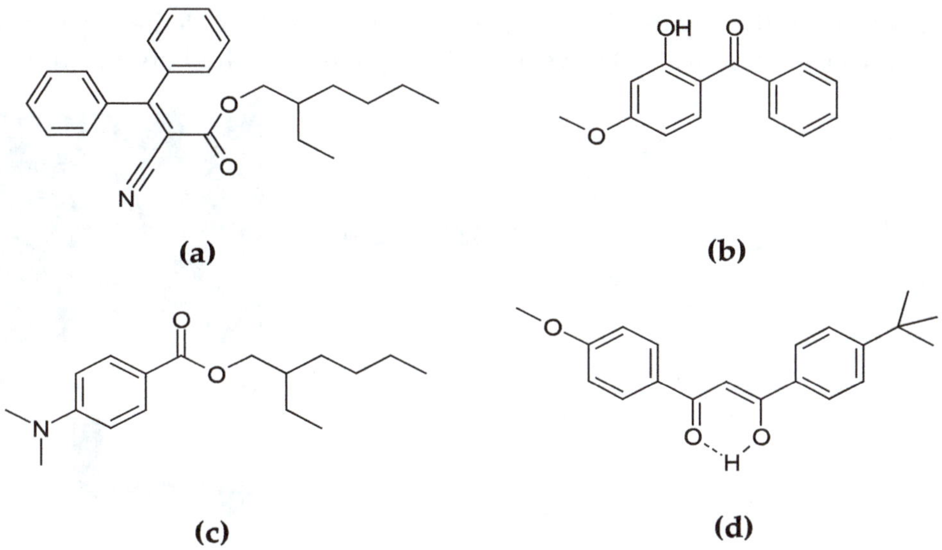

Figure 11. Chemical structure of immobilized chemical UV filter as GSPN: (**a**) octocrylene; (**b**) benzophenone-3 (BP-3); (**c**) 2-ethylhexyl-p-dimethylaminobenzoate (EHMAB); (**d**) avobenzone.

4.2. Hydrophobic Reservoirs for Drug Delivery

Other studies demonstrated the interest of colloidal dispersions of GLN for the encapsulation and delivery of lipophilic active compounds (Figure 12).

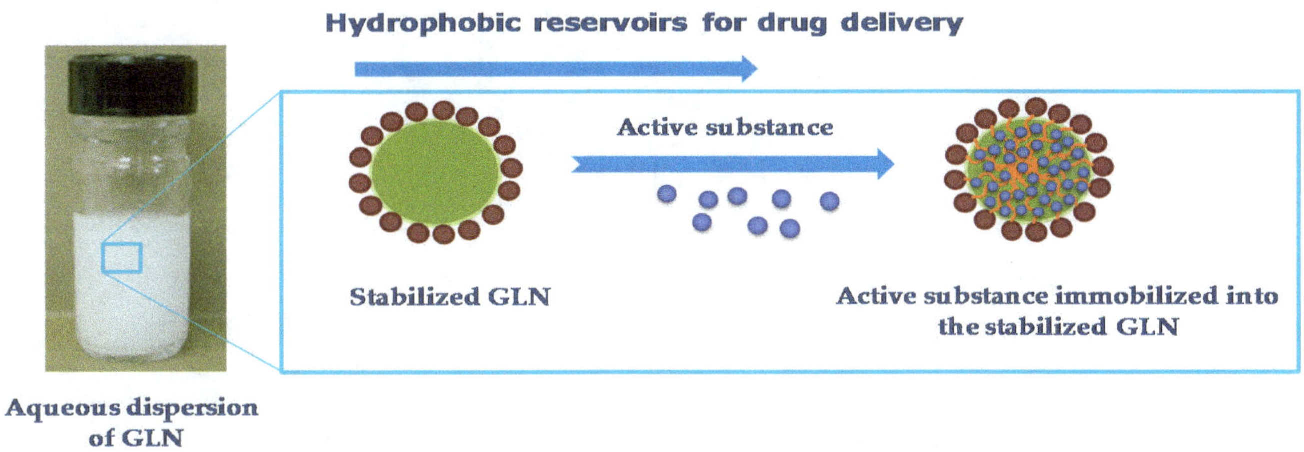

Figure 12. Representation of the drug delivery ability of GLN.

Sigueira-Mourra et al. prepared stable dispersions of soybean oil GLN with excellent chloroaluminum phthalocyanine (ClAlPc) encapsulation efficiency. The obtained ClAlPc-loaded GLN formulation could be a very useful system for photodynamic therapy (PDT) procedures (Figure 13).

Authors have efficiently encapsulated ClAlPc (Figure 14a)—a second-generation photosensitizer agent usually used in PDT—into soybean oil GLN dispersed in aqueous medium. Nanoparticles were jellified and stabilized by HSA and PEI, respectively. They were obtained by hot homogenization using ultrasound probe and then cooling at room temperature until obtaining a stable GLN dispersion. Nanoparticles showed mean diameter of 280 nm, narrow size distribution (PDI ~0.3), positive ζ potential value (~+50 mV), and spherical morphology (Table 2). Dispersions exhibited pronounced physical stability after a 6-month aging period. The PEI and HSA proportions allowed the GLN size, charge, and stability to be controlled, resulting from the complex electrostatic interactions between the positively-charged PEI and the negatively-charged HSA fibers present on the gelled particles' surface on the one hand and stearic repulsions between positively-charged GLN on the other hand.

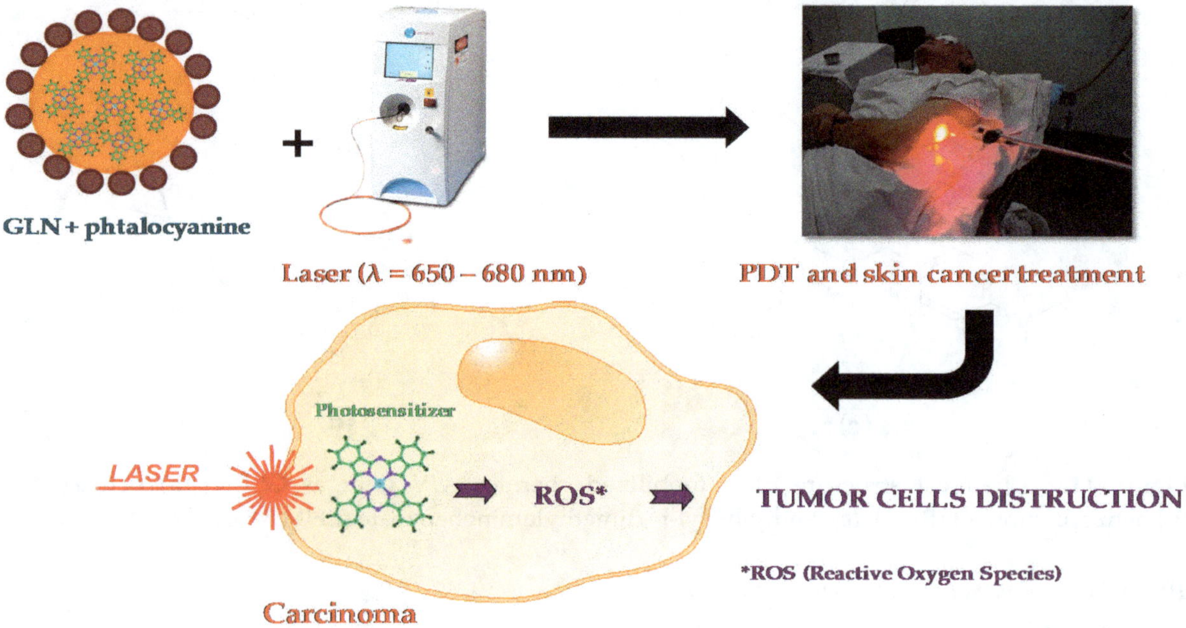

Figure 13. Representation of the medical treatment of skin cancers using photodynamic therapy (PDT).

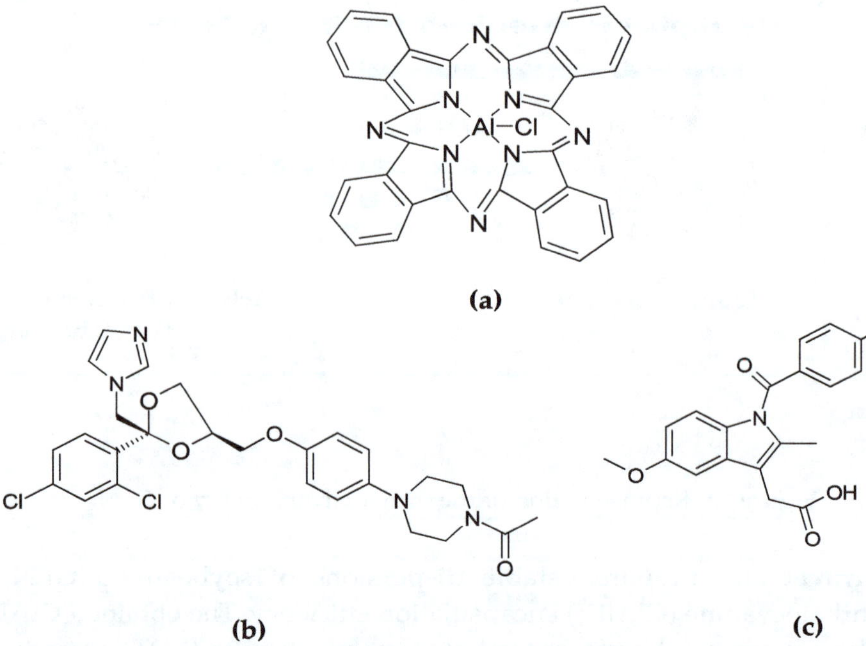

Figure 14. Chemical structure of immobilized active substances: **(a)** chloroaluminum phthalocyanine (ClAlPc); **(b)** ketoconazole; **(c)** indomethacin.

Table 2. Some experimental results from physicochemical characterization GLN containing active substances (n = 3), pH = 7. PDI: polydispersity index.

Active Substance	Mean Size (nm)	PDI	ζ-Potential (mV)	Encapsulation Efficiency (%)
ClAlPc	280 (±10)	0.3 (±0.04)	+50 (±1.6)	~100
indomethacin	244 (±13)	0.26 (±0.05)	−30 (±5.6)	91.9 (±0.2)
ketoconazole	266 (±16)	0.25 (±0.03)	−25.9 (±4.7)	97.4 (±0.1)

Martin et al. studied the release profile and efficiency of GLN obtained by the same preparation process as Siqueira-Mourra et al. [24]. In the first step, they used DLS and differential scanning calorimetry (DSC) to select the most suitable formulations by optimizing the proportion of ingredients (HSA, PVA, castor oil) to obtain particles of the smallest size and greatest stability. Two lipophilic drug models—indomethacin (Figure 14b) and ketoconazole (Figure 14c)—were entrapped in the GLS made of castor oil and gelled by HSA.

Thermal studies confirmed that there was no significant alteration of gelation phenomenon in presence of the entrapped drugs. Very stable dispersions were obtained (>3 months), with GLN presenting a mean diameter between 240 and 300 nm. High encapsulation efficiency (>98%) was measured for indomethacin and ketoconazole (Table 2). The release profile determined by in vitro dialysis showed an immediate release of the drug from the organogel nanoparticles, due to rapid diffusion.

All these studies demonstrate the interest in the colloidal dispersions of GLN for dermo-cosmetic applications as sun protection systems and for drug delivery for lipophilic active substances.

5. Conclusions

Since lipophilic drug formulation remains the major challenge in the pharmaceutical field, recent preparations of colloidal dispersions of GLN present very interesting advantages as drug delivery formulations as well as in the cosmetic industry. These dispersions have promising stability properties, are easily formulated, and can be produced with natural organogelators or simple synthetic polymers, giving them a great range of possible industrial applications. The basic knowledge of these colloidal GLN dispersions, their properties, and characterization methods will allow the optimization of the formulation with regards to obtaining more stable dispersions according to their intended applications. Furthermore, they are still a broad area of possibilities in the synthesis of new organogelators, formulations of better and more stable dispersions, and drug immobilization. The interest for these materials will continue to grow in the future years.

Acknowledgments: We gratefully thank the region Rhone-Alpes and European Region Action Scheme for the Mobility of University Students (ERASMUS) program for the financial support.

References

1. Gao, L.; Liu, G.; Ma, J.; Wang, X.; Zhou, L.; Li, X.; Wang, F. Applications of drug nanocrystal technologies on oral drug delivery of poorly soluble drugs. *Pharm. Res.* **2013**, *30*, 307–324. [CrossRef] [PubMed]
2. Boudier, A.; Kirilov, P.; Franceschi-Messant, S.; Belkhelfa, H.; Hadioui, L.; Roques, C.; Perez, E.; Rico-Lattes, I. Evaluation of biocompatible stabilized gelled soya bean oil nanoparticles as new hydrophobic reservoirs. *J. Microencapsul.* **2010**, *27*, 682–692. [CrossRef] [PubMed]
3. Kirilov, P.; Franceschi-Messant, S.; Rico-Lattes, I.; Perez, I.; Bordat, P. Novel organogel particles, method for the preparation thereof and cosmetic uses thereof. Patent WO2008037679 (A1), 3 April 2008.
4. Kirilov, P.; Rum, S.; Gilbert, E.; Roussel, L.; Salmon, D.; Abdayem, R.; Serre, C.; Villa, C.; Haftek, M.; Falson, F.; et al. Aqueous dispersions of organogel nanoparticles-potential systems for cosmetic and dermo-cosmetic applications. *Int. J. Cosmet. Sci.* **2014**, *36*, 336–346. [CrossRef] [PubMed]
5. Kirilov, P.; Le Cong, A.K.; Denis, A.; Rabehi, H.; Rum, S.; Villa, C.; Haftek, M.; Pirot, F. Organogels for cosmetic and dermo-cosmetic applications. Classification, preparation and characterization of organogel formulations Part I & II. *H&PC Today* **2015**, *10*, 15–21.
6. Kirilov, P. Solar Energy: New Long Chain Polypyridinic Ruthenium Complexes; Solar Protection: Preparation of Gelosomes for Cosmetic Applications. Ph.D. Thesis, University Paul Sabatier Toulouse III, Toulouse, France, 17 January 2007.
7. George, M.; Weiss, R.G. Low molecular mass-organic gelator. In *Molecular Gels. Materials with Self-Assembled Fibrillar Networks*, 1st ed.; Weiss, R.G., Terech, P., Eds.; Springer: Dordrecht, The Netherlands, 2006; Volume 14, pp. 449–477, ISBN 13-978-1-4020-3352-0-(HB).

8. Suzuki, M.; Hanabusa, K. Polymer organogelators that make supramolecular organogels through physical cross-linking and self-assembly. *Chem. Soc. Rev.* **2010**, *39*, 455–463. [CrossRef] [PubMed]
9. Bag, B.G.; Majumdar, R.; Dinda, S.K.; Dey, P.P.; Maity, G.C.; Mallia, V.A.; Weiss, R.G. Self-assembly of ketals of arjunolic acid into vesicles and fibers yielding gel-like dispersions. *Langmuir* **2013**, *29*, 1766–1778. [CrossRef] [PubMed]
10. Rogers, M.A.; Abraham, S.; Bodondics, F.; Weiss, R.G. Influence of the hydroxyl position in racemic hydroxyoctadecanoic acids on the crystallization kinetics and activation energies of gels and dispersions in mineral oil. *Cryst. Growth Des.* **2005**, *12*, 5497–5504. [CrossRef]
11. Desvergne, J.P. Organic gelators and hydrogelators. *Beilstein J. Org. Chem.* **2010**, *6*, 846–847. [CrossRef] [PubMed]
12. Kirilov, P.; Lukyanova, L.; Franceschi-Messant, S.; Perier, V.; Perez, E.; Rico-Lattes, I. A new type of colloidal dispersions based on nanoparticles of gelled oil. *Colloids Surf. A Physicochem. Eng. Asp.* **2008**, *328*, 1–7. [CrossRef]
13. Vintiloiu, A.; Leroux, J.-C. Organogels and their use in drug delivery—A review. *J. Control. Release* **2008**, *125*, 179–192. [CrossRef] [PubMed]
14. Das, D.; Roy, S.; Debnath, S.; Das, P.K. Surfactant-stabilized small hydrogel particles in oil: Hosts for remarkable activation of enzymes in organic solvents. *Chemistry* **2010**, *16*, 4911–4922. [CrossRef] [PubMed]
15. Hanabusa, K.; Suzuki, M. Development of low-molecular-weight gelators and polymer-based gelators. *Polym. J.* **2014**, *46*, 776–782. [CrossRef]
16. Elnaggar, Y.S.R.; El-Refaie, W.M.; El-Massik, M.A.; Abdallah, O.Y. Lecithin-based nanostructured gels for skin delivery: An update on state of art and recent applications. *J. Control. Release* **2014**, *180*, 10–24. [CrossRef] [PubMed]
17. Siqueira-Moura, M.P.; Franceschi-Messant, S.; Blanzat, M.; Ré, M.I.; Perez, E.; Rico-Lattes, I.; Lattes, A.; Tedesco, A.C. Gelled oil particles: A new approach to encapsulate a hydrophobic metallophthalocyanine. *J. Colloid Interface Sci.* **2013**, *401*, 155–160. [CrossRef] [PubMed]
18. Richards, J.J.; Weigandt, K.M.; Pozzo, D.C. Aqueous dispersions of colloidal poly(3-hexylthiophene) gel particles with high internal porosity. *J. Colloid Interface Sci.* **2011**, *364*, 341–350. [CrossRef] [PubMed]
19. Hadidi, N.; Nazari, N.; Aboofazeli, R. Formulation and optimization of microemulsion-based organogels containing propranolol hydrochloride using experimental design methods. *DARU J. Pharm. Sci.* **2009**, *17*, 217–224.
20. Kirilov, P.; Gauffre, F.; Franceschi-Messant, S.; Perez, E.; Rico-Lattes, I. Rheological characterization of a new type of colloidal dispersion based on nanoparticles of gelled oil. *J. Phys. Chem. B* **2009**, *113*, 11101–11108. [CrossRef] [PubMed]
21. Winyall, M.E. Process for Preparing Grade of Silica and Process of Using Name. Patent US4216113 (A), 5 August 1980.
22. Ribier, A.; Biatry, B. Cosmetic or Dermatologic Oil/Water Dispersion Stabilized with Cubic Gel Particles and Method of Preparation. Patent EP0711540 (A1), 3 October 2000.
23. Martin, B.; Brouillet, F.; Franceschi, S.; Perez, E. Evaluation of organogel nanoparticles as drug delivery systems for lipophilic compounds. *AAPS PharmSciTech* **2017**, *18*, 1261–1269. [CrossRef] [PubMed]
24. Chaouat, C.; Kirilov, P.; Franceschi-Messant, S.; Perez, E.; Giraud, I.; Rico-Lattes, I. Nanoparticles of gelled suncreen oil: Elaboration and physicochemical characterization. In Proceedings of the 42nd Journadas Anuales del Comite Espanol de la Detergencia, Barcelona, Spain, 28–29 March 2012; Volume 42, pp. 84–96.

2

Self-Assembly of Colloidal Nanocomposite Hydrogels using 1D Cellulose Nanocrystals and 2D Exfoliated Organoclay Layers

Takumi Okamoto [1,2], Avinash J. Patil [1,*], Tomi Nissinen [1] and Stephen Mann [1]

1 Centre for Organized Matter for Chemistry and Centre for Protolife Research, School of Chemistry, University of Bristol, Bristol BS8 1TS, UK; okamoto.bristol@gmail.com (T.O.); tomi.nissinen84@gmail.com (T.N.); s.mann@bristol.ac.uk (S.M.)

2 Denso Corporation, 1-1, Showa-cho, Kariya, Aichi 448-8661, Japan

* Correspondence: avinash.patil@bristol.ac.uk

Abstract: Stimuli-responsive colloidal nanocomposite hydrogels are prepared by exploiting non-covalent interactions between anionic cellulose nanocrystals and polycationic delaminated sheets of aminopropyl-functionalized magnesium phyllosilicate clays.

Keywords: cellulose nanocrystals; organoclay; nanocomposite hydrogels

---

1. Introduction

Hydrogels have emerged as an important class of soft materials that have found numerous applications ranging from cosmetics and personal care products to biotechnological and biomedical applications. A typical hydrogel comprises a physically or chemically cross-linked 3D network of natural or synthetic building blocks, which has the ability to encapsulate an extremely high percentage of water compared with their dry weight. Significantly, the capacity to retain water and degree of swelling can be regulated by tailoring the cross-linked network of building blocks in the gel matrix. As a result, the physico-chemical (porosity and hydrophilicity) and mechanical (viscoelastic) properties of hydrogels can be modulated for desired applications [1].

Traditionally, hydrogel assembly involves cross-linking of components such as synthetic or natural polymers [2] and biological molecules [3], or self-assembly of low molecular weight gelators [4–7], or both. Alternatively, recent studies have indicated that self-supporting nanocomposite hydrogels can be prepared by integrating nanoparticles or nanostructures into 3D hydrated cross-linked polymer or biopolymer networks [8,9]. This approach has opened up new opportunities to introduce novel physical, chemical, mechanical, electrical, magnetic, and optical properties into soft materials. For example, materials such as carbon nanotubes and graphene [10,11], polymeric nanoparticles [12,13], inorganic nanoparticles (hydroxyapatite, calcium phosphate, synthetic clays) [14–16], and metallic/metal-oxide nanoparticles (gold, silver, iron oxide, titania) [17–20] have been physically or chemically incorporated within polymeric networks to produce nanocomposite hydrogels with reinforced properties. In this study, we present a new type of stimuli-responsive nanocomposite hydrogel based on the cooperative assembly of two aqueous colloidal sols comprising 1D cellulose nanocrystals (CNCs) and 2D exfoliated sheets of an organically modified magnesium phyllosilicate clay (organoclay). Materials based on CNCs have received considerable attention in recent years due to their applicability in diverse areas such as composites [21], wound dressing and medical implants [22], and as chiral templates for synthesis of inorganic materials [23–25]. To the best of our knowledge, we demonstrate the first example of an organic–inorganic hybrid hydrogel in which non-covalent interactions between 1D and

2D nanoparticles spontaneously form a 3D cross-linked matrix, which expedites the entrapment of water to produce a self-supported colloidal nanocomposite hydrogel matrix. Notably, the CNC–organoclay nanocomposite hydrogels can be disassembled and reconstructed by exposing the soft materials to ammonia and carbon dioxide gases, respectively, suggesting that they could be developed as environmentally sensitive soft materials. We also show that the nanocomposite hydrogels are capable of storing and releasing small drug molecules such as ibuprofen. Significantly, introduction of ibuprofen facilitates co-assembly of the organoclay sheets to produce mesolamellar domains within the hydrogel matrix.

2. Results and Discussion

CNCs were prepared by sulfuric acid hydrolysis of microcrystalline cellulose powder (see Experimental section) [26]. Transmission electron microscopy images of uranyl acetate-stained CNCs revealed the presence of rod-like morphologies, with lengths of 100–300 nm and a thickness of 10–30 nm (Supporting Information, Figure S1a). Aminopropyl-functionalized magnesium phyllosilicate was synthesized using our previously reported studies (see Experimental section) [27]. For delamination, organoclay powders were dispersed in distilled water using ultrasonication. As a consequence, protonation of the aminopropyl-functional groups associated with the inorganic framework facilitated exfoliation of stacks of organoclay sheets and produced a clear suspension containing 50–300 nm sized delaminated organoclay particles (Supporting Information, Figure S1b). CNC–organoclay nanocomposite hydrogels were prepared by adding aqueous dispersions of exfoliated aminopropyl-functionalized magnesium phyllosilicate clay (1–10 wt %) to a 3 wt % colloidal sol of CNCs. The resulting mixtures showed gradual increase in viscosity with increased loadings of the exfoliated organoclay. TEM images of unstained CNC–organoclay hydrogel samples showed the presence of large aggregates comprising cross-linked networks of CNC nanoparticles (Figure 1a). Energy dispersive X-ray (EDX) analysis of the hybrid gel samples showed the presence of Si, Mg, and Cl (counter ion) associated with the organoclay particles and S from sulfonic functional groups of the CNC nanoparticles (Supporting Information, Figure S2). Typically, weight ratios of CNC:organoclay in the range of 1:0.03, 1:0.06. 1:0.13, 1:0.2, and 1:0.26 produced opaque self-supported hydrogels (Figure 1b), and showed no gravity-mediated flow upon inversion of the hydrogels. In contrast, mixing CNC and organoclay dispersions at weight ratios of 1:0.015 and 1:0.033 produced free-flowing liquid and viscous suspensions, respectively.

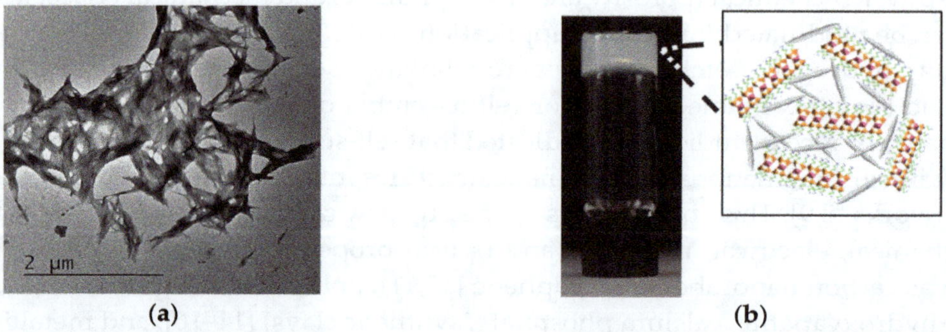

Figure 1. (**a**) Unstained TEM image of cellulose nanocrystal (CNC)-organoclay hydrogel; (**b**) photograph showing self-supported CNC:organoclay (1:0.13) colloidal nanocomposite hydrogel, inset showing schematic illustration of cross-linked network formed by non-covalent interactions between CNC (grey) and exfoliated organoclay sheets.

To probe the interactions between the exfoliated organoclay particles and CNCs, we characterized the colloidal nanocomposite hydrogels prepared at a CNC:organoclay weight ratio of 1:0.13 using zeta potential, powder X-ray diffraction (PXRD), and rheometry techniques. Zeta potential studies on the CNCs, exfoliated organoclay sheets, and CNC–organoclay nanocomposite hydrogel gave values of −43 mV, +20 mV, and −16 mV, respectively (Supporting Information, Figure S3). The significant decrease in the overall surface charge of the CNCs in the hydrogels was consistent with strong electrostatic interactions between the cationic aminopropyl-functionalities associated with the magnesium phyllosilicate framework and sulfonic functional groups of CNCs. Low-angle PXRD studies of the as-synthesized aminopropyl-functionalized magnesium phyllosilicate showed a broad reflection at $2\theta = 5.9°$, which was attributed to an interlayer spacing (d_{001}) of 1.5 nm, consistent with covalently anchored aminopropyl-functionalities present in the interlamellar regions [22]. In contrast, CNC–organoclay nanocomposite hydrogel samples displayed no reflections below $2\theta = 10°$, confirming that the organoclay sheets remained exfoliated within the hydrogel matrix (Supporting Information, Figure S4a). High-angle PXRD pattern showed that characteristic peaks associated with CNC nanoparticles were retained at $2\theta = 12.6°$ and 22.6° (Supporting Information, Figure S4b). Frequency sweep experiments showed a linear viscoelastic region where parallel storage ($G' = 240$ Pa) remained higher than loss ($G'' = 25$ Pa) moduli, which is a typical characteristic for solid-like viscoelastic hydrogels (Figure 2a). Oscillatory amplitude sweeps at a constant frequency of 1 Hz of hydrogels showed a linear viscoelastic region with parallel G' (290 Pa) and G'' (30 Pa) moduli up to a shear strain of 10%. Two crossover points observed above 10% shear strain were assigned to deformation of the hydrogel into a quasi-liquid state, indicating that the colloidal nanocomposite hydrogel exhibited responsive properties (Figure 2b).

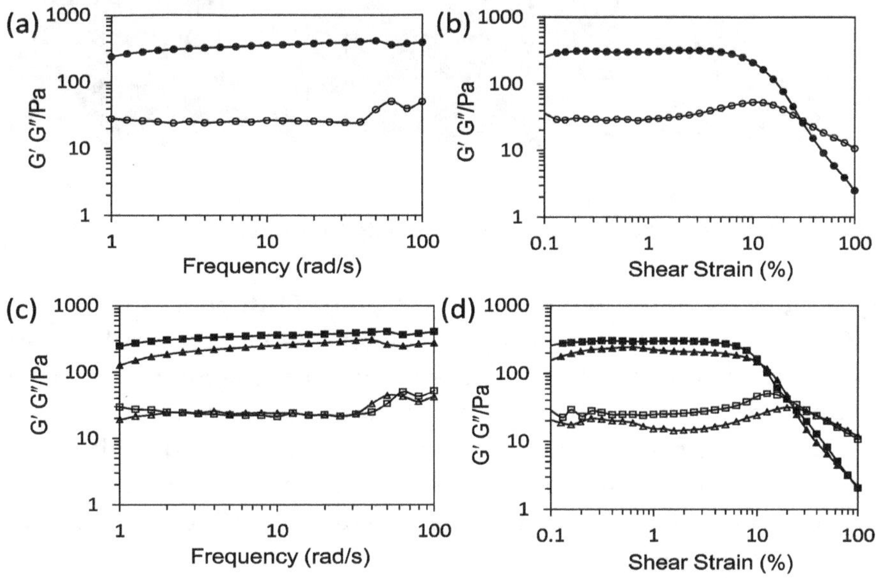

Figure 2. Rheometry studies showing (**a**) frequency sweep and (**b**) amplitude sweep curves for an as-prepared CNC–organoclay colloidal nanocomposite hydrogel (ratio 1:0.13); values for storage G' (filled circles) and loss G'' moduli (open circles); (**c**) and (**d**) show frequency and amplitude sweep profiles, respectively, for a CNC–organoclay nanocomposite hydrogel (ratio 1:0.13) after exposure to gaseous ammonia (G', filled triangles; G'', open triangles) and carbon dioxide (G', filled squares; G'', open squares).

We also investigated the gas-sensing properties of the colloidal nanocomposite hydrogels. For this, a hybrid hydrogel (CNC:organoclay, 1:0.13) was carefully exposed to ammonia gas to increase the pH of the bulk hydrogel from 9.4 to 10.4. Interestingly, the inversion tests carried out after change in the pH showed gravity-induced flow of the viscous hydrogel. Significantly, exposing the above viscous fluids to carbon dioxide gas lowered the pH value of the bulk hydrogel back to 9.4, which re-instigated cross-linking of the organic and inorganic building blocks to yield a self-supported hydrogel. Corresponding frequency sweep measurements on the ammonia-treated colloidal nanocomposite hydrogels at pH 10.4 showed a decrease in G' values to 127 Pa, whilst lowering the pH back to 9.4 using gaseous carbon dioxide increased the G' value to 240 Pa (Figure 2c), consistent with a decrease and restoration of the solid-like viscoelastic behavior of the hydrogel, respectively. In contrast, loss moduli (G'') values of the as-prepared hydrogel and hydrogels prepared at pH 10.4 or 9.4 remained unchanged. The oscillatory amplitude sweep experiments were also consistent with above observations (Figure 2d). Ammonia- and carbon dioxide-treated colloidal nanocomposite hydrogels maintained linear viscoelastic regions up to a shear strain of 10%, above which the samples became deformed.

We also employed cross-polarized (CP) optical microscopy to investigate the effect of gas-induced reassembly and disassembly of CNC–organoclay colloidal nanocomposite hydrogel networks (Figure 3). CP microscopy images of the as-prepared CNC:organoclay nanocomposite hydrogels showed birefringence, indicating the presence of anisotropic CNC–organoclay domains dispersed within an isotropic phase. Exposure to ammonia gas caused significant reduction in the birefringence of the hybrid gels, consistent with an increase in the isotropic phase due to loss of the interactions between the CNC and organoclay particles. Notably, birefringence was regained when the above samples were carefully exposed to the carbon dioxide gas, confirming that the non-covalent interactions between the organic and inorganic particles were restored.

Figure 3. Cross-polarized microscopy images of **(a)** as-prepared **(b)** ammonia- and **(c)** carbon dioxide-treated CNC–organoclay colloidal nanocomposite hydrogels, scale bar = 100 μm.

Taken together, the above observations suggest that pH-dependent electrostatic interactions between the CNC and exfoliated organoclay sheets were primarily responsible for the formation of the self-supported hydrogels. Gas-induced disassembly and reconstruction of the CNC–organoclay nanocomposite hydrogels was attributed to deprotonation and re-protonation of aminopropyl-functionalities (pKa of primary amine ~10.5) associated with the organoclay sheets. As a consequence, environmentally induced changes in pH are able to alter the columbic interactions between the CNC and organoclay layers, and, therefore, strongly influence the mechanical and optical properties of the colloidal nanocomposite hydrogels.

The potential use of CNC–organoclay colloidal nanocomposite hydrogels for the encapsulation and controlled release of functional small molecules such as the anti-inflammatory drug, ibuprofen, was investigated. Encapsulation of the drug was carried out by mixing an aqueous solution of ibuprofen (10 wt %) with a CNC sol (6 wt %), followed by addition of an aqueous dispersion of the exfoliated organoclay sheets. The resulting mixture produced a self-supported hydrogel with a CNC:organoclay:ibuprofen weight ratio of 1:0.13:1.6. Frequency sweeps revealed a linear viscoelastic region in which the parallel storage moduli were higher than the loss moduli. However, the storage moduli (G' = 121 Pa) and loss moduli (G'' = 13.8) values were significantly lower than that of parent CNC–organoclay hybrid hydrogel (Supporting Information, Figure S5). Oscillatory amplitude sweeps at a constant frequency of 1 Hz revealed that G' and G'' remained parallel and deformed above 10% shear strain (Supporting Information, Figure S6). The marked decrease in solid-like viscoelastic properties of the drug-loaded hybrid hydrogel was attributed to charge screening of the columbic interactions between the CNC and exfoliated clay layers due to the presence of the anionic drug molecules.

The concentration of ibuprofen released over time from the colloidal nanocomposite CNC–organoclay hydrogel was determined by using UV–vis spectroscopy. As a control sample, an ibuprofen–organoclay composite was prepared by adding an aqueous solution of ibuprofen to a freshly exfoliated suspension of organoclay. This allowed spontaneous restacking of the exfoliated sheets in the presence of the drug molecules. The resulting precipitates were isolated, dried, and pressed into pellet for the release studies. Both the ibuprofen-loaded colloidal nanocomposite hydrogel and organoclay–ibuprofen control nanocomposite showed a steady release of the drug molecules over a period of 6 h. Comparison of the initial release profiles obtained within an hour indicated that the extraction of ibuprofen molecules from the drug-loaded CNC–organoclay hydrogel was approximately 10% slower than that from the ibuprofen–organoclay nanocomposite pellet (Figure 4a). To further elucidate the origin of these differences, structural investigations on the nature of the ibuprofen–organoclay nanocomposite and CNC–organoclay–ibuprofen hydrogel were undertaken by using PXRD studies (Figure 4b). PXRD profiles recorded from the nanocomposite indicated that the exfoliated organoclay sheets were spontaneously restacked into a mesolamellar bulk phase in the presence of the drug molecules. As a result, the interlamellar spacing for the parent organoclay was increased from 1.5 nm ($2\theta = 5.9°$) to 2.3 nm ($2\theta = 3.8°$), which suggested that ibuprofen molecules were intercalated within the interlayer regions of the reconstituted organoclay. Significantly, PXRD patterns of the ibuprofen-containing CNC–organoclay hydrogels also exhibited a low-angle reflection at $2\theta = 3.8°$ corresponding to an interlayer distance of 2.3 nm, indicating that the colloidal nanocomposite hydrogels comprised mesolamellar domains containing intercalated drug molecules [22]. The slower displacement of the drug molecules observed from the nanocomposite hydrogels could be therefore attributed to a combination of charge-mediated and diffusion-limited processes associated with interactions between the aminopropyl side chains of the organoclay sheets and carboxylic acid moieties of ibuprofen, and physical immobilization of the drug molecules within the interconnected CNC network, respectively.

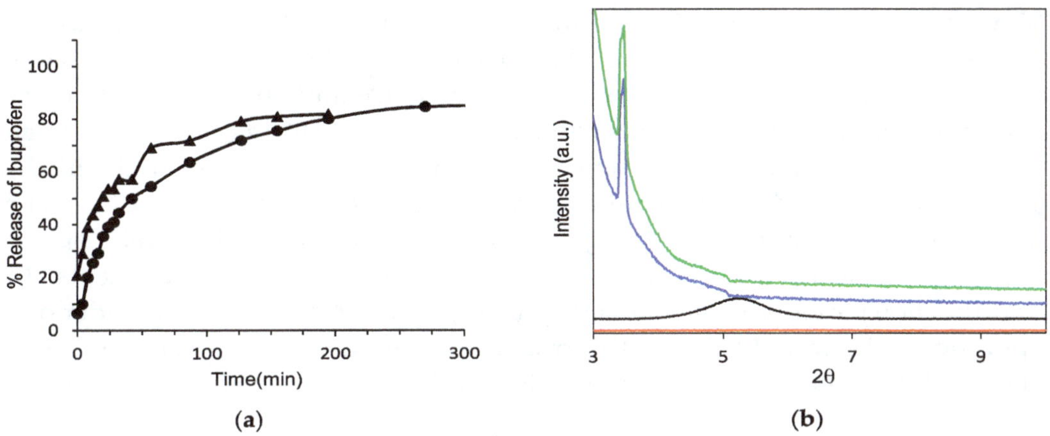

Figure 4. (**a**) Ibuprofen release profiles from organoclay–ibuprofen nanocomposite pellet (triangles) and CNC–organoclay–ibuprofen colloidal nanocomposite hydrogels (circles); (**b**) powder X-Ray diffraction (PXRD) pattern of CNC–organoclay nanocomposite hydrogel (red), as-synthesized organoclay (black), organoclay–ibuprofen nanocomposite (blue), and CNC–organoclay–ibuprofen colloidal nanocomposite hydrogel (green).

3. Conclusions

In summary, we have demonstrated a simple methodology for construction of colloidal nanocomposite hydrogels by using aqueous sols of 1D cellulose nanocrystals and 2D delaminated sheets of aminopropyl-functionalized magnesium phyllosilicate clay. The results indicate that electrostatic interactions between the negatively charged cellulose nanocrystals and polycationic exfoliated sheets of the organoclay spontaneously generate a 3D interconnected matrix that encapsulates water molecules to yield self-supported hydrogels. We also show the colloidal CNC–organoclay nanocomposite hydrogels are responsive to external stimuli such that gas-induced changes in pH-triggered reversible changes in viscoelastic properties. Moreover, inclusion of ibuprofen in the colloidal nanocomposite hydrogels facilitates reassembly of the cationic organoclay sheets to produce mesolamellar domains comprising intercalated drug molecules. As a consequence, the hydrogel matrices show slower drug release rates. Thus, the CNC–organoclay colloidal nanocomposite hydrogels should offer an excellent opportunity to fabricate new types of stimuli-responsive hydrogels that are bioactive and biocompatible, and of potential use in applications such as wound dressing and consumer care products.

4. Experimental Section

4.1. Materials and Methods

All chemicals were obtained from Sigma-Aldrich (St. Louis, MO, USA) and used without purification. Rheology experiments were performed by using Malvern Kinexus Pro Rheometer (Malvern Instruments Ltd, Malvern, UK) equipped with parallel plate (diameter of 20 mm). Hydrogel samples aged for 1 day were added to the base plate to minimize shear. The top plate was then lowered to a set gap width of 150 μm, and the normal force was then measured; once the normal force reached an equilibrium, the measurements were performed at room temperature. CNC–organoclay hydrogel samples obtained from different batches showed similar rheological behavior and the data were within ±10% error. We ascribe the differences to changes in hydration levels and shear-induced effects during sample loading. Cross-polarized optical microscopy images were obtained by using Leica DMI3000 B manual inverted fluorescence microscope (Leica Microsystems Ltd., UK). Transmission electron microscopy analysis was undertaken on JEOL1400 operating (JEOL Ltd., Tokyo, Japan) at 120 keV in bright field mode. Energy dispersive X-ray analysis (EDX) analysis was performed by using Oxford Instrument Aztec microanalysis SDD detector (Oxford Instruments, Abingdon, UK) attached to JEOL 2100F-STEM microscope (JEOL Ltd.). CNC samples were prepared by mounting 5 μL of the 1 wt % CNC dispersions onto carbon-coated grids and left to dry at room temperature. Negative staining was carried out using an aqueous solution of 1% uranyl acetate. Delaminated sheets of organoclay were imaged by mounting freshly exfoliated dispersion of organoclay (0.01 mg/mL). TEM micrographs of hydrogel samples were obtained by dispersing 5–10 μL of gel samples in 500 μL distilled water prior to mounting onto carbon-coated grids. Zeta-potential measurements for aqueous dispersions of CNCs (1 wt %), freshly exfoliated organoclay (1 mg/mL), and hydrogels (20 μL of gel dispersed in 5 mL distilled water) were performed using Malvern Zetasizer Nano ZS (Malvern Instruments Ltd., Malvern, UK). Powder diffraction patterns were obtained by using Bruker Advanced D8 powder diffractometer (Bruker Corporation, Billerica, MA, USA). Controlled release experiments were performed by using Lambda-35 Perkin-Elmer UV–vis spectrophotometer (PerkinElmer Inc., Waltham, MA, USA).

4.2. Synthesis and Exfoliation of Aminopropyl-Functionalized Magnesium Phyllosilicate (Organoclay)

Magnesium chloride hexahydrate (0.84 g) was dissolved in 20 g ethanol. To this solution, 1.3 mL of 3-aminopropyltriethoxysilane was added dropwise with continuous stirring. The white slurry obtained after 5 min was stirred overnight and the precipitate was isolated by centrifugation, washed with ethanol (50 mL), and dried at 40 °C. A stable clear aqueous dispersions containing exfoliated organoclay layers (1–10 wt %) were prepared by dispersing desired amounts of organoclay powders in distilled water followed by ultrasonication for 5 min.

4.3. Synthesis of Crystalline Nanocellulose (CNC)

Fibrous cellulose powder (10.0 g, Whatman–CF11, Sigma-Aldrich Co.) was hydrolyzed by sulfuric acid (87.5 mL, 64%) for 40 min at 45 °C with continuous stirring. The hydrolysis was quenched by adding a large amount of water (500 mL) to the reaction mixture. The resulting mixture was cooled to room temperature 25 °C and centrifuged (4000 rpm) for 10 min at room temperature. The supernatant was decanted. Distilled water (500 mL) was added to the precipitate and the mixture was then stirred vigorously to form a new suspension. This centrifugation process was repeated three times. The newly generated suspension was dialyzed with an ultrafiltration membrane (30,000 molecular weight cutoff) until the pH of the suspension reached a constant value (pH ~6). Finally, the suspension was collected into glass vessel, then evaporated at 40 °C to obtain the desired concentration of CNC.

4.4. Fabrication of Colloidal CNC–Organoclay Nanocomposite Hydrogels

Typically, nanocrystalline cellulose-containing nanocomposite hydrogels were prepared by adding an aqueous solution of exfoliated organoclay at low concentrations (1–10 wt %) to a colloidal sol of CNC with low concentration (3 wt %). The mixture was allowed to stand for several minutes to obtain a self-supported CNC–organoclay hybrid hydrogel.

4.5. Synthesis of CNC–Organoclay–Ibuprofen Nanocomposite Hydrogels

Typically, 0.1 mL of a colloidal sol of CNC at 6 wt % concentration was added to 0.1 mL of an aqueous ibuprofen (10 wt %) solution with stirring. Freshly exfoliated organoclay (5 wt %, 0.016 mL) dispersions were added and the reaction mixture was allowed to stand to obtain the CNC–organoclay–ibuprofen hydrogel.

4.6. Synthesis of Organoclay–Ibuprofen Composites

Ibuprofen solution (0.1 mg/mL) was added to a freshly prepared exfoliated dispersion of organoclay (0.05 mg/mL). The resulting precipitates were centrifuged, dried, and pressed into pellets for drug-release experiments.

4.7. Controlled Release Studies

Organoclay–ibuprofen pellets or CNC–organoclay–ibuprofen hydrogels were immersed in a known volume of distilled water. Aliquots were removed at regular time intervals and time-dependent release of ibuprofen was measured by monitoring changes in the intensity of the absorption peak at 264 nm using UV–vis spectroscopy. Release profiles were plotted by averaging absorption values obtained over three runs.

Acknowledgments: Takumi Okamoto and Stephen Mann thank Denso Corporation for research funds and Avinash J. Patil is grateful to University of Bristol for financial support. Authors would like to thank J. Jones for assistance with TEM.

Author Contributions: Avinash J. Patil conceived and designed the experiments; Takumi Okamoto and Tomi Nissinen performed the experiments; Takumi Okamoto, Tomi Nissinen and Avinash J. Patil analyzed the data; Avinash J. Patil and Stephen Mann wrote the paper.

References

1. Loh, X.J.; Scherman, O.A. *Polymeric and Self-Assembled Hydrogels: From Fundamental Understanding to Applications*; RSC Publishing: Cambridge, UK, 2013.
2. Haraguchi, K. Nanocomposite, Hydrogels. *Curr. Opin. Solid State Mater. Sci.* **2007**, *11*, 47–54. [CrossRef]
3. Jonker, A.M.; Lowik, D.W.P.M.; van Hest, J.C.M. Peptide and Protein-Based Hydrogels. *Chem. Mater.* **2012**, *24*, 759–773. [CrossRef]
4. Kopecek, J.; Yang, J. Smart Self-Assembled Hybrid Hydrogel Biomaterials. *Angew. Chem. Int. Ed.* **2012**, *51*, 7396–7417. [CrossRef] [PubMed]
5. Raeburn, J.; Cardoso, A.Z.; Adams, D.J. The Importance of the Self-Assembly Process to Control Mechanical Properties of Low Molecular Weight Hydrogels. *Chem. Soc. Rev.* **2013**, *42*, 5143–5156. [CrossRef] [PubMed]
6. Du, X.; Zhou, J.; Shi, J.; Xu, B. Supramolecular Hydrogelators and Hydrogels: From Soft Matter to Molecular Biomaterials. *Chem. Rev.* **2015**, *115*, 13165–13307. [CrossRef] [PubMed]

7. Gaharwar, A.K.; Peppas, N.A.; Khademhosseini, A. Nanocomposite Hydrogels for Biomedical Applications. *Biotechnol. Bioeng.* **2014**, *111*, 441–453. [CrossRef] [PubMed]
8. Merino, S.; Martin, C.; Kostarelos, K.; Prato, M.; Vazquez, E. Nanocomposite Hydrogels: 3D Polymer-Nanoparticle Synergies for On-Demand Drug Delivery. *ACS Nano* **2015**, *9*, 4686–4697. [CrossRef] [PubMed]
9. Shin, S.R.; Jung, S.M.; Zalabany, M.; Kim, K.; Zorlutuna, P.; Kim, S.B.; Nikkhah, M.; Khabiry, M.; Azize, M.; Kong, J.; et al. Carbon Nanotube-Embedded Hydrogel Sheets for Engineering Cardiac Constructs and Bioactuators. *ACS Nano* **2013**, *7*, 2369–2380. [CrossRef] [PubMed]
10. Giri, A.; Bhowmick, M.; Pal, S.; Bandopadhyay, A. Polymer Hydrogel from Caboxymethyl Guar Gum and Carbon Nanotube for Sustained Trans-dermal Release of Diclofenac Sodium. *Int. J. Biol. Macromol.* **2011**, *49*, 885–893. [CrossRef] [PubMed]
11. Goenka, S.; Sant, V.; Sant, S. Graphene-based Nanomaterials for Drug Delivery and Tissue Engineering. *J. Control Release* **2014**, *173*, 75–88. [CrossRef] [PubMed]
12. Zhong, S.P.; Yung, L.Y.L. Enhanced Biological Stability of Collagen with Incorporation of PAMAM Dendrimer. *J. Biomed. Mater. Res. A* **2009**, *91A*, 114–122. [CrossRef] [PubMed]
13. Appel, E.A.; Tibbitt, M.W.; Webber, M.J.; Mattix, B.A.; Veiseh, O.; Langer, R. Self-Assembled Hydrogels Utilizing Polymer-Nanoparticle Interactions. *Nat. Commun.* **2015**. [CrossRef] [PubMed]
14. Schexnailder, P.; Schmidt, G. Nanocomposite Polymer Hydrogels Colloid. *Polym. Sci.* **2009**, *287*, 1–11.
15. Wang, Q.; Mynar, J.L.; Yoshida, M.; Lee, E.; Lee, M.; Okura, K.; Kinbara, K.; Aida, T. High-Water content Mouldable Hydrogels by Mixing Clay and a Dendritic Molecular Binder. *Nature* **2010**, *463*, 339–343. [CrossRef] [PubMed]
16. Martin, J.E.; Patil, A.J.; Butler, M.F.; Mann, S. Guest-Molecule Directed Assembly of Mesostructured Nanocomposite Polymer/Organoclay Hydrogels. *Adv. Funct. Mater.* **2011**, *21*, 674–681. [CrossRef]
17. Liu, T.Y.; Hu, S.H.; Liu, D.M.; Chen, S.Y. Magnetic-sensitive Behaviour of Intelligent Ferrogels for Controlled Release of Drug. *Langmuir* **2006**, *22*, 5974–5978. [CrossRef] [PubMed]
18. Dvir, T.; Timko, B.P.; Brigham, M.D.; Naik, S.R.; Karajangai, S.S.; Levy, O.; Jin, H.; Parker, K.K.; Langer, R.; Kohane, D.S. Nanowired Three-Dimensional Cardiac Patches. *Nat. Nanotechnol.* **2011**, *6*, 720–725. [CrossRef] [PubMed]
19. Liu, Y.; Ma, W.; Liu, W.; Li, C.; Liu, Y.; Jiang, X.; Tang, Z. Silver (I)-Glutathione Biocoordination Polymer Hydrogel: Effective Antibacterial Activity and Improved Cytocompatibility. *J. Mater. Chem.* **2011**, *21*, 19214–19218. [CrossRef]
20. Thoniyot, P.; Tan, M.J.; Karim, A.A.; Young, D.J.; Loh, X.J. Nanoparticle-Hydrogel Composites: Concept, Design, and Applications of These Promising, Multi-Functional Materials. *Adv. Sci.* **2015**. [CrossRef] [PubMed]
21. Dufresne, A. Comparing the Mechanical Properties of High Performance Polymer Nancomposites from Biological Resources. *J. Nanosci. Nanotechnol.* **2006**, *6*, 322–330. [CrossRef] [PubMed]
22. Thomas, S. A Review of the Physical Biological and Clinical Properties of a Bacterial Cellulose Wound. *J. Wound Care* **2008**, *17*, 349–352. [CrossRef] [PubMed]
23. Shopsowitz, K.E.; Qi, H.; Hamad, W.Y.; MacLachlan, M.J. Free-Standing Mesoporous Silica Films with Tuneable Chiral Nematic Structures. *Nature* **2010**, *468*, 422–425. [CrossRef] [PubMed]
24. Shopsowitz, K.E.; Hamad, W.Y.; MacLachlan, M.J. Chiral Nematic Mesoporous Carbon Derived from Nanocrystalline Cellulose. *Angew. Chem. Int. Ed.* **2011**, *50*, 10991–10995. [CrossRef] [PubMed]
25. Shopsowitz, K.E.; Stahl, A.; Hamad, W.Y.; MacLachlan, M.J. Hard Templating of Nanocrystalline Titanium Dioxide with Chiral Nematic Ordering. *Angew. Chem. Int. Ed.* **2012**, *51*, 6886–6890. [CrossRef] [PubMed]
26. Dujardin, E.; Blaseby, M.; Mann, S. Synthesis of Mesoporous Silica by Sol-Gel Mineralisation of Cellulose Nanorod Nematic Suspensions. *J. Mater. Chem.* **2003**, *13*, 696–699. [CrossRef]
27. Patil, A.J.; Muthusamy, E.; Mann, S. Synthesis and Self-Assembly of Organoclay-Wrapped Biomolecul. *Angew. Chem. Int. Ed.* **2004**, *43*, 4928–4933. [CrossRef] [PubMed]

3

Hydrogels for Biomedical Applications: Cellulose, Chitosan and Protein/Peptide Derivatives

Luís J. del Valle, Angélica Díaz and Jordi Puiggalí *

Barcelona Research Center for Multiscale Science and Engineering, Universitat Politècnica de Catalunya, Escola d'Enginyeria de Barcelona Est-EEBE, c/Eduard Maristany 10-14, Barcelona 08019, Spain; luis.javier.del.valle@upc.edu (L.J.d.V.); angelicadiaz07@gmail.com (A.D.)

* Correspondence: Jordi.Puiggali@upc.edu

Abstract: Hydrogels based on polysaccharide and protein natural polymers are of great interest in biomedical applications and more specifically for tissue regeneration and drug delivery. Cellulose, chitosan (a chitin derivative), and collagen are probably the most important components since they are the most abundant natural polymers on earth (cellulose and chitin) and in the human body (collagen). Peptides also merit attention because their self-assembling properties mimic the proteins that are present in the extracellular matrix. The present review is mainly focused on explaining the recent advances on hydrogels derived from the indicated polymers or their combinations. Attention has also been paid to the development of hydrogels for innovative biomedical uses. Therefore, smart materials displaying stimuli responsiveness and having shape memory properties are considered. The use of micro- and nanogels for drug delivery applications is also discussed, as well as the high potential of protein-based hydrogels in the production of bioactive matrices with recognition ability (molecular imprinting). Finally, mention is also given to the development of 3D bioprinting technologies.

Keywords: cellulose; chitosan; collagen; gelatin; peptides; self-assembling; nanogels; shape memory; molecularly imprinting; 3D printing

1. Introduction

Hydrogels are three-dimensional polymer matrices able to retain large amounts of water in a swollen state, a feature that makes them similar to biological tissues. In fact, biomedical applications of hydrogels have been explored continuously since the 1960s, when they were first discovered [1]. Physical and chemical crosslinks are fundamental to building a hydrophilic network in which chemical agents can also be incorporated, giving rise to drug delivery systems and even to new functional materials. In fact, research on hydrogels is nowadays mainly focused on developing stimuli-responsive smart materials and hydrogels with shape memory properties that could be applied for innovative biomedical uses [2–5]. Different reviews have addressed natural and synthetic formulations as well as the corresponding general applications [6–9].

Great effort is directed towards designing molecules capable of promoting molecular aggregation (gelators) [10,11]. The incorporation of units able to establish one-dimensional hydrogen bonding interactions such as amides and saccharides merits attention. In fact, derived low molecular weight compounds can be heated in an appropriate solvent to form a supersaturated solution, which after cooling to room temperature can give rise to a gel through an aggregation process [12–14]. Small molecules can be organized into polymer-like fibers that become entangled and constitute a continuous matrix that entraps the solvent by surface tension. A heterogeneous "solid" matrix is obtained with a hierarchical superstructure aggregation that creates dimensions from the nanometric to the micrometric scale.

Development of in situ gelling polymeric matrices is also of great interest in tissue regeneration since these materials can be used as injectable hydrogels. These can act as cell vehicles that have the ability to take the shape of the corresponding tissue cavity. Furthermore, problems related to cell adhesion can be minimized since cells can directly be incorporated into the injectable solution [15,16]. A suitable in situ hydrogel for biomedical applications should be soluble in aqueous media and have a fast sol-gel transition under physiological conditions without releasing toxic byproducts or harming surrounding tissue [17]. In general, injectable hydrogels are designed with functional groups sensitive to external stimuli such as pH, temperature, and light [18].

The present review is basically focused on the study of hydrogels for biomedical applications, mainly restricted to cellulose and chitosan (a chitin derivative) as the most abundant natural polymers on earth and collagen as the most abundant protein in the human body. Attention is also paid to the use of peptides due to their abovementioned self-assembling properties. Different subsections also introduce the most relevant topics that can nowadays be considered: (a) responsiveness of hydrogels to external stimuli; (b) development of micro and nanogels (i.e., micro or nanoscopic three-dimensional networks comprising cross-linked polymer molecules dispersed in a proper solvent); (c) shape memory hydrogels; (d) molecularly imprinted hydrogels and (e) protein-based hydrogels for 3D printing.

2. Hydrogels Derived from Cellulose

Cellulose is the most abundant biopolymer, mainly as a consequence of its properties that make it an essential structural component of green plants, marine animals, algae, and bacteria. Sustainability, biodegradability, biocompatibility, and low cytotoxicity are other characteristics that justify the development of a great number of applications in the biomedical field that concern cellulosic materials. From a chemical point of view, cellulose is defined by the connection through β-(1→4) glycosidic bonds between D-glucose units (Figure 1), which gives rise to a linear syndiotactic polymer with hydroxyl groups arranged in an equatorial disposition. The molecular chain can be visualized as a stiff rod-like conformation that can be arranged, giving rise to crystalline fibrous materials. Strong intra- and intermolecular hydrogen bonding interactions can be found, with different crystalline structures of cellulose reported depending on its origin (e.g., cellulose I_α and cellulose I_β for polymers produced by bacteria and plants, respectively) and chemical treatments (e.g., cellulose II for regenerated fibers).

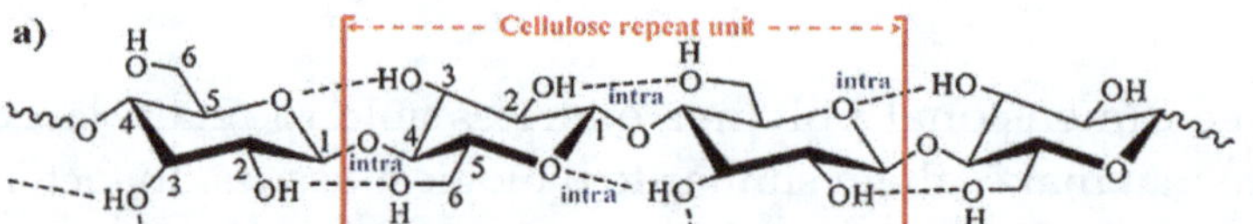

Figure 1. (**a**) Scheme of the linear molecular chain (green box), the syndiotactic repeat unit (garnet), the establishment of glycosylic bonds between glucose rings (violet ellipsoid) and intra and intermolecular hydrogen bonding interactions; (**b**) TEM micrograph of cellulose nanowhiskers (left), scheme and SEM micrograph of nanofibers derived from a fiber of cellulose (middle) and TEM micrograph of bacterial cellulose (right).

Cellulose can be employed to produce hydrogels for biomedical applications according to two differentiated methodologies: (a) the use of cellulose-based matrices and (b) the use of composites incorporating nanocellulose.

Cellulose based-hydrogel matrices are nowadays ideal materials for tissue engineering applications [21] due to intrinsic properties like non-toxicity, biocompatibility, tunable and porous microstructure, and good mechanical properties [22]. Applications of cellulose derivatives can be enhanced when blends or hybrids with other components like chitosan are considered [23].

Nevertheless, cellulose-based hydrogels have as a main limitation the low solubility of cellulose in both water and most organic solvents due to the hydrogen-bonded structure [24]. This problem can be avoided via chemical modification and specifically by the conversion of the hydroxyl pendant groups into ether and cationic groups, although it should be taken into account that final properties can be negatively altered. Research has also been focused on directly dissolving cellulose in appropriate non-toxic solvents. In this sense, alkali/urea aqueous systems appear highly promising [25,26] since inclusion complexes can be formed at low temperatures, which demonstrates the possibility of preparing membranes and hydrogels from these media [27,28].

Hydrogels derived from cellulose can be prepared by the crosslinking of aqueous solutions of cellulose ethers (e.g., methylcellulose (MC), ethylcellulose (EC), sodium carboxymethylcellulose (NaCMC), or hydroxypropyl methylcellulose (HPMC)). MC is ideal for the preparation of thermoresponsive hydrogels due to its hydrophobic–hydrophilic equilibrium, which gives rise to a collapse or an expansion of molecular chains by small temperature changes around its critical value [29].

Nanocellulose is a general term that defines a nanostructured material that comprises cellulose nanocrystals (CNC), cellulose nanofibers (NFC), and bacterial cellulose (BC). Nanocrystals with a whisker morphology (Figure 1b) can be easily produced by the treatment of cellulose with strong acids [30] that cause the degradation of amorphous regions and produce whisker-like crystals to be used as fillers in bio-based matrices [31]. Cellulose nanofibers (NFC) (Figure 1b) are produced by mechanical treatments of natural fibers (e.g., high pressure and ultrasonic homogenization, grinding, and microfluidization). Clinical applications of NFCs have been justified considering their cytocompatibility and also the tolerogenic potential in the immune system [32]. Finally, bacterial or microbial cellulose is directly produced by bacteria (e.g., *Acetobacter* strain) from glucose residues. The polymerized material is secreted from the cell and crystallizes, giving rise to nanofibers with diameters smaller than 100 nm (Figure 1b) [33]. These nanofibers can be connected, forming a 3D networked structure.

CNCs can display interesting effects on the gelation mechanism of hydrogels and can improve mechanical properties and dimensional stability and even favor the drug release [34–36]. CNCs have also been employed as nanofillers to improve, for example, the compression modulus (up to 92 kPa for a load of 20 wt %) of hybrid hydrogels based on gelatin and alginate [37]. Hydrogels composed of an interpenetrating network of sodium alginate and gelatin reinforced with 50 wt % of CNCs have also been developed for cartilage applications. Specifically, modulus, strength, and strain values of 0.5 GPa, 14.4 MPa, and 15.2%, respectively, were attained, with the modulus clearly higher than that determined for natural cartilage. The double cross-linked system was prepared by a freeze-drying process, with the carboxyl surface groups of CNC contributing to the enhancement of mechanical properties and structural stability [38].

Surface modification of CNCs can improve their performance since they cannot only be used as fillers but also as cross-linking agents that increase the adhesion of the filler with the polymeric matrix [39]. Thus, aldehyde-functionalized CNCs have been employed as cross-linkers for carboxymethylcellulose hydrogels [40]. Maleimide-functionalized CNCs have also been proposed as effective cross-linkers for the formation of hydrogels based on gelatin and chondroitin sulfate, which are stiffer networks with lower swelling ratios [41].

NFC hydrogels can be prepared from suspensions of NFC fibrils produced by a simple mechanical treatment, but nowadays better results have been described when fibrils with a high negative charge on their surface are employed. Thus, oxidation processes such as the treatment with 2,2,6,6-tetramethylpiperidine-1-oxyl radical (TEMPO) produced more stable dispersions/suspensions than hydrogels formed under limiting values of polymer concentration and charge density [42,43]. Great efforts have also been developed to control the porosity and network microstructure to ensure the transportation of nutrients and waste products when tissue engineering applications are considered [44]. Thus, hydrogels with controllable swelling degree, porosity, and surface area can be prepared by tuning the charge density of fibers, the conditions of the swelling media, and the processing methodology [45]. In general, NFC hydrogels have low mechanical properties (e.g., a storage modulus close to 10 Pa was reported for a native NFC hydrogel at 0.5 wt % [46]) and therefore applications are usually limited to soft tissues [47,48]. The use of reinforcing agents has been proposed in order to improve mechanical properties since they can easily be incorporated into NFC hydrogels due to their high degree of swelling. Furthermore, the addition of molecules such as hemicellulose can provide additional benefits like anticancer and antioxidative properties [49,50]. Different hemicelluloses (e.g., galactoglucomannan, xyloglucan, and xylan) have been studied as crosslinkers to tune the structural and mechanical properties of NFC hydrogels, as well as to study their effect on cell behavior (adhesion, growth, and proliferation) during wound healing processes [51]. Nanocellulose charge density was found to be a determining factor for the incorporation of hemicellulose, the derived surface (topography and roughness) and the mechanical and biological properties of the composite hydrogels.

BCs are currently used as promising hydrogels for the development of functional nano-biocomposites [52–54]. Furthermore, derived materials can display appropriate mechanical properties when being used as membranes. Composite systems with collagen were, for example, evaluated in order to promote cell adhesion and viability. Specifically, BC membrane surfaces were coated with collagen and alginate on each side to favor cell adhesion and protect transplanted cells from immune rejection, respectively [55]. It was also demonstrated that cells were able to release dopamine through the BC composite membrane, a promising feature for its use as a material for cell encapsulation.

The high purity and hydrophilicity of bacterial cellulose make it a promising material for wound-healing applications [56], with different products already commercialized for dressings (e.g., XCell, Biofill, or Dermafill) [57,58].

3. Hydrogels Derived from Chitosan

Chitin (poly-(1→4)-*N*-acetyl-glucosamine) is one of the most abundant natural polymers since its facility to produce microfibrils makes it an essential structural component of cell walls (e.g., fungi and yeast) and of the exoskeleton of many invertebrates (e.g., shrimps and crabs). This polysaccharide is characterized, like cellulose, by a β-(1→4) glycosidic bond and can be transformed into the water-soluble chitosan (CS) upon deacetylation in strong alkaline solutions. The high availability of chitosan, its biodegradability and biocompatibility have enhanced interest in its use as a hydrogel with improved structural stability and high capability to absorb water. The physical properties of chitosan can be controlled by changing the molecular weight of the precursor, the degree of depolymerization, and deacetylation, and finally by modifying the interactions that can be established with both hydroxyl and amine groups that are present in the molecular backbone. The cationic character of CS favors the formation of gel particles through electrostatic interactions [59,60], with, e.g., sodium sulfate employed as a precipitant [61]. CS molecules can also interact with hydrophobic components, giving rise to amphiphilic particles with great self-assembly and encapsulation ability. Interactions established between chitosan and drugs should be appropriate to produce the expected pharmacological effect at the target site.

Chitin has also been extensively considered for tissue engineering and drug delivery applications; specifically, it has been processed in the form of hydrogels and nanogels despite the disadvantages associated with its water insolubility [62,63]. Nevertheless, it should be taken into account that biological properties (e.g., antimicrobial, hemostatic, or mucoadhesion, effects) decrease with the degree of acetylation. Several attempts have been proposed to get three-dimensional sponge-like materials from chitin [64,65]. These scaffolds favor the deposition of the extracellular matrix on chondrocytes and have applications in cartilage tissue engineering [66]. Preparation of chitin hydrogels is feasible by employing a mild medium based on a supersaturated $CaCl_2$–methanol solution [67]. Hydrophilic chains are able to aggregate through intermolecular interactions and form a network-like structure [68]. Chitin nanogels have been employed to release anti-cancer drugs (e.g., doxorubicin [69]), anti-fungal drugs (e.g., fluconazole [70]), and proteins (e.g., bovine serum albumine (BSA) [71]) [72].

The mucoadhesion property of chitosan gels has enhanced interest in using them as drug delivery systems since the bioavailability of loaded bioactive drugs can be increased [59]. Furthermore, CS hydrogels can be processed in the form of micro-/nano-sized spherical particles or beads where bioactive compounds can be encapsulated. These beads swell in acidic media, facilitating drug release. An additional advantage of chitosan beads is their relative high hydrophobicity, which facilitates intestinal absorption [73]. Delivery systems based on chitosan have been developed for the treatment of colon [74,75] and hepatic [76,77] diseases. Several hydrogels based on chitosan have also been designed to encapsulate radioisotope drugs for site-specific cancer therapy. These chitosan hydrogels can display photo-responsiveness and a thermoreversible gelling capacity [78,79].

Hydrogels composed of chitosan have the capacity to adsorb both anionic and cationic molecules if hydrogen bond interactions can still be established. This feature confers on CS-based hydrogels a great potential in fields as diverse as water purification and protein encapsulation [80].

Most injectable hydrogels for biomedical applications are based on CS, it being possible to modify the composition in order to undertake chemical or physical gelling by UV irradiation or the increase of temperature or pH [81,82]. Probably, the main systems studied are those based on the addition of a glycerophosphate salt that initiated the sol-gel transition at body temperature [83] and also the additional incorporation of genipin as a crosslinking agent [84]. Incorporation of hydroxyapatite has also been revealed to be useful for bone tissue regeneration since it enhanced cell adhesion and proliferation and improved osteogenic properties [85,86]. A pH-responsive CS-based injectable hydrogel incorporating hydroxyapatite has been prepared using sodium bicarbonate ($NaHCO_3$) as the gelling agent (Figure 2). This system provided a neutral environment suitable for cell encapsulation and allowed non-cytotoxic, fast gelation (e.g., 4 min). Physical crosslinks were the consequence of glucosamine deprotonation and produced materials with good resistance to applied shear deformation [87].

Quaternized chitosan-*g*-polyaniline copolymers have been synthesized [88] in order to enhance the antibacterial activity and cytocompatibility of CS. The grafted copolymer was postulated as an idoneous injectable hydrogel dressing [89] due to its good biocompatibility, bactericidal effect, conductivity, and good free radical scavenging capacity [90]. Specifically, injectable hydrogel dressings were prepared at physiological conditions by mixing solutions of quaternized chitosan-*g*-polyaniline with a poly(ethylene glycol)-*co*-poly(glycerol sebacate) copolymer having benzaldehyde functional groups (PEGS-FA) (Figure 3). Soft and flexible hydrogels were derived as a consequence of the chain flexibility given by PEGS-FA and the dynamic network of chemical bonds. The final conductivity was a result of ionic conductivity from amino groups and ammonium groups, and electronic conductivity from doped polyaniline. Conductivity varied between 3.13 mS/cm to 2.25 mS/cm as the crosslinker concentration increased from 0.5 to 2 wt %. A crosslinker concentration of 1.5 wt % was found to be optimal for enhancing blood clotting capacity and the in vivo wound healing process.

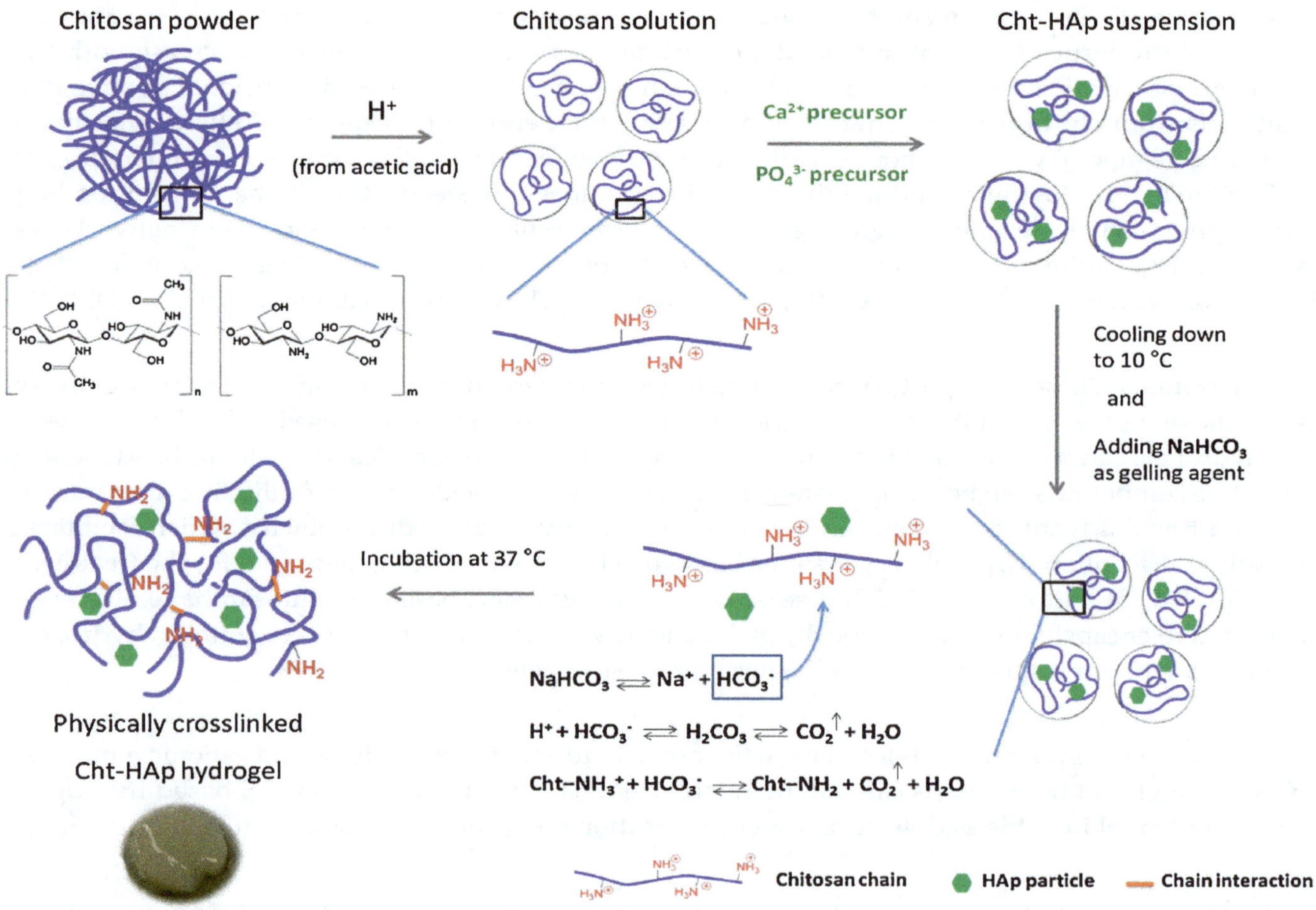

Figure 2. Preparation of a physically crosslinkinked injectable hydrogel based on chitosan and hydroxyapatite.

pH-Sensitive CS hydrogels reinforced with CNCs were prepared using glutaraldehyde as a crosslinker due to its high reactivity with chitosan amine groups [91]. CNCs were incorporated in the preformed polymer network, giving rise to hydrogels characterized by a combination of amorphous and crystalline phases and the increase of compression modulus from 25.9 ± 1 to 50.8 ± 3 kPa when the nanocellulose content was 2.5 wt %. The maximum swelling ratio was found for a pH of 4.01, where chitosan amine groups were protonated and hydrogen bonds became consequently dissociated. The CS/CNC hydrogel was also interesting for drug release and specifically for the delivery of curcumin [92].

Blending cellulose with chitosan may give rise to hydrogels with improved properties, but it is problematic to get a homogeneous aqueous solution of both components since they need alkaline (cellulose) and acidic (chitosan) aqueous media. Nevertheless, a water-soluble chitosan derivative (i.e., hydroxyethyl chitosan) has successfully been employed to make porous scaffolds using silicon dioxide particles as a porogen and following a freeze-drying process [93]. Good overall performance and an ability to reach the equilibrium swelling state were observed.

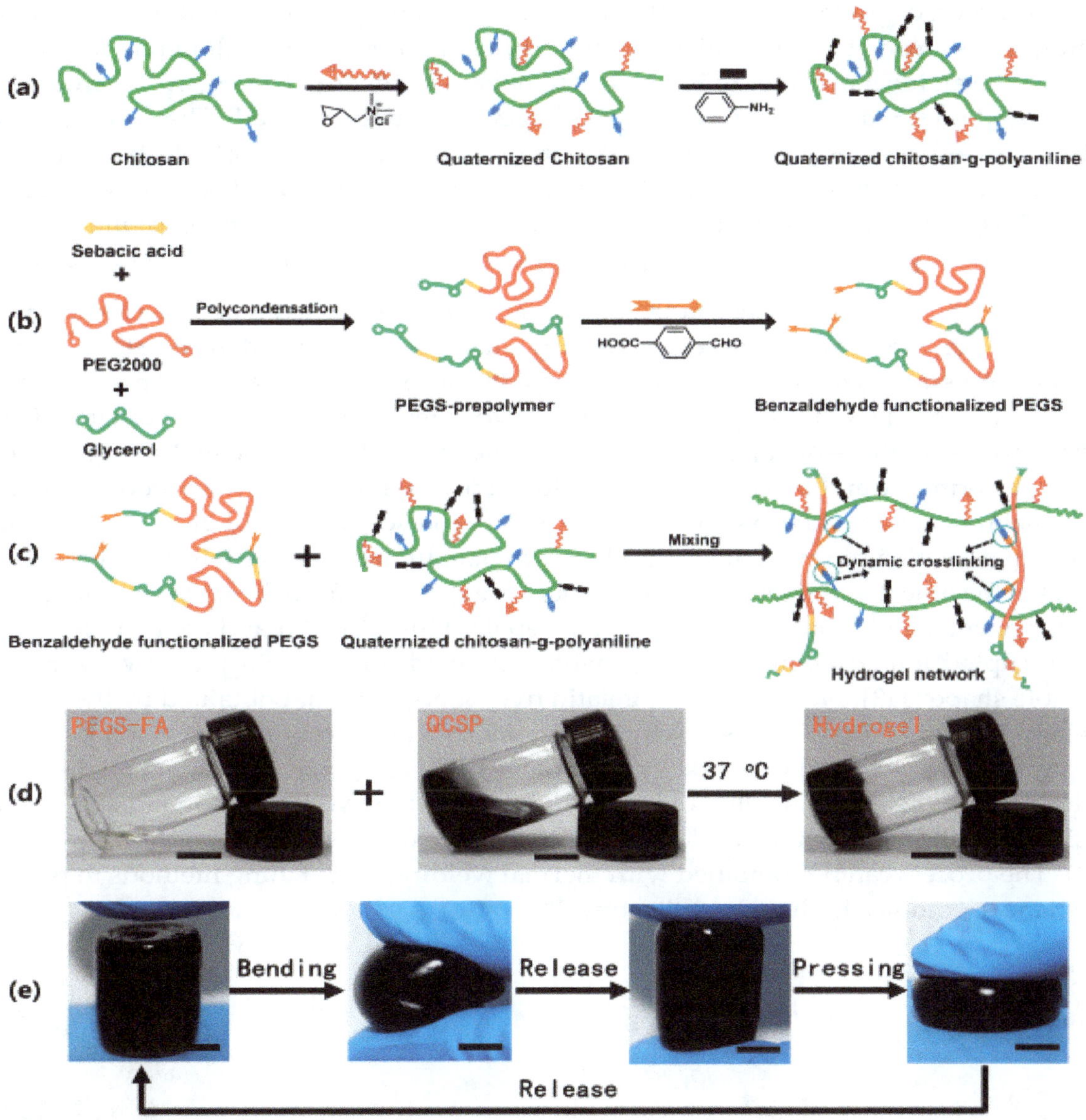

Figure 3. Scheme showing the synthesis of chitosan-*g*-aniline (**a**), PEGS-FA copolymers (**b**) and the structure of the hydrogel derived from both copolymers (**c**). Photographs showing the corresponding solutions (**d**) and flexible behavior of the hydrogel under bending and pressing efforts (**e**). Reproduced with permission from [89].

4. Hydrogels Derived from Collagen and Gelatin

Collagen is the most important protein that forms part of the extracellular matrices. It can be obtained from skin and other tissues by enzymatic and acidic treatments. A hydrogel can be produced after neutralization of the acid solution and subsequent heating to body temperature. Gelatin is derived when the typical triple helix of collagen is broken into a single molecule that could undergo a reversible sol-gel transition at room temperature [94]. Gelatin exhibits biocompatibility but its applications are hindered by its low mechanical properties, which usually make the establishment of additional crosslinks necessary. Non-toxic enzymes (e.g., transglutaminase and tyrosinase) are usually preferred [95,96] as new cross-linking agents. Suitable hydrogels, mainly used as vehicles for cell transplantation (e.g., mesenchymal stem and stromal cells), can therefore be prepared [96,97].

Multiple studies have demonstrated that collagen can play a highly positive role in tissue regeneration [98–100], but its use in biomedical applications is somehow limited by its poor mechanical properties and high degradation rate [101]. Therefore, blending collagen with other polymers has been postulated as an alternative solution to improve final performance [102]. In this way, protonated CS appears to be an ideal polymer to interact with negatively charged collagen. Thermoresponsive hydrogels based on CS and different types of collagen have been studied. They have potential for biomedical uses as matrices for the encapsulation of cells, repair of bone defects, and promotion of in vivo cell differentiation [103–105]. Hydrogels containing CS, acid-soluble collagen (ASC), and glycerophosphates have recently been prepared [106] for tissue regeneration, having demonstrated good biocompatibility and the ability to support the survival and proliferation of encapsulated cells.

High-strength hydrogels based on BC and gelatin and with good biocompatibility have been studied [107]. The preparation process was difficult since it involved three steps, and it was consequently necessary to develop similar systems with easier processing than the two-step method applied for dual-crosslinked chitin and cellulose hydrogels [108,109]. Development of multiple crosslinked structures therefore appears to be a suitable option for producing hydrogels with good mechanical performance and biocompatibility. Efforts are also focused on new processes that ensure shape designability and good formability. Namely, we try to avoid the use of injectable hydrogels that require the modification of raw materials, which has usually led to materials with low mechanical properties. On the other hand, tubular hydrogels hold great interest for delivery applications, exchange channels for oxygen and nutrients, and vascular repair [110–112]. Wu and collaborators proposed a mild interrupted ion crosslinked process, which allowed for obtaining hollow hydrogels with a controllable shape [113]. Non-stable CS/gelatin hydrogels were first obtained by the aggregation of the gelatin helix domains, but a subsequent treatment with a sodium citrate solution produced a stable, physically crosslinked network. The efficiency of this step was dependent on the immersing time in the solution (i.e., the capability of ions to diffuse into the hydrogel). Therefore, at low times it was possible to melt/dissolve the non-ionic crosslinked core by exposure to deionized water at 37 °C (Figure 4). The process can be combined with thermal welding and etching methods to program the external shape of complex hydrogel architectures.

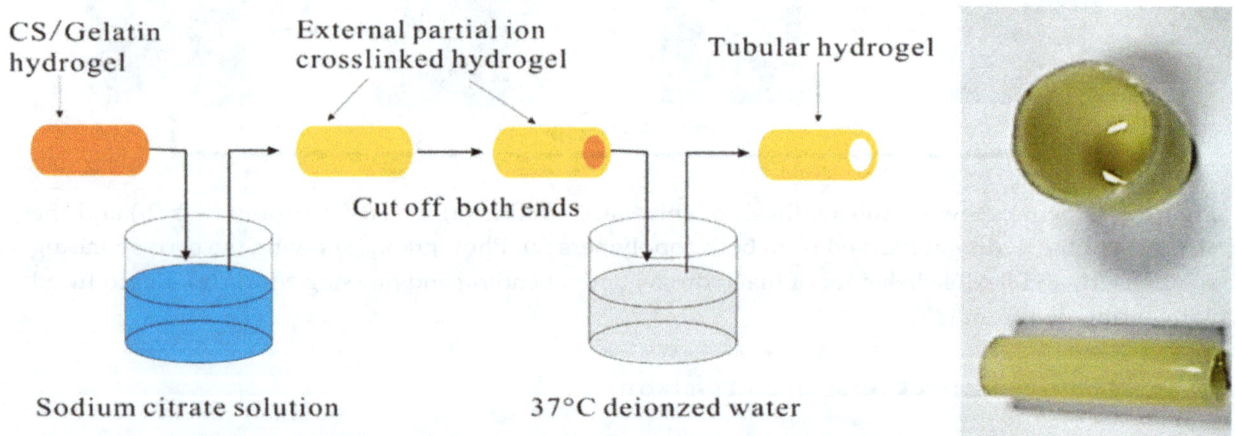

Figure 4. Preparation of hollow structures (e.g., cup and tube) from CS/gelatin hydrogels based on a controllable ion crosslinking process. Reproduced with permission from [113].

Development of conductive hydrogels is highly interesting for cardiac regeneration and repair. Natural hydrogels derived from collagen and gelatin [114,115] have usually been considered to support cardiac cell functions despite having an insulating character. This shortcoming could be avoided by using electrically conductive nanomaterials. Gold nanostructures have several advantages like high conductivity, easy modification and fabrication, minimum cytotoxicity, and varied architecture (e.g., nanowires, nanorods, and nanoparticles). In any case, hydrogel matrices should be designed to encourage good cell adhesion, a feature that can be attained with gelatin-based hydrogels. Specifically, hybrid hydrogels composed of a UV-crosslinkable gelatin methacrylate and incorporating gold nanorods have recently been found to be appropriate for cardiac tissue engineering [116].

Carbon nanotubes (CNTs) [117] are also considered for enhancing conductivity, although several limitations concerning cytotoxicity and a complex fabrication procedure have been indicated [118,119]. In any case, considerable research has been carried out to develop conductive hydrogels based on CNTs. For example, the bulk electrical properties of collagen can be increased by the addition of single-walled carbon nanotubes (SWCNT). These clearly influenced neurite extension and enhanced neurite outgrowth, leading to suitable hydrogels for nerve regeneration [120].

5. Peptide Hydrogels

Natural fibrillar proteins of the extracellular matrix (ECM) can be mimicked by the self-assembling of fully synthetic peptides (SAPs). A porous network having cell-binding sites or functional motifs can be formed and used to induce the growth and differentiation of host cells or alternatively as carriers for transplanted cells. Peptidic sequences can be assembled, giving rise to a variety of morphologies (e.g., nanofibers, nanotubes, nanovesicles, or nanoparticles), it being possible in some cases to trigger the assembly by modifying pH or temperature or by the presence of external cations. Different types of self-assembly can be considered (Figure 5): (a) alternate disposition of charged hydrophilic and hydrophobic residues (e.g., peptides based on the Arg-Ala-Asp-Ala sequence named RADA-like SAPs); (b) complementary co-assembling peptides (CAPs); (c) peptide amphiphiles; (d) cyclic peptides; and (e) functionalized peptides.

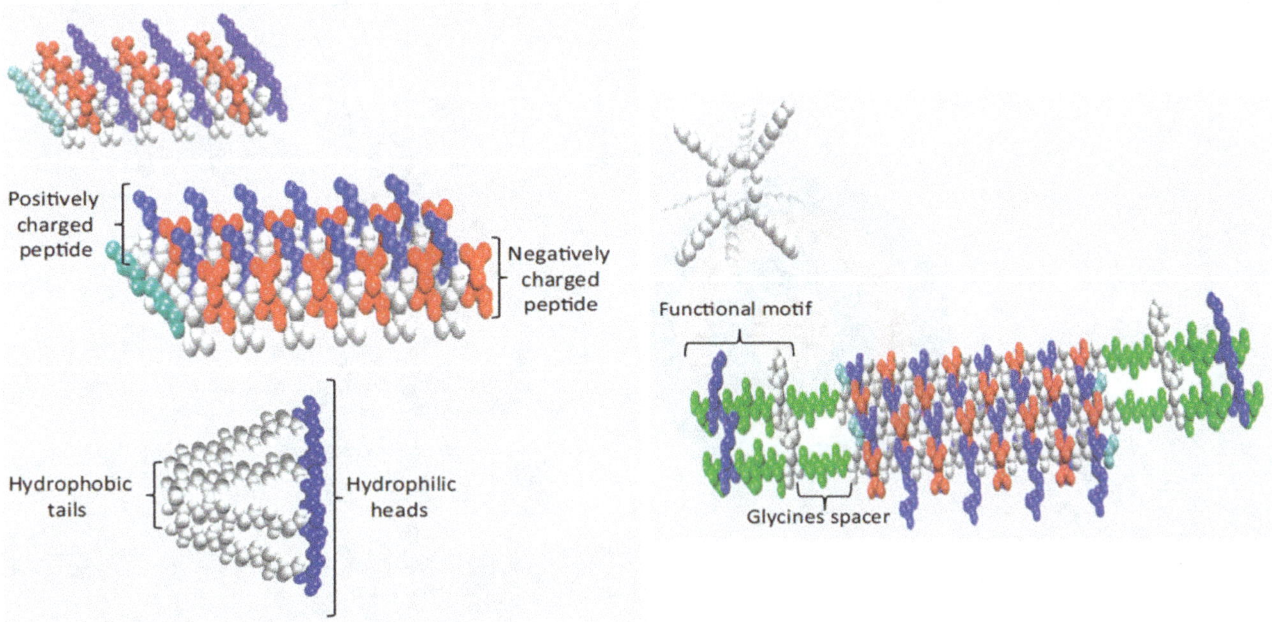

Figure 5. Typical structures of different self-assembled peptides: RADA-like SAPs, complementary coassembling peptides, peptide amphiphiles, cyclo-SAPs, and functionalized SAPs. Reproduced with permission from [121].

Organized stable β-sheet superstructures are characteristic of RADA and similar sequences. Specifically, Ac-(RARADADA)$_2$-CONH$_2$ [122] and Ac-(RADA)$_4$-CONH$_2$ [123] are composed of 16 amino acids. They form structures in aqueous media with charged amino acid side groups oriented on one side of the sheet and the hydrophobic side groups on the other side (i.e., a hydrophobic inner pocket was derived). The resultant scaffolds are appropriate for 3D cell culture, wound healing, and synapse growth. In fact, the indicated sequence has great similarity with the well-known RGD cell adhesion motif.

CAPs are based on the attraction and self-repulsion of two peptides having opposite electric charges [124]. For example, mixing of positive Ac-(LKLH)$_3$-CONH$_2$ and negative (Ac-(LDLD)$_3$CONH$_2$ spontaneously formed double layers of β-sheets and gave rise to 3D molecular assemblies composed of nanofibrillar structures, as is also typical for the above indicated SAPs. Derived hydrogels from CAPs can retain 95–99% of water and had pores between 5 and 200 nm [125]. CAPs avoid the use of pH shift as a triggering stimulus, which may be problematic in some biological conditions.

Peptide amphiphiles are similar to the phospholipids existing in membranes since they are composed of hydrophobic tails and hydrophilic heads. The former are based on nonpolar amino acids with different degrees of hydrophobicity, while the latter can be based on positively or alternatively negatively charged amino acids.

Self-assembled tubular structures are usually derived from the stacking of cyclic peptides. These structures are stabilized by hydrogen bonds that are established between carbonyl and amide groups. These interactions are directed perpendicularly to the ring, while the amino acid side groups are directed outward [126].

Amino acids and peptides may contain hydrophobic groups, which may provide hydrophobic interactions and create a synergistic effect with the more hydrophilic hydrogen bonding interactions. This point is illustrated in Figure 6, where a hydrogelator with good self-assembling ability was prepared by coupling three amino acids with hydrophobic side chains to a hydrophobic cyclohexane molecule. One-dimensional hydrogen-bonded stacks were derived, also taking advantage of the hydrophobic interactions established between the central cores and the shielding effect of the side chains over potential interactions between the amides and water molecules of the solvent [127].

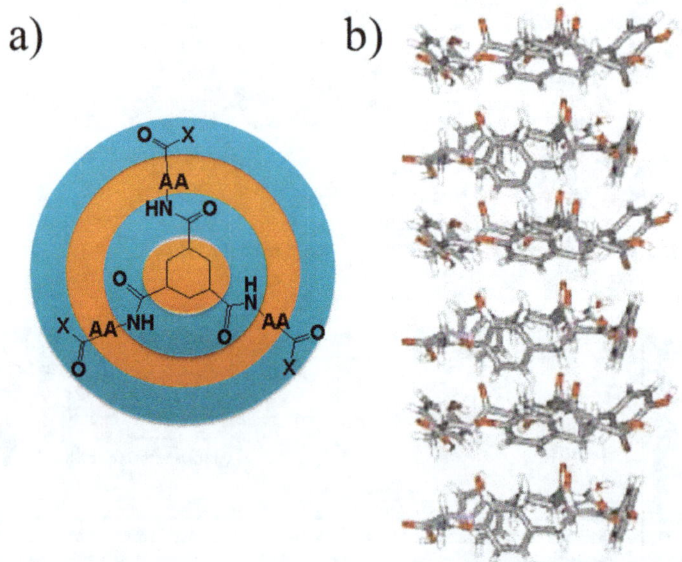

Figure 6. (**a**) Scheme showing the hydrophobic (blue) and hydrophilic (orange) regions of a cyclohexane-based hydrogelator having amino acids (AA) with hydrophobic side groups; (**b**) a single strand formed through the multiple hydrogen bonds that each single molecule can establish. Reproduced with permission from [127].

In general, an ideal functionalized self-assembling structure should have: long alkyl chains that constitute the hydrophobic domain; a peptide sequence promoting β-sheet structures; charged amino acids to favor water solubility; and a bioactive epitope [128]. The hydrophobic interactions favor the development of self-assembled structures. Nevertheless, the final morphology can be influenced by the size of the hydrophobic moiety, there being, for example, a report of a morphological change from simple sheet stacking to long nanotubes and finally to short nanorods as the length increased [129].

6. Responsiveness to External Stimuli of Peptide-Derived Hydrogels for 3D Cell Culture

Control of supramolecular interactions that lead to the formation of assembled supramolecular structures may facilitate the in situ encapsulation of cells and even the subsequent delivery of the proliferated cells by switching the hydrogel to the sol state by means of proper external stimuli.

The self-assembly of ionic peptides can be controlled through the pH of the solution. Thus, assembly is possible when the solution pH leads to a practically zero net charge of the peptide molecules. Ionic force also has a great influence in this assembling process since the counter-ions can decrease the charge distribution. Multiple examples can be mentioned concerning the encapsulation of cells (e.g., chondrocytes [130,131], human mesenchymal cells [132], or murine embryonic pluripotent stem cells [133]).

The switching between the hydrogel and solution phases can be controlled by the protonation and deprotonation of their amino and carboxylic terminal groups [134]. Short dipeptides with their amino terminal group blocked with the aromatic 9-fluorenylmethoxy-carbonyl group (Fmoc) were able to be dissolved in alkali solutions and form 3D hydrogels at neutral pH through hydrogen bonds and aromatic stacking interactions. In this way, cell dispersion could be mixed with a peptide solution before lowering the pH to physiological conditions. Cells like dermal fibroblasts were successfully encapsulated following the indicated procedure [135]. Peptides derived from serine and mainly from phenylalanine were also reported to produce such pH switchable hydrogels [136–139].

A method based on the change of polarity of the solvent has been proposed for the autoassembling of peptides scarcely soluble in aqueous solutions. Usually dimethylsulfoxide is considered as the ideal organic solvent due to its compatibility with water and low cytotoxicity. In this way, assembly and cell encapsulation can be produced when the peptide solution in the organic polar solvent is mixed with the cell culture medium [140,141].

Efforts have also been focused to develop light-sensitive hydrogels despite the fact that sol-gel transitions under UV irradiation may be harmful to cells. Therefore, nowadays systems mainly use UV before introducing the cells to the hydrogel. For instance, UV can be employed to produce channels in a formed hydrogel that could then be filled by the sol containing the cells [142]. Formation of a hydrogel can also be triggered by the action of enzymes able to form a gelator molecule from a given precursor. Usually this is characterized by the presence of a hydrophilic moiety that is susceptible to enzyme-catalyzed cleavage [143].

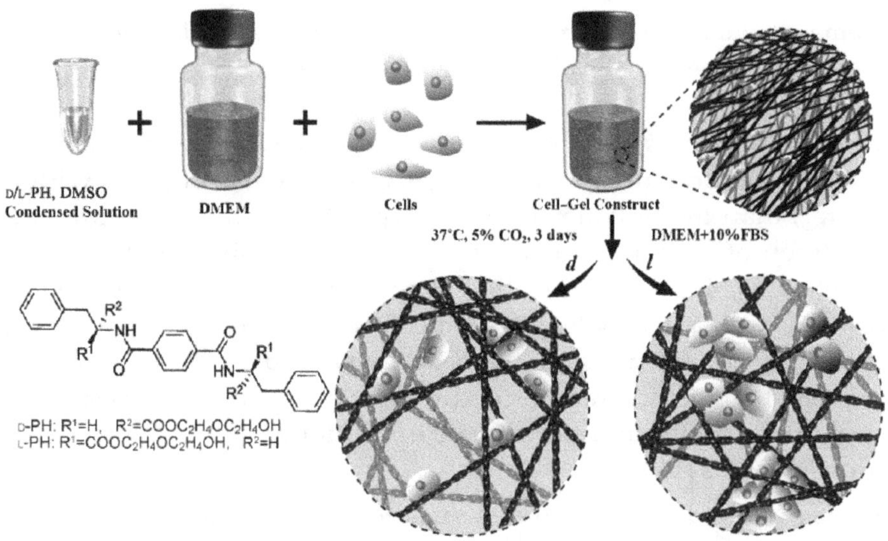

Figure 7. Schematic representation of the culture of fibroblast or endothelial cells in enantiomeric nanofibrous hydrogels (*d*, *l* right-handed and left-handed helices, respectively). Reproduced with permission from [144].

The use of peptides with chiral centers allows for obtaining chiral nanofibrous hydrogels that are influenced by adhesion and cell proliferation events. Specifically, two enantiomers of a 1,4-benzenedicarboxamide phenylalanine derivative have been employed as supramolecular gelators that produced left- and right-handed helices (Figure 7) [144]. The cell densities of fibroblast and endothelial cells in left-handed hydrogels were found to be twice those determined for hydrogels based on right-handed helices. Stereospecific interactions between chiral nanofibers and fibronectin were responsible for the observed effect.

7. Micro- and Nanogels

Gelling biopolymers as proteins and polysaccharides can be intramolecularly crosslinked to produce small particles [145] with a globular structure when the concentration of the polymer is moderate and specifically lower than that employed to produce macroscopic gels [60]. These particles can be classified into microgels and nanogels when their diameter is lower than 0.5 μm or between 0.5–5 μm, respectively. These hydrogel particles are receiving increasing attention for rheology control, drug encapsulation, and targeted delivery [146]. Furthermore, particles can swell and deflate in response to external stimuli (e.g., pH, temperature, ionic strength) and consequently are appropriate for use in controllable and responsive systems. Cellulose, chitosan, gluten, soy protein, corn zein, casein, whey protein, gelatin, and collagen are the natural polymers most often employed for the preparation of these micro/nanogels [147].

Different techniques have been applied to get small gel particles, as indicated in Figure 8. These involve a typical phase separation caused by coacervation and desolvation, as well as processes induced by mechanical methods.

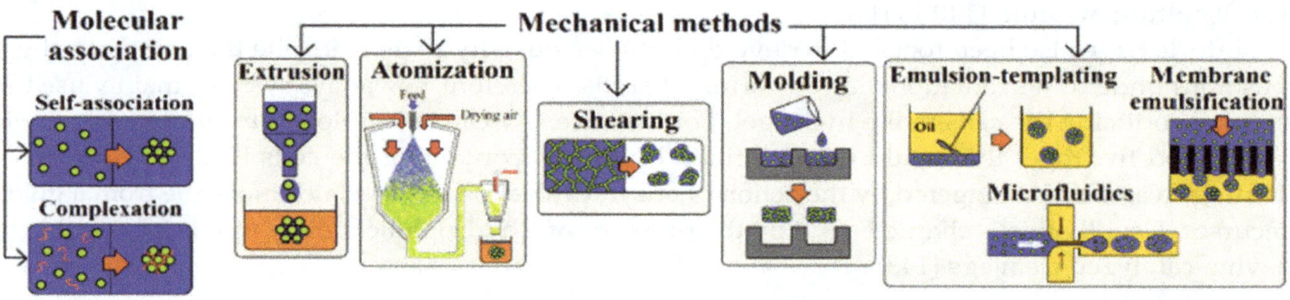

Figure 8. Scheme showing the different preparation methods applied for the production of micro/nanogel particles. Reproduced with permission from [147].

A simple coacervation process implies only one polymer, whose molecular association may be favored, for example, by temperature or pH changes or by the use of a salt with higher affinity to the solvent (e.g., water) than the polymer itself. A typical example is the preparation of chitosan microgels by the addition of sodium hydroxide to a polymer solution also having a crosslinking agent (e.g., glutaraldehyde) [148]. Self-association of proteins can also be obtained by typical thermal denaturation processes as well as by cooling preheated protein solutions. In this case, unfolding took place in the heating step, giving rise to protein filaments that are able to be associated during the cooling step under appropriate pH and ionic strength conditions. Microgels based on soy protein (28–179 nm) were, for example, prepared by the addition of calcium cations, which favored the establishment of bridges between protein chains [149].

Protein microgels prepared by thermal denaturation are mainly based on the establishment of hydrophobic interactions. Basically, spherical particles (with diameters close to 50 nm) and strands (diameter less than 10 nm and length up to tens of microns) can be formed during denaturation, with the second morphology being favored when strong electrostatic repulsions exist. In a second aggregation step, fine strands or particulate gels can be derived, as has been reviewed by Nicolai and

Durant [150]. pH and ionic strength were found to have a high influence on final morphology [151], as well as the protein concentration, temperature, and heating time.

Two polymers with opposite charges are required for the named complex coacervation process or associative phase separation. This process appears ideal for the preparation of gel particles composed of proteins and negatively charged polysaccharides. For example, ovalbumin is a protein that can be gelled by heat treatment and gives rise to nanogel particles where CS could subsequently be entrapped through electrostatic interactions [152].

A mechanical device can also be employed to form particles that will then be gelled through a physicochemical process. These mechanical methods were well explained by Farjami et al. [147] and include extrusion, atomization, shearing, emulsion, and micromolding processes.

The extrusion method is based on the flow of a polymer solution through a syringe needle, with the formed droplets hardened in a solution having crosslinking agents (e.g., enzymes and glutaraldehyde), by a temperature change (as in the case or gelatin), or by forming a complex with other polymers [153].

An aqueous solution of the selected biopolymers can also be atomized into a stream of hot air that evaporates the solvent. The resultant spray-dried particles can subsequently be rehydrated to form the corresponding gel particles. Particle sizes can be modified by using different gelling polymers and agents as well as controlling the speed of the process (gel particles or a continuous gel can be obtained under fast and slow gelation conditions, respectively). Collagen and carrageenan are good examples of microgels prepared under this atomization technology [154].

Application of a shear to a biopolymer gel system may induce its breakage and form irregular gel particles. The process has some advantages related to the facility of modulating the final properties (i.e., by controlling the shear and thermal history) [155].

Water-in-oil emulsions can be employed as templates where polymers dissolved in the oil phase are subsequently cross-linked. This technique has been successfully applied for the preparation of microgels able to encapsulate microorganisms and drugs, as is the case of those based on casein [156] and whey proteins [157], chitosan [158], and protein/sugar conjugates [159]. In the case of chitosan, the cross-linking process can also be completed easily by the addition of anionic agents such as sulfates or citrates, but in general the derived morphologies were irregular. The incorporation of gelatin allowed an improvement in the ionic crosslinking process and gave rise to regular, spherical and smooth microspheres with diameters in the micrometer range [160].

Preparation of anisotropic particles with complex architectures is also interesting for innovative applications like Janus motors [161]. The use of capillary flow-based approaches (e.g., microfluidics) has been revealed to be highly effective for producing polysaccharide hydrogel hetero Janus microparticles [162]. By contrast, drug delivery applications require the use of uniform particles in order to guarantee good repeatable release behavior. The use of a membrane is a methodology that allows for getting uniformly sized particles since the polymer aqueous solution should permeate through a membrane with a well-controlled pore size before being emulsified in the oil phase. This technique has been applied to produce uniform chitosan microspheres for drug delivery, taking advantage of its excellent mucoadhesive character and good permeability through biological surfaces. Thus, particles with a highly uniform diameter close to 0.4 mm have been obtained and applied to encapsulate proteins such as bovine serum albumin [163] and peptides like insulin [164].

Figure 9 shows three methods that are usually applied to form networks from the nano/microgel particles: (a) direct cross-linking via physical or chemical interactions [165]; (b) reaction of functionalized microgel particles (e.g., having peroxy groups) with reactive polymers or even by grafting polymerization of monomers [166]; and (c) physical entrapment of hydrogel microparticles [167]. Microgels can, for example, be immobilized within the network produced by an in situ gelling hydrogel. Therefore, administration of the hydrogel is facilitated, while sustained or prolonged drug release can be achieved when the targeted drug is loaded in the entrapped particles.

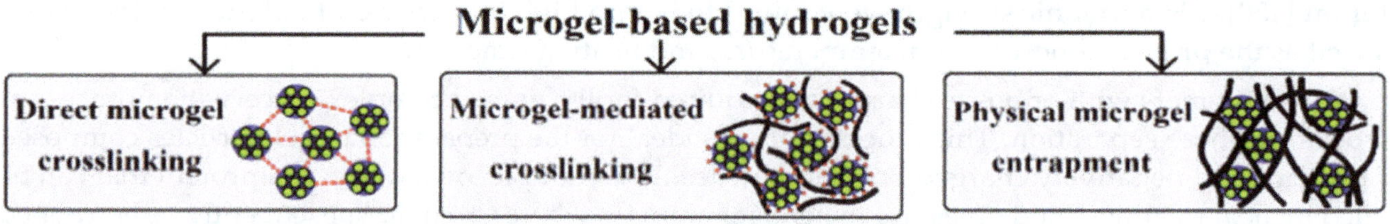

Figure 9. Scheme showing the different preparation methods applied for the production microgel networks. Reproduced with permission from [147].

8. Shape Memory Hydrogels

Shape memory polymers (SMPs) are extensively studied due to their capability to fix one or more temporary shapes and recover their permanent shape under the effect of an external stimuli [168–170]. These properties (i.e., dual and multi shape memory effects) pave the way for the development of attractive new materials for applications in the biomedical field, aerospace, textiles, etc. [171,172]. Thermo-responsive SMPSs are probably the most usual systems and are specifically based on crosslinked polymers that adopt a temporary shape by vitrification or crystallization, recovering their permanent shape after heating [168,169].

Efforts are nowadays also focused on developing shape memory materials based on the establishment of supramolecular interactions and dynamic hydrogen bonds [173] (Figure 10). These supramolecular shape memory hydrogels (SSMHs) have the advantage of providing materials responsive to a wide range of external stimuli at body or ambient temperature since reversible interactions are usually multiresponsive [174].

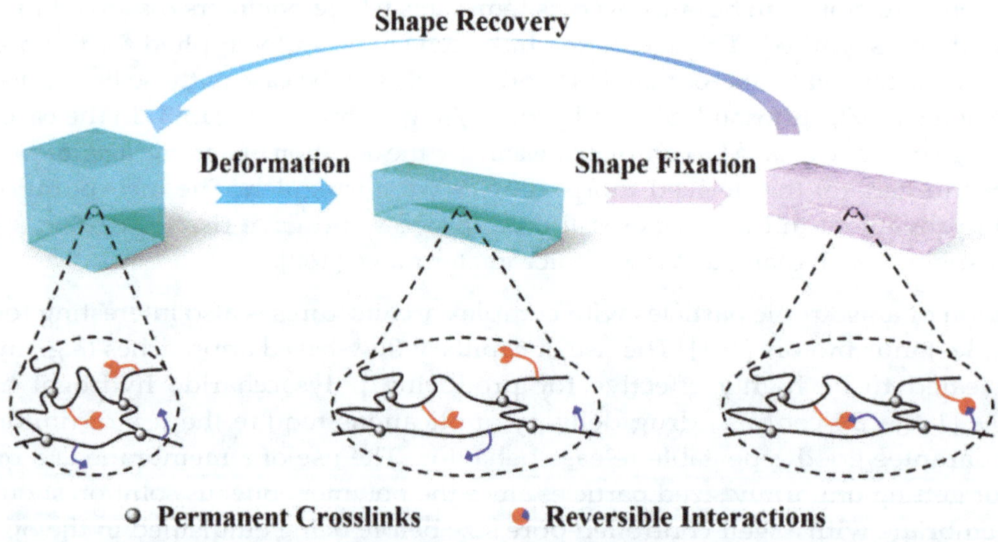

Figure 10. Mechanism of SSMHs: A crosslinked hydrogel can be deformed under an external stress and the temporary shape fixed by an external stimulus that induces the establishment of reversible interactions. A second stimulus may break the interactions and the material reverts to its permanent shape. Reproduced with permission from [173].

A β-cyclodextrine modified CS has been employed as a SSMHs material taking advantage of the capability of the hydrophobic internal cavity of cyclodextrine to accommodate guest molecules such as ferrocene. Specifically, this supramolecular hydrogel was formed by the crosslinked CS derivative and a ferrocene-modified hyperbranched poly(ethylene imine) [175]. The derived host–guest interactions were redox-sensitive, being the temporary and permanent shapes achieved in the reduced and oxidized states, respectively.

Photoactive gels have also been developed since light is a non-destructive and easily controllable stimuli. Therefore, photochromic groups (e.g., azobenzene) have been incorporated into functional gels by means of the supramolecular approach [176,177]. Host–guest inclusion complexes can be formed between *trans*-azobenzene groups and cyclodextrines, whereas steric repulsions are dominant when a bulky *cis* conformation is preferred. Therefore, systems based on an amphiphilic dendron with three L-glutamic acid units and an azobenzene moiety have been considered to form hydrogels susceptible to photo-triggered changes [178].

9. Molecularly Imprinted Hydrogels

Bioactive scaffolds for tissue engineering applications can be prepared by molecular imprinting, as recently reviewed by Neves et al. (2017) [179]. This technology has the great advantage of providing molecular memory effects when appropriate intelligent materials are selected [180,181]. Basically, a template molecule is combined with a functional monomer and finally cross-linked to create a polymer network. After removal of the template, the final material has ideal functionalization, cavity size, and shape to act specifically towards specific molecules (Figure 11). This high bioactivity and recognition ability make molecular imprinting a highly promising tool for tissue engineering.

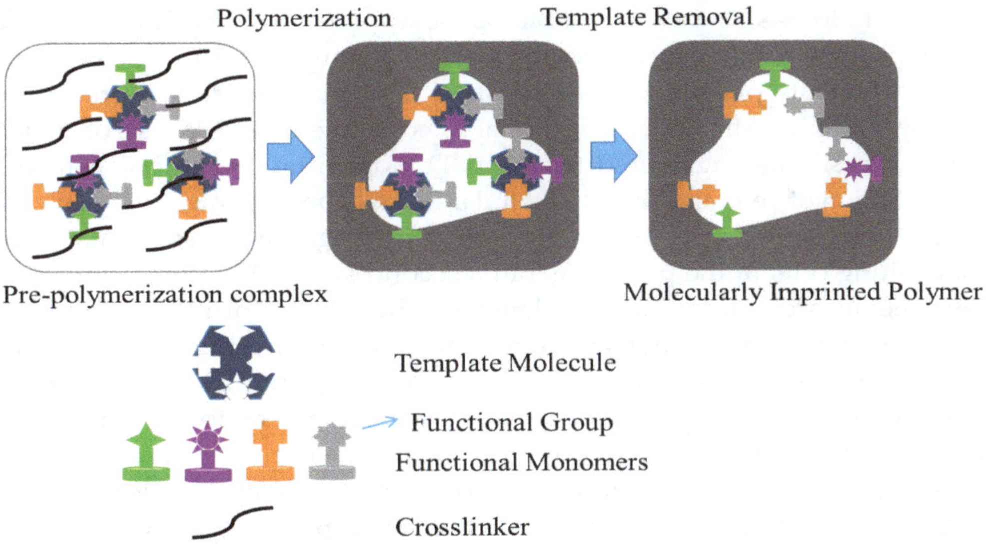

Figure 11. Scheme showing the different steps involved in the molecular imprinting process: Mixing of the appropriate template molecule and the selected functional monomer(s) and cross-linker(s) in a solvent; the polymerization of the formed complex; and finally the removal of the template, unreacted monomer, and cross-linker molecules. Adapted from [179].

Hydrogels have great potential for macromolecular imprinting since they facilitate both the movement of high molecular weight templates and the production of reversible systems sensitive to external stimuli (e.g., pH or temperature). In addition, they can be easily processed in different forms (e.g., sheets, coatings, or capsules). The main problems with such imprinting hydrogels concern the distortion of binding sites as a consequence of their easy expansion and contraction. Efficiency is consequently affected but different works have been focused on the development of tissue engineering applications that extend their most common use as drug delivery systems.

BSA [182,183] and fibronectin (Fn) [184,185], a high molecular weight protein with wound healing and tissue repair activities, have been successfully employed as template molecules for alginate-based hydrogels.

CS has also been considered to create pH-, temperature- and ionic-strength-sensitive hydrogels able to recognize, for example, the dipeptide of carnosine [186], an antioxidant that is found in muscle and brain tissue. Most commonly, chitosan has been combined with acrylamide monomers to form hydrogels that can also be used in biosensing applications such as the recognition of hemoglobine [187]. Systems based on a mixture of alginates and CS have also received attention due to the combination of oppositely charged functional groups (e.g., carboxyl and amine for alginate and CS, respectively), which support the interaction of the hydrogel with differently charged domains of proteins. The imprinting effect of the alginate/CS combination has effectively been demonstrated for BSA, lysozime, hemoglobin, and ovoalbumine proteins [188].

Cell adhesion on a particular substrate can also be improved by imprinting morphological and topographic features of cells (i.e., cells are considered the templates, as an extension of the abovementioned macromolecules). Despite its high potential, the technique is still in development and therefore mainly the more elemental acrylamide hydrogels have been assayed. Promising results have been attained with both epithelial and fibroblast cell lines [189].

10. Protein-Based Hydrogels for 3D Printing

Three-dimensional printing technologies are a focus of increasing interest in the preparation of scaffolds based on the most important structural proteins (e.g., collagen, fibrin, silk, and even their composites with hydroxyapatite). Excellent reviews detail the printing parameters and physical properties of bioinks as well as the biological applications of printed scaffolds [190–195]. Three-dimensional printing technology allows for preparing complex scaffolds where the distribution of the different components plays a key role in their final performance. Furthermore, scaffolds can be designed in complex shapes at a reasonable cost, avoiding the use of expensive molds. Artificial tissues can easily be engineered through 3D bioprinting by precise control over spatial and temporal distribution of cells and the extracellular matrix components. Cells can be incorporated into the scaffolds following two approaches: printing a hydrogel precursor containing the selected cells [196] or depositing cells in the printed gel in a second step [197]. Processing conditions must be carefully selected in order to avoid cell damage when direct printing of embedded cells is performed [198]. Attention is nowadays given to three main bioprinting methodologies/strategies (Figure 12): inkjet-based, laser-assisted, and extrusion-based bioprinting [199]. The first methodology is based on the deposition of ink drops on successive layers, it being possible to place cells and proteins onto targeted spatial positions. This layer-by-layer technique is characterized by high resolution, reproducibility, and inexpensiveness [200]. In the second methodology, a pulsed laser source is able to transfer heat and eject a cell suspension from a coated absorption layer toward a substrate. Small volumes (from 10 to 7000 pL) of cell suspension can be printed with high resolution [201]. Cells or proteins are encapsulated in a hydrogel when the extrusion-based bioprinting is applied. In this case, a syringe, micronozzle, and pressure system are required to apply the hydrogel onto the substrate according to a specific 3D design. The process allows obtaining constructions with a relevant shape and size [202].

Water-soluble polymers able to form hydrogels are ideal components for the bioinks used in bioprinting technologies. The main advantages correspond to the possibility of controlling the gelation process and a favorable environment for cell growth [193], with poor mechanical properties and low dimensional stability as the main problems [199]. In any case, materials should contribute to: (a) an acceptable printing time, which is mainly influenced by the cross-linking kinetics and the rheological properties; and (b) a balanced cross-linking density that guarantees mechanical stability as well as the movement and proliferation of cells [203].

Collagen has been processed by the three aforementioned technologies, concerning in general the main problem of the mechanical stability of the constructions. Therefore, the use of salts and fillers has been proposed as well as the development of core/shell structures. In this case, alginate has been employed as the shell component since it can provide mechanical support to the collagen core after being cross-linked by the application of a $CaCl_2$ aerosol solution [204]. Treatment of printed scaffolds with bicarbonate solutions increased the gelation rate while retaining the original shape and avoiding shrinkage or swelling effects [205].

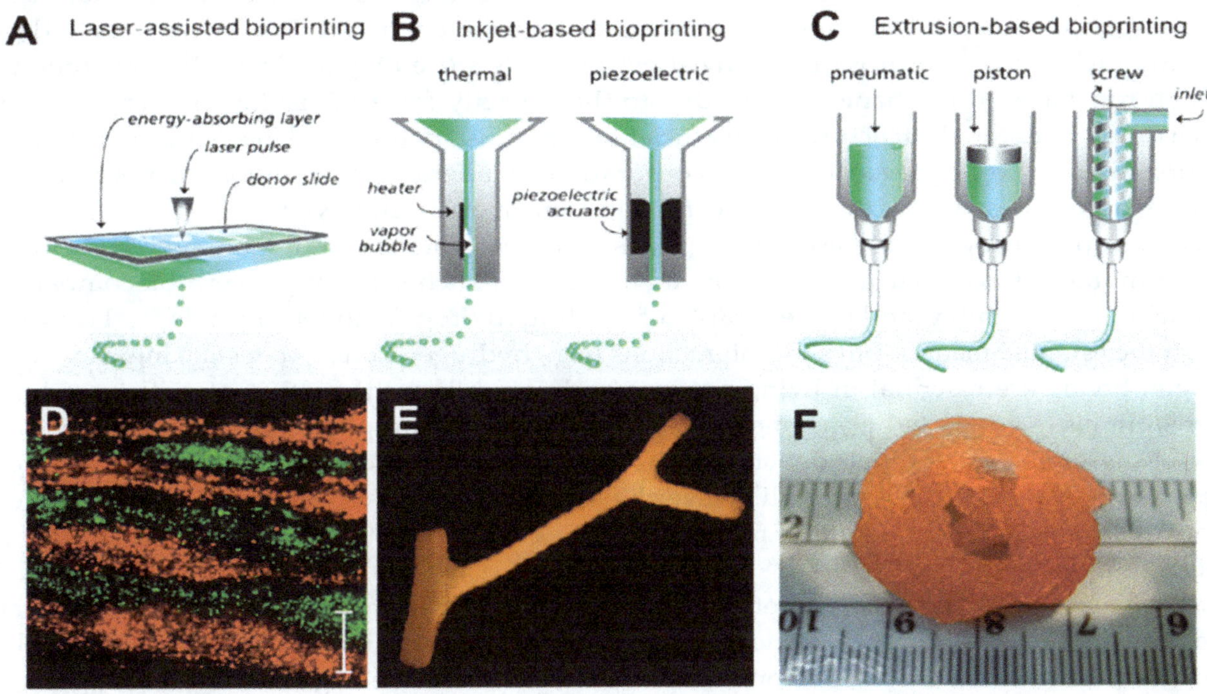

Figure 12. Main 3D bioprinting strategies: (**A**) laser-assisted; (**B**) injet-based and (**C**) extrusion-based. Examples of bioprinting tissues correspond to skin prepared by laser printing (**D**); branched vasculature obtained by inkjet printing (**E**) and heart aortic valve by extrusion bioprinting (**F**). Reproduced with permission from [194].

Fibrin is another well-studied structural protein for preparing bioprinted scaffolds. Fibrin matrices display good mechanical properties and fast gelation and promote tissue regeneration and cell proliferation [206,207]. A layer-by-layer printing process was, for example, applied to produce scaffolds with embedded neurons. Specifically, layers of gelled fibrin were prepared from fibrinogen and thrombin and then neurons were printed on top of the layer. The 3D neural scaffold was attained after repeating the alternate bilayer printing five times. The final scaffold showed a modulus as high as 2.92 MPa, tensile strength of 1.7 MPa, and neurite overgrowth after 12 days of culturing [208].

Collagen/fibrin mixtures are also considered since they combine the high biocompatibility of collagen with the fast gelation rate of fibrin. The derived printed scaffolds showed faster wound closure and vascularization than hydrogels based only on collagen [209].

11. Conclusions

Cellulose and chitin are the two most abundant natural polymers on earth and consequently the use of them or their derivatives (e.g., chitosan) is widely studied. Biodegradability, biocompatibility, and renewability are characteristics that enhance their interest for biomedical applications and specifically for the development of hydrogels useful for tissue engineering, wound dressing, and drug delivery systems. Advantages related to processability, the presence of functional groups, and bioactivity have also shifted the research focus to improving solubility and basic properties and also to developing new physical and chemical cross-linkers. Hydrogels based on proteins are of great interest for use in physiological environments, taking into account the maintenance of their bioactive properties (e.g., the capability of collagen to enhance cell proliferation and tissue reconstruction). Hybrid materials based on their mixtures and the combination with other natural proteins such as collagen and gelatin may provide materials with considerable advantages over the individual components. Opportunities to create natural hydrogels are clear, although effort is still necessary to improve the final performance and get effective nontoxic cross-linkers. In this sense, the development of injectable hydrogels able to form in the body without requiring surgery or the development of stimuli-responsive smart materials is a clear example of the present trends. Double network hydrogels are highly promising to improve mechanical performance since they intrinsically have toughness and high resistance to mechanical stress due to the capacity for establishing entanglements that involve different network structures with efficient energy dissipation. The mechanical performance of hydrogels can be improved by the incorporation of fillers, with cellulose nanocrystals and hydroxyapatite being some of the natural compounds most often employed.

The use of hydrogelators based on peptides has great potential in the biomedical field since they can mimic extracellular matrices. The advantages of such systems concern biocompatibility, controllable self-assembly, and the feasibility of creating hydrogels under physiological conditions through the establishment of physical interactions (e.g., hydrogen bonds, π–π stacking, etc.). In fact, hydrogen bonds are essential in living systems and may allow the culture of cells by means of proper stimulus (e.g., pH, light, enzymes, etc.). Furthermore, the properties of the derived hydrogels can be easily varied due to their dependence on the amino acid sequence, it being possible to get complex responses to external stimuli. Nevertheless, research must continue to address different challenges such as limited mechanical properties or the control over the porosity and morphology of the derived hydrogel.

Micro- and nanogels have great potential as drug delivery systems and can be used in other areas as bioimaging, sensors, and even tissue engineering. The main applications are based on the ability to tune the particle size from the micrometric to the nanometric scale, together with the large surface area, which makes bioconjugation feasible, and the inner network that favors efficient encapsulation of molecules. In this sense, hydrogel particles show clear advantages over typical systems based on micelles and liposomes.

Shape memory hydrogels can be designed to switch their form in response to external stimuli as heat, pH, light, or ions. Temporary cross-links can be modified and dual or even more complex effects can be derived. Recent advances are focused on achieving the appropriate functionality to satisfy the increasing demand for materials for the biomedical field and even for their use as smart actuators.

Hydrogels appear to be ideal systems for the molecular imprinting of biomacromolecules and the production of bioactive matrices with recognition ability. Promising results as cell culture systems have specifically been attained using hydrogels based on natural polymers.

Three-dimensional printing technologies are a major area of development, given that they are probably the most promising devices focused on biomaterial deposition. In addition, applications in tissue engineering and medicine are well documented, and the possibility of producing artificial organs justifies the great research efforts. Different examples emphasize the potential of the technology for bioprinting natural proteins despite the difficulty of preserving their structure and functionality. Recent efforts have demonstrated the feasibility of employing ink that incorporates cells as well as the possibility of culturing cells within the printed scaffolds.

Acknowledgments: The authors acknowledge support from MINECO and FEDER (MAT2015-69547-R and MAT2015-69367-R), and the Generalitat de Catalunya (2014SGR188).

Author Contributions: Angélica Díaz, Luís J. del Valle, and Jordi Puiggalí contributed equally to the development of the review.

References

1. Wichterle, O.; Lím, D. Hydrophilic gels for biological use. *Nature* **1960**, *185*, 117–118. [CrossRef]
2. Buwalda, S.; Boere, J.K.; Dijksra, P.; Fiejen, J.; Vermoden, T.; Hennink, W. Hydrogels in an historical perspective: From simple networks to smart materials. *J. Control. Release* **2014**, *190*, 254–273. [CrossRef] [PubMed]

3. Wang, T.; Chen, L.; Shen, T.; Wu, D. Preparation and properties of a novel thermo-sensitive hydrogel based on chitosan/hydroxypropylmethylcellulose/glicerol. *Int. J. Biol. Macromol.* **2016**, *93*, 775–782. [CrossRef] [PubMed]
4. Hennink, W.E.; van Nostrum, C.F. Novel crosslinking methods to design hydrogels. *Adv. Drug Deliv. Rev.* **2002**, *54*, 13–36. [CrossRef]
5. Moreira Teixeira, L.S.; Feijen, J.; van Blitterswijk, C.A.; Dijkstra, P.J.; Karperien, M. Enzyme-catalyzed crosslinkable hydrogels: Emerging strategies for tissue engineering. *Biomaterials* **2012**, *33*, 1281–1290. [CrossRef] [PubMed]
6. Gyles, D.A.; Castro, L.D.; Carréra Silva, J.O.; Ribeiro-Costa, R.M. A review of the designs and prominent biomedical advances of natural and synthetic hydrogel formulations. *Eur. Polym. J.* **2017**, *88*, 373–392. [CrossRef]
7. Das, N. Preparation methods and properties of hydrogel: A review. *Int. J. Pharm. Pharm. Sci.* **2013**, *5*, 112–117.
8. Hoffman, A.S. Hydrogels for biomedical applications. *Adv. Drug Deliv. Rev.* **2002**, *54*, 3–12. [CrossRef]
9. Bae, K.H.; Wang, L.S.; Kurisawa, M. Injectable biodegradable hydrogels: Progress and challenges. *J. Mater. Chem. B* **2013**, *1*, 5371–5388. [CrossRef]
10. George, M.; Weiss, R.G. Molecular organogels. Soft matter comprised of low-molecular-mass organic gelators and organic liquids. *Acc. Chem. Res.* **2006**, *39*, 489–497. [CrossRef] [PubMed]
11. Sangeetha, N.M.; Maitra, U. Supramolecular gels: Function and uses. *Chem. Soc. Rev.* **2005**, *34*, 821–836. [CrossRef] [PubMed]
12. Yoshida, R.; Uchida, K.; Kaneko, Y.; Sakai, K.; Kikuchi, A.; Sakurai, Y.; Okano, T. Comb-type grafted hydrogels with rapid deswelling response to temperature changes. *Nature* **1995**, *374*, 240–242. [CrossRef]
13. Du, X.W.; Zhou, J.; Xu, B. Supramolecular hydrogels made of basic biological building blocks. *Chem. Asian J.* **2014**, *9*, 1446–1472. [CrossRef] [PubMed]
14. Bakota, E.L.; Sensoy, O.; Ozgur, B.; Sayar, M.; Hartgerink, J.D. Self-assembling multidomain peptide fibers with aromatic cores. *Biomacromolecules* **2013**, *14*, 1370–1378. [CrossRef] [PubMed]
15. Bi, L.; Cheng, W.; Fan, H.; Pei, G. Reconstruction of goat tibial defects usingan injectable tricalcium phosphate/chitosan in combination with autologousplatelet-rich plasma. *Biomaterials* **2010**, *31*, 3201–3211. [CrossRef] [PubMed]
16. Li, H.; Ji, Q.; Chen, X.; Sun, Y.; Xu, Q.; Denq, P.; Hu, F.; Yanq, J. Accelerated bony defect healing based on chitosan thermosensitive hydrogel scaffolds embedded with chitosan nanoparticles for the delivery of BMP2 plasmid DNA. *J. Biomed. Mater. Res. A* **2017**, *105A*, 265–273. [CrossRef] [PubMed]
17. Tan, H.; Marra, K.G. Injectable, biodegradable hydrogels for tissue engineering applications. *Materials* **2010**, *3*, 1746–1767. [CrossRef]
18. Amini, A.A.; Nair, L.S. Injectable hydrogels for bone and cartilage repair. *Biomed. Mater.* **2012**, *7*, 24105–24118. [CrossRef] [PubMed]
19. Czaja, W.K.; Young, D.J.; Kawecki, M.; Brown, R.M. The future prospects of microbial cellulose in biomedical applications. *Biomacromolecules* **2007**, *8*, 1–17. [CrossRef] [PubMed]
20. Araki, J.; Mishima, S. Steric stabilization of "charge-free" cellulose nanowhiskers by grafting of poly(ethylene glycol). *Molecules* **2015**, *20*, 169–184. [CrossRef] [PubMed]
21. Joshi, M.K.; Pant, H.R.; Tiwari, A.P.; Maharjan, B.; Liao, N.; kim, H.J.; Park, C.H.; Kim, C.S. Three-dimensional cellulose sponge: Fabrication, characterization, biomimetic mineralization, and in vitro cell infiltration. *Carbohydr. Polym.* **2016**, *136*, 154–162. [CrossRef] [PubMed]
22. Shelke, N.B.; James, R.; Laurencin, C.T.; Kumbar, S.G. Polysaccharide biomaterials for drug delivery and regenerative engineering. *Polym. Adv. Technol.* **2014**, *25*, 448–460. [CrossRef]
23. Ko, H.F.; Sfeir, C.; Kumta, P.N. Novel synthesis strategies for natural polymer and composite biomaterials as potential scaffolds for tissue engineering. *Philos. Trans. A Math. Phys. Eng. Sci.* **2010**, *368*, 1981–1997. [CrossRef] [PubMed]
24. Park, S.; Baker, J.O.; Himmel, M.E.; Parilla, P.A.; Johnson, D.K. Cellulose crystallinity index: Measurement techniques and their impact on interpreting cellulase performance. *Biotechnol. Biofuels* **2010**, *3*, 10. [CrossRef] [PubMed]

25. Cai, J.; Zhang, L.N.; Chang, C.Y.; Cheng, G.Z.; Chen, X.M.; Chu, B.J. Hydrogen-bond-induced inclusion complex in aqueous cellulose/LiOH/urea solution at low temperature. *Chem. Phys. Chem.* **2007**, *8*, 1572–1579. [CrossRef] [PubMed]
26. Cai, J.; Zhang, L.N.; Liu, S.L.; Liu, Y.T.; Xu, X.J.; Chen, X.M.; Chu, B.; Guo, X.; Cheng, H.; Han, C.C.; et al. Dynamic self-assembly induced rapid dissolution of cellulose at low temperatures. *Macromolecules* **2008**, *41*, 9345–9351. [CrossRef]
27. Sun, B.Z.; Duan, L.; Peng, G.G.; Li, X.X.; Xu, A.H. Efficient production of glucose by microwave-assisted acid hydrolysis of cellulose hydrogel. *Bioresour. Technol.* **2015**, *192*, 253–256. [CrossRef] [PubMed]
28. Ruan, D.; Zhang, L.N.; Mao, Y.; Zeng, M.; Li, X.B. Microporous membranes prepared from cellulose in NaOH/thiourea aqueous solution. *J. Membr. Sci.* **2004**, *241*, 265–274. [CrossRef]
29. Takigami, M.; Amada, H.; Nagasawa, N.; Yagi, T.; Kasahara, T.; Takigami, S.; Tamada, M. Preparation and properties of CMC gel. *Trans. Mater. Res. Soc. Jpn.* **2007**, *32*, 713–716.
30. Peng, B.L.; Dhar, N.; Liu, H.L.; Tam, K.C. Chemistry and applications of nanocrystalline cellulose and its derivatives: A nanotechnology perspective. *Can. J. Chem. Eng.* **2011**, *89*, 1191–1206. [CrossRef]
31. Pranger, L.; Tannenbaum, R. Chemistry and applications of nanocrystalline cellulose and its derivatives: A nanotechnology perspective. *Macromolecules* **2008**, *41*, 8682–8687. [CrossRef]
32. Čolić, M.; Mihajlović, D.; Mathew, A.; Naser, N.; Kokol, V. Cytocompatibility and immunomodulatory properties of wood based nanofibrillated cellulose. *Cellulose* **2015**, *22*, 763–778. [CrossRef]
33. Grimm, S.; Giesa, R.; Sklarek, K.; Langner, A.; Gosele, U.; Schmidt, H.W.; Steinhart, M. Nondestructive replication of self-ordered nanoporous alumina membranes via cross-linked polyacrylate nanofiber arrays. *Nano Lett.* **2008**, *8*, 1954–1959. [CrossRef] [PubMed]
34. Liu, J.; Korpinen, R.; Mikkonen, K.; Willför, S.; Xu, C. Nanofibrillated cellulose originated from birch sawdust after sequential extractions: A promising polymeric material from waste to films. *Cellulose* **2014**, *21*, 2587–2598. [CrossRef]
35. Syverud, K.; Pettersen, S.; Draget, K.; Chinga-Carrasco, G. Controlling the elastic modulus of cellulose nanofibril hydrogels—Scaffolds with potential in tissue engineering. *Cellulose* **2015**, *22*, 473–481. [CrossRef]
36. Zhang, X.; Huang, J.; Chang, P.R.; Li, J.; Chen, Y.; Wang, D.; Yu, J.; Chen, J. Structure and properties of polysaccharide nanocrystal-doped supramolecular hydrogels based on cyclodextrin inclusion. *Polymer* **2010**, *51*, 4398–4407. [CrossRef]
37. Lin, N.; Bruzzese, C.; Dufresne, A. Tempo-oxidized nanocellulose participating as crosslinking aid for alginate-based sponges. *ACS Appl. Mater. Interfaces* **2012**, *4*, 4948–4959. [CrossRef] [PubMed]
38. Domingues, R.M.; Gomes, M.E.; Reis, R.L. The potencial of cellulose nanocrystals in tissue engineering strategies. *Biomacromolecules* **2014**, *15*, 2327–2346. [CrossRef] [PubMed]
39. Wang, K.; Nune, K.; Misra, R. The functional response of alginate-gelatin-nanocrystalline cellulose injectable hydrogels toward delivery of cells and bioactive molecules. *Acta Biomater.* **2016**, *36*, 143–151. [CrossRef] [PubMed]
40. Naseri, N.; Deepa, B.; Mathew, A.P.; Oksman, K.; Girandon, L. Nanocellulose-based interpenetrating polymer network (IPN) hydrogels for cartilage applications. *Biomacromolecules* **2016**, *17*, 3714–3723. [CrossRef] [PubMed]
41. Chen, J.; Lin, N.; Huang, J.; Dufresne, A. Highly alkynyl-functionalization ofcellulose nanocrystals and advanced nanocomposites thereof via click chemistry. *Polym. Chem.* **2015**, *6*, 4385–4395. [CrossRef]
42. Domingues, R.M.A.; Silva, M.; Gershovich, P.; Betta, S.; Babo, P.; Caridade, S.G.; Mano, J.F.; Motta, A.; Reis, R.L.; Gomes, M.E. Development of injectable hyaluronic acid/cellulose nanocrystals bionanocomposite hydrogels for tissue engineering applications. *Bioconjugate Chem.* **2015**, *26*, 1571–1581. [CrossRef] [PubMed]
43. García-Astrain, C.; González, K.; Gurrea, T.; Guaresti, O.; Algar, I.; Eceiza, A.; Gabilondo, N. Maleimide-grafted cellulose nanocrystals as cross-linkers for bionanocomposite hydrogels. *Carbohydr. Polym.* **2016**, *149*, 94–101. [CrossRef] [PubMed]
44. Stella, J.A.; D'Amore, A.; Wagner, W.R.; Sacks, M.S. On the biomechanical function of scaffolds for engineering load-bearing soft tissues. *Acta Biomater.* **2010**, *6*, 2365–2381. [CrossRef] [PubMed]
45. Liu, J.; Cheng, F.; Grénman, H.; Spoljaric, S.; Seppälä, J.; Eriksson, J.E.; Willför, S.; Xu, C. Development of nanocellulose scaffolds with tunable structures to support 3D cell culture. *Carbohydr. Polym.* **2016**, *148*, 259–271. [CrossRef] [PubMed]

46. Bhattacharya, M.; Malinen, M.M.; Lauren, P.; Lou, Y.R.; Kuisma, S.W.; Kanninen, L.; Lille, M.; Corlu, A.; GuGuen-Guillouzo, C.; Ikkala, O.; et al. Nanofibrillar cellulose hidrogel promotes three-dimensional liver cell culture. *J. Control. Release* **2012**, *164*, 291–298. [CrossRef] [PubMed]
47. Malinen, M.M.; Kanninen, L.K.; Corlu, A.; Isoniemi, H.M.; Lou, Y.-R.; Yliperttula, M.L.; Urtti, A.O. Differentiation of liver progenitor cell line to functional organotypic cultures in 3D nanofibrillar cellulose and hyaluronan-gelatin hydrogels. *Biomaterials* **2014**, *35*, 5110–5121. [CrossRef] [PubMed]
48. Mertaniemi, H.; Escobedo-Lucea, C.; Sanz-Garcia, A.; Gandía, C.; Mäkitie, A.; Partanen, J.; Ikkala, O.; Yliperttula, M. Human stem cell decorated nanocellulose threads for biomedical applications. *Biomaterials* **2016**, *82*, 208–220. [CrossRef] [PubMed]
49. Alexandrescu, L.; Syverud, K.; Gatti, A.; Chinga-Carrasco, G. Cytotoxicity tests of cellulose nanofibril-based structures. *Cellulose* **2013**, *20*, 1765–1775. [CrossRef]
50. Liu, J.; Willför, S.; Xu, C. A review of bioactive plant polysaccharides: Biological activities, functionalization, and biomedical applications. *Bioact. Carbohydr. Diet Fibre* **2015**, *5*, 31–61. [CrossRef]
51. Liu, J.; Chinga-Carrasco, G.; Cheng, F.; Xu, W.; Willför, S.; Syverud, K.; Xu, C. Hemicellulose-reinforced nanocellulose hydrogels for wound healing application. *Cellulose* **2016**, *23*, 3129–3143. [CrossRef]
52. Park, M.; Chang, H.; Jeong, D.H.; Hyun, J. Spatial deformation of nanocellulose hydrogel enhances SERS. *Biochip J.* **2013**, *7*, 234–241. [CrossRef]
53. Shah, N.; Ul-Islam, M.; Khattak, W.A.; Park, J.K. Overview of bacterialcellulose composites: A multipurpose advanced material. *Carbohydr. Polym.* **2013**, *98*, 1585–1598. [CrossRef] [PubMed]
54. Ul-Islam, M.; Khan, S.; Ullah, M.W.; Park, J.K. Bacterial cellulosecomposites: Synthetic strategies and multiple applications in bio-medical and electroconductive fields. *Biotechnol. J.* **2015**, *10*, 1847–1861. [CrossRef] [PubMed]
55. Park, M.; Shin, S.; Cheng, J.; Hyun, J. Nanocellulose based asymmetric composite membrane for the multiple functions in cell encapsulation. *Carbohydr. Polym.* **2017**, *158*, 133–140. [CrossRef] [PubMed]
56. Kusuma, S.; Shen, Y.-I.; Hanjaya-Putra, D.; Mali, P.; Cheng, L.; Gerecht, S. Self-organized vascular networks from human pluripotent stem cells in a synthetic matrix. *Proc. Natl. Acad. Sci. USA* **2013**, *110*, 12601–12606. [CrossRef] [PubMed]
57. Petersen, N.; Gatenholm, P. Bacterial cellulose-based materials and medical devices: Current state and perspectives. *Appl. Microbiol. Biotechnol.* **2011**, *91*, 1277–1286. [CrossRef] [PubMed]
58. Portal, O.; Clark, W.A.; Levinson, D.J. Microbial cellulose wound dressing in the treatment of nonhealing lower extremity ulcers. *Wounds* **2009**, *21*, 1–3. [PubMed]
59. Joye, I.J.; McClements, D.J. Biopolymer-based nanoparticles and microparticles: Fabrication, characterization, and application. *Curr. Opin. Colloid Int. Sci.* **2014**, *19*, 417–427. [CrossRef]
60. McClements, D.J. *Nanoparticle- and Microparticle-Based Delivery Systems*; CRC Press: Boca Raton, FL, USA, 2015.
61. Berthold, A.; Cremer, K.; Kreuter, J. Preparation and characterization of chitosan microspheres as drug carrier for prednisolone sodium phosphate as model for anti-inflammatory drugs. *J. Control. Release* **1996**, *39*, 17–25. [CrossRef]
62. Anitha, A.; Sowmya, S.; Sudheesh Kumar, P.T.; Deepthai, S.; Chennazhi, K.P.; Ehrlich, H.; Tsurkan, M.; Jayakumar, R. Chitin and chitosan in selected biomedical applications. *Prog. Polym. Sci.* **2014**, *39*, 1644–1667. [CrossRef]
63. Shen, X.; Shamshina, J.L.; Berton, P.; Gurau, G.; Rogers, R.D. Hydrogels based on cellulose and chtin: Fabrication, properties and applications. *Green Chem.* **2016**, *18*, 53–75. [CrossRef]
64. Muramatsu, K.; Masuda, S.; Yoshihara, S.; Fujisawa, A. In vitro degradation behavior of freeze-dried carboxymethyl-chitin sponges processed by vacuum-heating and gamma irradiation. *Polym. Degrad. Stabil.* **2003**, *81*, 327–332. [CrossRef]
65. Suzuki, D.; Takahashi, M.; Abe, M.; Sarukawa, J.; Tamura, H.; Tokura, S.; Kurahashi, Y.; Nagano, A. Comparison of various mixtures of β-chitin and chitosan as a scaffold for three-dimensional culture of rabbit chondrocytes. *J. Mater. Sci. Mater. Med.* **2008**, *19*, 1307–1315. [CrossRef] [PubMed]
66. Hausmann, R.; Vitello, M.P.; Leitermann, F.; Syldatk, C. Advances in the production of sponge biomass *Aplysina aerophoba*—A model sponge for ex situ sponge biomass production. *J. Biotechnol.* **2006**, *124*, 117–127. [CrossRef] [PubMed]
67. Tamura, H.; Nagahama, H.; Tokura, S. Preparation of chitin hydrogel under mild conditions. *Cellulose* **2006**, *13*, 357–364. [CrossRef]

68. Kabanov, A.V.; Vinogradov, S.V. Nanogels as pharmaceutical carriers: Finite networks of infinite capabilities. *Angew. Chem. Int. Ed.* **2009**, *48*, 5418–5429. [CrossRef] [PubMed]
69. Arunraj, T.; Rejinold, N.S.; Kumar, N.A.; Jayakumar, R. Doxorubicin–chitin–poly(caprolactone) composite nanogel for drug delivery. *Int. J. Biol. Macromol.* **2013**, *62*, 35–43. [CrossRef] [PubMed]
70. Mohammed, N.; Rejinold, N.S.; Mangalathillam, S.; Biswas, R.; Nair, S.V.; Jayakumar, R. Fluconazole loaded chitin nanogels as a topical ocular drug delivery agent for corneal fungal infections. *J. Biomed. Nanotech.* **2013**, *9*, 1521–1531. [CrossRef]
71. Rejinold, N.S.; Chennazhi, K.P.; Tamura, H.; Nair, S.V.; Jayakumar, R. Multifunctional chitin nanogels for simultaneous drug delivery, bioimaging and biosensing. *ACS Appl. Mater. Interfaces* **2011**, *3*, 3654–3665. [CrossRef] [PubMed]
72. Vishnu Priya, M.; Sabitha, M.; Jayakumar, R. Colloidal chitin nanogels: A pletora of applications under one shell. *Carbohydr. Polym.* **2016**, *136*, 609–617. [CrossRef] [PubMed]
73. Rani, M.; Agarwal, A.; Negi, Y.S. Review: Chitosan based hydrogel polymeric beads—As drug delivery system. *BioResources* **2010**, *5*, 2765–2807.
74. Park, J.H.; Saravanakumar, G.; Kim, K.; Kwon, I.C. Targeted delivery of low molecular drugs using chitosan and its derivatives. *Adv. Drug Deliv. Rev.* **2010**, *63*, 28–41. [CrossRef] [PubMed]
75. Chourasia, M.K.; Jain, S.K. Polysaccharides for colon targeted drug delivery. *Drug Deliv.* **2004**, *11*, 129–148. [CrossRef] [PubMed]
76. Ogawara, K.; Yoshida, M.; Higaki, K.; Kimura, T.; Shiraishi, K.; Nishikawa, M.; Takakura, Y.; Hashida, M. Hepatic uptake of polystyrene microspheres in rats: Effect of particle size on intrahepatic distribution. *J. Control. Release* **1999**, *59*, 15–22. [CrossRef]
77. Kato, Y.; Onishi, H.; Machida, Y. Biological characteristics of lactosaminated N-succinyl-chitosan as a liver-specific drug carrier in mice. *J. Control. Release* **2001**, *70*, 295–307. [CrossRef]
78. Ruel-Gariepy, E.; Shive, M.; Bichara, A.; Berrada, M.; Le Garrec, D.; Chenite, A.; Leroux, J.C. A thermosensitive chitosan-based hydrogel for the local delivery of paclitaxel. *Eur. J. Pharm. Biopharm.* **2004**, *57*, 53–63. [CrossRef]
79. Obara, K.; Ishihara, M.; Ozeki, Y.; Ishizuka, T.; Hayashi, T.; Nakamura, S.; Saito, Y.; Yura, H.; Matsui, T.; Hattori, H.; et al. Controlled release of paclitaxel from photocrosslinked chitosan hydrogels and its subsequent effect on subcutaneous tumor growth in mice. *J. Control. Release* **2005**, *110*, 79–89. [CrossRef] [PubMed]
80. Boardman, S.J.; Lad, R.; Green, D.C.; Thornton, P.D. Chitosan hydrogels for targeted dye and protein adsorption. *J. Appl. Polym. Sci.* **2017**, *134*, 44846. [CrossRef]
81. Cao, L.; Cao, B.; Lu, C.; Wang, G.; Yu, L.; Ding, J. An injectable hydrogelformed by in situ cross-linking of glycol chitosan and multi-benzaldehydefunctionalized PEG analogues for cartilage tissue engineering. *J. Mater. Chem. B* **2015**, *3*, 1268–1280. [CrossRef]
82. Yasmeen, S.; Lo, M.K.; Bajracharya, S.; Roldo, M. Injectable scaffolds for bone regeneration. *Langmuir* **2014**, *30*, 12977–12985. [CrossRef] [PubMed]
83. Chenite, A.; Buschmann, M.; Wang, D.; Chaput, C.; Kandani, N. Rheological characterisation of thermogelling chitosan/glycerol-phosphate solutions. *Carbohydr. Polym.* **2001**, *46*, 39–47. [CrossRef]
84. Songkroh, T.; Xie, H.; Yu, W.; Liu, X.; Sun, G.; Xu, X.; Ma, X. Injectable in situ forming chitosan-based hydrogels for curcumin delivery. *Macromol. Res.* **2015**, *23*, 53–559. [CrossRef]
85. Frohbergh, M.E.; Katsman, A.; Botta, G.P.; Lazarovici, P.; Schauer, C.L.; Wegst, U.G.K.; Lelkes, P.I. Electrospun hydroxyapatite-containing chitosan nanofibers crosslinked with genipin for bone tissue engineering. *Biomaterials* **2012**, *33*, 9167–9178. [CrossRef] [PubMed]
86. Peter, M.; Ganesh, N.; Selvamurugan, N.; Nair, S.V.; Furuike, T.; Tamura, H.; Jayakumar, R. Preparation and characterization ofchitosan-gelatin/nanohydroxyapatite composite scaffolds for tissue engineering applications. *Carbohydr. Polym.* **2010**, *80*, 687–694. [CrossRef]
87. Rogina, A.; Ressler, A.; Matíc, I.; Gallego Ferrer, G.; Marijanovic, I.; Ivankovic, M.; Ivankovic, H. Cellular hydrogels based on pH-responsive chitosan-hydroxyapatite system. *Carbohydr. Polym.* **2017**, *166*, 173–182. [CrossRef] [PubMed]
88. Zhao, X.; Li, P.; Guo, B.; Ma, P.X. Antibacterial and conductive injectable hydrogels based on quaternized chitosan-graft-polyaniline/oxidized dextran for tissue engineering. *Acta Biomater.* **2015**, *26*, 236–248. [CrossRef] [PubMed]

89. Zhao, X.; Wu, H.; Guo, B.; Dong, R.; Qiu, Y.; Ma, P.X. Antibacterial anti-oxidant electroactive injectable hydrogel as self-healing wound dressing with hemostasis and adhesiveness for cutaneous wound healing. *Biomaterials* **2017**, *122*, 34–47. [CrossRef] [PubMed]
90. Kilmartin, P.; Gizdavic-Nikolaidis, M.; Zujovic, Z.; Travas-Sejdic, J.; Bowmaker, G.; Cooney, R. Free radical scavenging and antioxidant properties of conducting polymers examined using EPR and NMR spectroscopies. *Synth. Met.* **2005**, *153*, 153–156. [CrossRef]
91. Gunathilake, T.M.S.U.; Ching, Y.C.; Chuah, C.H.; Singh, R.; Lin, P.-C. Preparation and characterization of nanocellulose reinforced semi-interpenetrating polymer network of chitosan hidrogel. *Cellulose* **2017**, *24*, 2215–2228.
92. Gunathilake, T.M.S.U.; Ching, Y.C.; Chuah, C.H. Enhancement of curcumin bioavailability using nanocellulose reinforced chitosan hydrogel. *Polymers* **2017**, *9*, 64. [CrossRef]
93. Wang, Y.; Qian, J.; Zhao, N.; Liu, T.; Xu, W.; Suo, A. Novel hydroxyethyl chitosan/cellulose scaffolds with bubble-like porous structure for bone tissue engineering. *Carbohydr. Polym.* **2017**, *167*, 44–51. [CrossRef] [PubMed]
94. Gasperini, L.; Mano, J.F.; Reis, R.L. Natural polymers for the microencapsulation of cells. *J. R. Soc. Interface* **2014**, *11*. [CrossRef] [PubMed]
95. Radhakrishnan, J.; Krishnan, U.M.; Sethuraman, S. Hydrogel based injectable scaffolds for cardiac tissue regeneration. *Biotechnol. Adv.* **2014**, *32*, 449–461. [CrossRef] [PubMed]
96. Yang, G.; Xiao, Z.; Ren, X.; Long, H.; Qian, H.; Ma, K.; Guo, Y. Enzymatically crosslinked gelatin hydrogel promotes the proliferation of adipose tissue derived stromal cells. *Peer J* **2016**, *4*, e2497. [CrossRef] [PubMed]
97. Gan, Y.; Li, S.; Li, P.; Xu, Y.; Wang, L.; Zhao, C.; Ouyang, B.; Tu, B.; Zhang, C.; Luo, L.; et al. A controlled release codelivery system of MSCs encapsulated in dextran/gelatin hydrogel with TGF-β3-loaded nanoparticles for nucleus pulposus regeneration. *Stem Cells Int.* **2016**, *2016*, 9042019. [CrossRef] [PubMed]
98. Ding, L.; Li, X.; Sun, H.; Su, J.; Lin, N.; Péault, B.; Song, T.; Yanq, J.; Dai, J.; Hu, Y. Transplantation of bone marrow mesenchymal stem cells on collagen scaffolds for the functional regeneration of injured rat uterus. *Biomaterials* **2014**, *35*, 4888–4900. [CrossRef] [PubMed]
99. Jia, W.; Tang, H.; Wu, J.; Hou, X.; Chen, B.; Chen, W.; Zhao, Y.; Shi, C.; Zhou, F.; Yu, W.; et al. Urethral tissue regeneration using collagen scaffold modified with collagen binding VEGF in a beagle model. *Biomaterials* **2015**, *69*, 45–55. [CrossRef] [PubMed]
100. Xiao, W.; Hu, X.Y.; Zeng, W.; Huang, J.H.; Zhang, Y.G.; Luo, Z.J. Rapid sciatic nerve regeneration of rats by a surface modified collagen-chitosan scaffold. *Injury* **2013**, *44*, 941–946. [CrossRef] [PubMed]
101. Martínez, A.; Blanco, M.D.; Davidenko, N.; Cameron, R.E. Tailoring chitosan/collagen scaffolds for tissue engineering: Effect of composition and different crosslinking agents on scaffold properties. *Carbohydr. Polym.* **2015**, *132*, 606–619. [CrossRef] [PubMed]
102. Zuber, M.; Zia, F.; Zia, K.M.; Tabasum, S.; Salman, M.; Sultan, N. Collagen based polyurethanes: A review of recent advances and perspective. *Int. J. Biol. Macromol.* **2015**, *80*, 366–374. [CrossRef] [PubMed]
103. Wang, L.; Stegemann, J.P. Thermogelling chitosan and collagen composite hydrogels initiated with β-glycerophosphate for bone tissueengineering. *Biomaterials* **2010**, *31*, 3976–3985. [CrossRef] [PubMed]
104. Wang, L.; Stegemann, J.P. Glyoxal crosslinking of cell-seeded chitosan/collagen hydrogels for bone regeneration. *Acta Biomater.* **2011**, *7*, 2410–2417. [CrossRef] [PubMed]
105. Song, K.; Li, L.; Yan, X.; Zhang, W.; Zhang, Y.; Wang, Y.; Liu, T. Characterizationof human adipose tissue-derived stem cells in vitro culture and in vivo differentiation in a temperature-sensitive chitosan/β-glycerophosphate/collagen hybrid hydrogel. *Mater. Sci. Eng. C* **2017**, *70*, 231–240. [CrossRef] [PubMed]
106. Dang, Q.; Liu, K.; Zhang, Z.; Liu, C.; Liu, X.; Xin, Y.; Cheng, X.; Xu, T.; Cha, D.; Fan, B. Fabrication and evaluation of thermosensitive chitosan/collagen/β-glycerophosphate hydrogels for tissue regeneration. *Carbohydr. Polym.* **2017**, *167*, 145–157. [CrossRef] [PubMed]
107. Nakayama, A.; Kakugo, A.; Gong, J.P.; Osada, Y.; Takai, M.; Erata, T.; Kawano, S. High mechanical strength double-network hydrogel with bacterial cellulose. *Adv. Funct Mater.* **2004**, *14*, 1124–1128. [CrossRef]
108. Xu, D.; Huang, J.; Zhao, D.; Ding, B.; Zhang, L.; Cai, J. High-flexibility, high-toughness double-cross-linked chitin hydrogels by sequential chemicaland physical cross-linkings. *Adv. Mater.* **2016**, *28*, 5844–5849. [CrossRef] [PubMed]

109. Zhao, D.; Huang, J.; Zhong, Y.; Li, K.; Zhang, L.; Cai, J. High-strength and high-toughness double-cross-linked cellulose hydrogels: A new strategy usingsequential chemical and physical cross-linking. *Adv. Funct. Mater.* **2016**, *26*, 6279–6287. [CrossRef]
110. Lim, H.-S.; Kwon, E.; Lee, M.; Lee, Y.M.; Suh, K.-D. One-pot template-free synthesis of monodisperse hollow hydrogel microspheres and their resulting properties. *Macromol. Rapid Comm.* **2013**, *34*, 1243–1248. [CrossRef] [PubMed]
111. Kageyama, T.; Kakegawa, T.; Osaki, T.; Enomoto, J.; Ito, T.; Nittami, T.; Fukuda, J. Rapid engineering of endothelial cell-lined vascular-like structures in situ crosslinkable hydrogels. *Biofabrication* **2014**, *6*, 025006. [CrossRef] [PubMed]
112. McClendon, M.T.; Stupp, S.I. Tubular hydrogels of circumferentially aligned nanofibers to encapsulate and orient vascular cells? *Biomaterials* **2012**, *33*, 5713–5722. [CrossRef] [PubMed]
113. Wu, S.; Dong, H.; Li, Q.; Wang, G.; Cao, X. High strength, biocompatible hydrogels with designable shapes and special hollow-formed character using chitosan and gelatin. *Carbohydr. Polym.* **2017**, *168*, 147–152. [CrossRef] [PubMed]
114. Pedron, S.; van Lierop, S.; Horstman, P.; Penterman, R.; Broer, D.J.; Peeters, E. Stimuli responsive delivery vehicles for cardiac microtissue transplantation. *Adv. Funct. Mater.* **2011**, *21*, 1624–1630. [CrossRef]
115. Saini, H.; Navaei, A.; van Putten, A.; Nikkhah, M. 3D cardiac microtissues encapsulated with the co-culture of cardiomyocytes and cardiac fibroblasts. *Adv. Healthc. Mater.* **2015**, *4*, 1961–1971. [CrossRef] [PubMed]
116. Navaei, A.; Saini, H.; Christenson, W.; Sulliva, R.T.; Ros, R.; Nikkhah, M. Gold nanorod-incorporated gelatin-based conductive hydrogels for engineering cardiac tissue constructs. *Acta Biomater.* **2016**, *41*, 133–146. [CrossRef] [PubMed]
117. Martinelli, V.; Cellot, G.; Toma, F.M.; Long, C.S.; Caldwell, J.H.; Zentilin, L.; Giacca, M.; Turco, A.; Prato, M.; Ballerini, L.; et al. Carbon nanotubes promote growth and spontaneous electrical activity in cultured cardiac myocytes. *Nano Lett.* **2012**, *12*, 1831–1838. [CrossRef] [PubMed]
118. Firme, C.P.; Bandaru, P.R. Toxicity issues in the application of carbon nanotubes to biological systems. *Nanomed. Nanotechnol. Biol. Med.* **2010**, *6*, 245–256.
119. Vardharajula, S.; Ali, S.Z.; Tiwari, P.M.; Eroglu, E.; Vig, K.; Dennis, V.A.; Singh, S.R. Functionalized carbon nanotubes: Biomedical applications. *Int. J. Nanomed.* **2012**, *7*, 5361–5374.
120. Koppes, A.N.; Keating, K.W.; McGregor, A.L.; Koppes, R.A.; Kearns, K.R.K.; Ziemba, A.M.; McKay, C.A.; Zuidema, J.M.; Rivet, C.J.; Gilbert, C.J.; et al. Robust neurite extension following exogenous electrical stimulation within single walled carbon nanotube-composite hydrogels. *Acta Biomater.* **2016**, *39*, 34–43. [CrossRef] [PubMed]
121. Pugliese, R.; Gelain, F. Peptidic Biomaterials: From self-assembling to regenerative medicine. *Trends Biotechnol.* **2017**, *35*, 145–158. [CrossRef] [PubMed]
122. Zhang, S.; Holmes, T.C.; DiPersio, C.M.; Hynes, R.O.; Su, X.; Rich, A. Self-complementary oligopeptide matrices support mammalian cell attachment. *Biomaterials* **1995**, *16*, 1385–1393. [CrossRef]
123. Zhang, S.; Altman, M. Peptide self-assembly in functional polymer science and engineering. *React. Funct. Polym.* **1999**, *41*, 91–102. [CrossRef]
124. Ramachandran, S.; Tseng, Y.; Ye, Y.B. Repeated Rapid Shear-Responsiveness of Peptide Hydrogels with Tunable Shear Modulus. *Biomacromolecules* **2005**, *6*, 1316–1321. [CrossRef] [PubMed]
125. Raspa, A.; Saracino, G.A.A.; Pugliese, R.; Silva, D.; Cigognini, D.; Vescovi, A.; Gelain, F. Complementary Co-assembling Peptides: From In Silico Studies to In Vivo Application. *Adv. Funct. Mater.* **2014**, *24*, 6317–6328. [CrossRef]
126. Tarek, M. Molecular dynamics investigation of an oriented cyclic peptide nanotube in DMPC bilayers. *Biophys. J.* **2003**, *85*, 2287–2298. [CrossRef]
127. Van Bommel, K.J.C.; van der Pol, C.; Kuizebelt, I.; Friggeri, A.; Heeres, A.; Meetsma, A.; Feringa, B.L.; van Esch, J. Responsive cyclohexane-based low-molecular-weight hydrogelators with modular architecture. *Angew. Chem. Int. Ed.* **2004**, *43*, 1663. [CrossRef] [PubMed]
128. Cui, H.; Webber, M.J.; Stupp, S.I. Self-assembly of peptide amphiphiles: From molecules to nanostructures to biomaterials. *Pept. Sci.* **2010**, *94*, 1–18. [CrossRef] [PubMed]
129. Xu, H. Hydrophobic-region-induced transitions in self-assembled peptide nanostructures. *Langmuir* **2009**, *25*, 4115–4123. [CrossRef] [PubMed]

130. Kisiday, J.; Jin, M.; Kurz, B.; Hung, H.; Semino, C.; Zhang, S.G.; Grodzinsky, A.J. Self-assembling peptide hydrogel fodters chondrocyte extracellular matrix production and cell division: Implications for cartilage tissue repair. *Proc. Natl. Acad. Sci. USA* **2002**, *99*, 9996–10001. [CrossRef] [PubMed]
131. Mujeeb, A.; Miller, A.F.; Saiani, A.; Gough, J.E. Self-assembled octapeptide scaffolds for in vitro chondrocyte culture. *Acta Biomater.* **2013**, *9*, 4609–4617. [CrossRef] [PubMed]
132. Zhang, S.M.; Greenfied, M.A.; Mata, A.; Palmer, L.C.; Bitton, R.; Mantei, J.R.; Aparicio, C.; de la Cruz, M.O.; Stupp, S.I. A self-assembly pathway to aligned monodomain gels. *Nat. Mater.* **2010**, *9*, 594–601. [CrossRef] [PubMed]
133. Tian, Y.F.; Devgun, J.M.; Collier, J.H. Fibrillized peptide microgels for cell ancapsulation and 3D cell culture. *Soft Matter* **2011**, *7*, 6005–6011. [CrossRef] [PubMed]
134. Zhao, Y.; Yokoi, H.; Tanaka, M.; Kinoshita, T.; Tan, T. Self-assembled pH-responsive hydrogels composed of the RATEA16 peptide. *Biomacromolecules* **2008**, *9*, 1511–1518. [CrossRef] [PubMed]
135. Zhou, M.; Smith, A.M.; Das, A.K.; Hodson, N.W.; Collins, R.F.; Ulijn, R.V.; Gough, J.E. Self-assembled peptide-based hydrogels as scaffolds for anchorage-dependent cells. *Biomaterials* **2009**, *30*, 2523–2530. [CrossRef] [PubMed]
136. Dou, X.-Q.; Feng, C.-L. Amino Acids and Peptide-Based Supramolecular Hydrogels for Three-Dimensional Cell Culture. *Adv. Mater.* **2017**, *29*, 1604062. [CrossRef] [PubMed]
137. Lin, B.F.; Megley, K.A.; Viswanathan, N.; Krogstad, D.V.; Drews, L.B.; Kade, M.J.; Qian, Y.C.; Tirrell, M.V. pH-Responsive Branched Peptide Amphiphile Hydrogel Designed for Applications in Regenerative Medicine with Potential as Injectable Tissue Scaffolds. *J. Mater. Chem.* **2012**, *22*, 19447–19454. [CrossRef]
138. Panda, J.J.; Dua, R.; Mishra, A.; Mittra, B.; Chauhan, V.S. 3D cell growth and proliferation on RGD functionalized nanofibrillar hydrogel based on a conformationally restricted residue containing dipeptide. *ACS Appl. Mater. Interfaces* **2010**, *2*, 2839–2848. [CrossRef] [PubMed]
139. Marchesan, S.; Qu, Y.; Waddington, L.J.; Easton, C.D.; Glattauer, V.; Lithgow, T.J.; McLean, K.M.; Forsythe, J.S.; Hartley, P.G. Self-assembly of ciprofloxacin and a tripeptide into an antimicrobial nanostructured hydrogel. *Biomaterials* **2013**, *34*, 3678–3687. [CrossRef] [PubMed]
140. Zhao, F.; Gao, Y.; Shi, J.F.; Browdy, H.M.; Xu, B. Novel anisotropic supramolecular hydrogel with high stability over a wide pH range. *Langmuir* **2011**, *27*, 1510–1512. [CrossRef] [PubMed]
141. Orbach, R.; Adler-Abramovich, L.; Zigerson, S.; Mironi-Harpaz, I.; Seliktar, D.; Gazit, E. Self-assembled Fmoc-peptides as a platform for the formation of nanostructures and hydrogel. *Biomacromolecules* **2009**, *10*, 2646–2651. [CrossRef] [PubMed]
142. Komatsu, H.; Tsukiji, S.; Ikeda, M.; Hamachi, I. Stiff, multistimuli-responsive supramolecular hydrogels as unique molds for 2D/3D microarchitectures of live cells. *Chem. Asian J.* **2011**, *6*, 2368–2375. [CrossRef] [PubMed]
143. Zhou, J.; Du, X.W.; Gao, Y.; Shi, J.F.; Xu, B. Aromatic-aromatic interactions enhance interfiber contacts for enzymatic formation of a spontaneously aligned supramolecular hydrogel. *J. Am. Chem. Soc.* **2014**, *136*, 2970–2973. [CrossRef] [PubMed]
144. Liu, G.F.; Zhang, D.; Feng, C.L. Control of three-dimensional cell adhesion by the chirality of nanofibers in hydrogels. *Angew. Chem. Int. Ed.* **2014**, *53*, 7789–7793. [CrossRef] [PubMed]
145. Funke, W.; Okay, O.; Joos-Muller, B. Microgels—Intramolecularly crosslinked macromolecules with a globular structure. *Adv. Polym. Sci.* **1998**, *136*, 139–234.
146. Shewan, H.M.; Stokes, J.R. Review of techniques to manufacture micro-hydrogel particles for the food industry and their applications. *J. Food Eng.* **2013**, *119*, 781–792. [CrossRef]
147. Farjami, T.; Madadlou, A. Fabrication methods of biopolymeric microgels and microgel-based hydrogels. *Food Hydrocoll.* **2017**, *62*, 262–272. [CrossRef]
148. Gonçalves, V.L.; Laranjeira, M.C.M.; Fávere, V.T.; Pedrosa, R.C. Effect of crosslinking agents on chitosan microspheres in controlled release of diclofenac sodium. *Polímeros* **2005**, *15*, 6–12. [CrossRef]
149. Zhang, J.; Liang, L.; Tian, Z.; Chen, L.; Subirade, M. Preparation and in vitro evaluation of calcium-induced soy protein isolate nanoparticles and their formation mechanism study. *Food Chem.* **2012**, *133*, 390–399. [CrossRef] [PubMed]
150. Nicolai, T.; Durand, D. Controlled food protein aggregation for new functionality. *Curr. Opin. Colloid Interface Sci.* **2013**, *18*, 249–256. [CrossRef]

151. Nicolai, T. Formation and functionality of self-assembled whey protein microgels. *Colloids Surf. B* **2016**, *137*, 32–38. [CrossRef] [PubMed]
152. Yu, S.; Hu, J.; Pan, X.; Yao, P.; Jiang, M. Stable and pH-sensitive nanogels prepared by self-assembly of chitosan and ovalbumin. *Langmuir* **2006**, *22*, 2754–2759. [CrossRef] [PubMed]
153. Murakami, R.; Takashima, R. Mechanical properties of the capsules of chitosan–soy globulin polyelectrolyte complex. *Food Hydrocoll.* **2003**, *17*, 885–888. [CrossRef]
154. Thomas, W.R. Carrageenan. In *Thickenning and Gelling Agents for Food*; Imeso, A.P., Ed.; Springer: Hong Kong, China, 1997; pp. 45–59.
155. Gabriele, A.; Spyropoulos, F.; Norton, I.T. Kinetic study of fluid gel formation and viscoelastic response with kappa-carrageen. *Food Hydrocoll.* **2009**, *23*, 2054–2061. [CrossRef]
156. Heidebach, T.; Först, P.; Kulozik, U. Transglutaminase-induced caseinate gelation for the microencapsulation of probiotic cells. *Int. Dairy J.* **2009**, *19*, 77–84. [CrossRef]
157. Madadlou, A.; Jaberipour, S.; Eskandari, M.H. Nanoparticulation of enzymatically cross-linked whey proteins to encapsulate caffeine via microemulsification/heat gelation procedure. *LWT—Food Sci. Technol.* **2014**, *57*, 725–730. [CrossRef]
158. Kumbar, S.G.; Kulkarni, A.R.; Aminabhavi, T.M. Crosslinked chitosan microspheres for encapsulation of diclofenac sodium: Effect of crosslinking agent. *J. Microencapsul.* **2002**, *19*, 173–180. [CrossRef] [PubMed]
159. Gazme, B.; Madadlou, A. Fabrication of whey protein–pectin conjugate particles through laccase-induced gelation of microemulsified nanodroplets. *Food Hydrocoll.* **2014**, *40*, 189–195. [CrossRef]
160. Shu, X.Z.; Zhu, K.J. Chitosan/gelatin microspheres prepared by modified emulsification and ionotropic gelation. *J. Microencapsul.* **2001**, *18*, 237–245. [PubMed]
161. Jiang, S.; Chen, Q.; Tripathy, M.; Luijten, E.; Schweizer, K.S.; Granick, S. Janus particle synthesis and assembly. *Adv. Mater.* **2010**, *22*, 1060–1071. [CrossRef] [PubMed]
162. Marquis, M.; Davy, J.; Cathala, B.; Fang, A.; Renard, D. Microfluidics assisted generation of innovative polysaccharide hydrogel microparticles. *Carbohydr. Polym.* **2015**, *116*, 189–199. [CrossRef] [PubMed]
163. Wang, L.Y.; Ma, G.H.; Su, Z.G. Preparation of uniform sized chitosan microspheres by membrane emulsification technique and application as a carrier of protein drug. *J. Control. Release* **2005**, *106*, 62–75. [CrossRef] [PubMed]
164. Wang, L.Y.; Gu, Y.H.; Zhou, Q.Z.; Ma, G.H.; Wan, Y.H.; Su, Z.G. Preparation and characterization of uniform-sized chitosan microspheres containing insulin by membrane emulsification and a two-step solidification process. *Colloids Surf. B* **2006**, *50*, 126–135. [CrossRef] [PubMed]
165. Farjami, T.; Madadlou, A.; Labbafi, M. Characteristics of the bulk hydrogels made of the citric acid cross-linked whey protein microgels. *Food Hydrocoll.* **2015**, *50*, 159–165. [CrossRef]
166. Zheng, C.; Huang, Z. Microgel reinforced composite hydrogels with pH-responsive, self-healing properties. *Colloids Surf. A* **2015**, *468*, 327–332. [CrossRef]
167. Sivakumaran, D.; Maitland, D.; Hoare, T. Injectable microgel-hydrogel composites for prolonged small-molecule drug delivery. *Biomacromolecules* **2011**, *12*, 4112–4120. [CrossRef] [PubMed]
168. Xie, T. Tunable polymer multi-shape memory effect. *Nature* **2010**, *464*, 267–270. [CrossRef] [PubMed]
169. Zhao, Q.; Qi, H.J.; Xie, T. Recent progress in shape memory polymer: New behavior, enabling materials, and mechanistic understanding. *Prog. Polym. Sci.* **2015**, *49–50*, 79–120. [CrossRef]
170. Habault, D.; Zhang, H.J.; Zhao, Y. Light-triggered self-healing and shape-memory polymers. *Chem. Soc. Rev.* **2013**, *42*, 7244–7256. [CrossRef] [PubMed]
171. Xu, H.X.; Yu, C.J.; Wang, S.D.; Malyarchuk, V.; Xie, T.; Rogers, J.A. Deformable, programmable, and shape-memorizing micro-optics. *Adv. Funct. Mater.* **2013**, *23*, 3299–3306. [CrossRef]
172. Small, W.; Singhal, P.; Wilson, T.S.; Maitland, D.J. Thermo-moisture responsive polyurethane shape-memory polymer and composites: A review. *J. Mater. Chem.* **2010**, *20*, 3356–3366. [PubMed]
173. Lu, W.; Le, X.; Zhang, J.; Huang, Y.; Chen, T. Supramolecular shape memory hydrogels: A new bridge between stimuli-responsive polymers and supramolecular chemistry. *Chem. Soc. Rev.* **2017**, *46*, 1284–1294. [CrossRef] [PubMed]
174. Sun, Z.F.; Lv, F.C.; Cao, L.J.; Liu, L.; Zhang, Y.; Lu, Z.G. Multistimuli-Responsive, moldable supramolecular hydrogels croos-linked by ultrafast complexation of metal ions and biopolymers. *Angew. Chem. Int. Ed.* **2015**, *54*, 7944–7948. [CrossRef] [PubMed]

175. Dong, Z.Q.; Cao, Y.; Yuan, Q.J.; Wang, Y.F.; Li, J.H.; Li, B.J.; Zhang, S. Redox- and glucose-induced shape-memory polymers. *Macromol. Rapid Commun.* **2013**, *34*, 867–872. [CrossRef] [PubMed]
176. Dong, M.X.; Babalhavaeji, A.; Samanta, S.; Beharry, A.A.; Woolley, G.A. Red-shifting azobenzene photoswitches for in vivo use. *Acc. Chem. Res.* **2015**, *48*, 2662–2670. [CrossRef] [PubMed]
177. Draper, E.R.; Adams, D.J. Photoresponsive gelators. *Chem. Commun.* **2016**, *52*, 8196–8206. [CrossRef] [PubMed]
178. Xie, F.; Ouyang, G.; Qin, L.; Liu, M. Supra-dendron gelator based on azobenzene–cyclodextrin host–guest interactions: Photoswitched optical and chiroptical reversibility. *Chem. Eur. J.* **2016**, *22*, 18208–18214. [CrossRef] [PubMed]
179. Neves, M.I.; Wechsler, M.E.; Gomes, M.E.; Reis, R.L.; Granja, P.L.; Peppas, N.A. Molecularly imprinted intelligent scaffolds for tissue engineering applications. *Tissue Eng. B* **2017**, *23*, 27–43. [CrossRef] [PubMed]
180. Kryscio, D.R.; Peppas, N.A. Critical review and perspective of macromolecularly imprinted polymers. *Acta Biomater.* **2012**, *8*, 461–473. [CrossRef] [PubMed]
181. Culver, H.R.; Daily, A.D.; Khademhosseini, A.; Peppas, N.A. Intelligent recognitive systems in nanomedicine. *Curr. Opin. Chem. Eng.* **2014**, *4*, 105–113. [CrossRef] [PubMed]
182. Zhao, K.; Cheng, G.; Huang, J.; Ying, X. Rebinding and recognition properties of protein-macromolecularly imprinted calcium phosphate/alginate hybrid polymer microspheres. *React. Funct. Polym.* **2008**, *68*, 732–741. [CrossRef]
183. Zhao, K.Y.; Wei, J.F.; Zhou, J.Y.; Zhao, Y.P.; Cheng, G.X. The rebinding properties of bovine serum albumin imprinted calcium phosphate/polyacrylate/alginate hybrid polymer microspheres. *Adv. Mater. Res.* **2010**, *152*, 1636–1644. [CrossRef]
184. Zhu, D.W.; Chen, Z.; Zhao, K.Y.; Kan, B.H.; Liu, L.X.; Dong, X.; Wang, H.; Zhang, C.; Leng, X.G.; Zhang, L.H. Polypropylene non-woven supported fibronectin molecular imprinted calcium alginate/polyacrylamide hydrogel filmfor cell adhesion. *Chin. Chem. Lett.* **2015**, *26*, 807–992. [CrossRef]
185. Fukazawa, K.; Ishihara, K. Fabrication of a cell adhesive protein imprinting surface with an artificial cell membrane structure for cell capturing. *Biosens. Bioelectron.* **2009**, *25*, 609–614. [CrossRef] [PubMed]
186. Singh, L.K.; Singh, M.; Singh, M. Biopolymeric receptor for peptide recognition by molecular imprinting approach—Synthesis, characterization and application. *Mater. Sci. Eng. C Mater. Biol. Appl.* **2014**, *45*, 383–394. [CrossRef] [PubMed]
187. Xia, Y.Q.; Guo, T.Y.; Song, M.D.; Zhang, B.H.; Zhang, B.L. Hemoglobin recognition by imprinting in semiinterpenetrating polymer network hydrogel based on polyacrylamide and chitosan. *Biomacromolecules* **2005**, *6*, 2601–2606. [CrossRef] [PubMed]
188. Gao, R.; Zhao, S.; Hao, Y.; Zhang, L.; Cui, X.; Liu, D.; Tang, Y. Facile and green synthesis of polysaccharide-based magnetic molecularly imprinted nanoparticles for protein recognition. *RSC Adv.* **2015**, *5*, 88436–88444. [CrossRef]
189. DePorter, S.M.; Lui, I.; McNaughton, B.R. Programmed cell adhesion and growth on cell-imprinted polyacrylamide hydrogels. *Soft Matter* **2012**, *8*, 10403–10408. [CrossRef]
190. Włodarczyk-Biegun, M.K.; del Campo, A. 3D Bioprinting of structural proteins. *Biomaterials* **2017**, *134*, 180–201. [CrossRef] [PubMed]
191. Arslan-Yildiz, A.; El Assal, R.; Chen, P.; Guven, S.; Inci, F.; Demirci, U. Towards artificial tissue models: Past, present, and future of 3D bioprinting. *Biofabrication* **2016**, *8*, 014103. [CrossRef] [PubMed]
192. Sears, N.A.; Seshadri, D.R.; Dhavalikar, P.S.; Cosgriff-Hernandez, E. A review of three-dimensional printing in tissue engineering. *Tissue Eng. B* **2016**, *22*, 298–310. [CrossRef] [PubMed]
193. Lei, M.; Wang, X. Biodegradable polymers and stem cells for bioprinting. *Molecules* **2016**, *21*, 539. [CrossRef] [PubMed]
194. Pati, F.; Gantelius, J.; Svahn, H.A. 3D Bioprinting of tissue/organ models. *Angew. Chem. Int. Ed. Engl.* **2016**, *55*, 4650–4665. [CrossRef] [PubMed]
195. Chia, H.N.; Wu, B.M. Recent advances in 3D printing of biomaterials. *J. Biol. Eng.* **2015**, *9*, 4. [CrossRef] [PubMed]
196. Xu, F.; Moon, S.; Emre, A.E.; Lien, C.; Turali, E.S.; Demirci, U. Cell Bioprinting as a Potential High-Throughput Method for Fabricating Cell-Based Biosensors (CBBs). In Proceedings of the IEEE Sensors, Christchurch, New Zealand, 25–28 October 2009; pp. 387–391.

197. Lee, W.; Debasitis, J.C.; Lee, V.K.; Lee, J.H.; Fischer, K.; Edminster, K.; Park, J.K.; Yoo, S.S. Multi-layered culture of human skin fibroblasts and keratinocytes through three-dimensional freeform fabrication. *Biomaterials* **2009**, *30*, 1587–1595. [CrossRef] [PubMed]
198. Lee, W.; Pinckney, J.; Lee, V.; Lee, J.H.; Fischer, K.; Polio, S.; Park, J.K.; Yoo, S.S. Three-dimensional bioprinting of rat embryonic neural cells. *Neuroreport* **2009**, *20*, 798–803. [CrossRef] [PubMed]
199. Malda, J.; Visser, J.; Melchels, F.P.; Jüngst, T.; Hennink, W.E.; Dhert, W.J.A.; Groll, J.; Hutmacher, D.W. Engineering Hydrogels for Biofabrication. *Adv. Mater.* **2013**, *25*, 5011–5028. [CrossRef] [PubMed]
200. Roth, E.A.; Xu, T.; Das, M.; Gregory, C.; Hickman, J.J.; Boland, T. Inkjet printing for high-throughput cell patterning. *Biomaterials* **2004**, *25*, 3707–3715. [CrossRef] [PubMed]
201. Gruene, M.; Unger, C.; Koch, L.; Deiwick, A.; Chichkov, B. Dispensing pico to nanolitre of a natural hydrogel by laser-assisted bioprinting. *Biomed. Eng. Online* **2011**, *10*, 19. [CrossRef] [PubMed]
202. Melchels, F.P.W.; Domingos, M.A.N.; Klein, T.J.; Malda, J.; Bartolo, P.J.; Hutmacher, D.W. Additive manufacturing of tissues and organs. *Prog. Polym. Sci.* **2012**, *37*, 1079–1104. [CrossRef]
203. Murphy, S.V.; Skardal, A.; Atala, A. Evaluation of hydrogels for bio-printing applications. *J. Biomed. Mater. Res. A* **2013**, *101*, 272–284. [CrossRef] [PubMed]
204. Yeo, M.; Lee, J.-S.; Chun, W.; Kim, G.H. An innovative collagen-based cell-printing method for obtaining human adipose stem cell-laden structures consisting of core–sheath structures for tissue engineering. *Biomacromolecules* **2016**, *17*, 1365–1375. [CrossRef] [PubMed]
205. Lee, V.; Singh, G.; Trasatti, J.P.; Bjornsson, C.; Xu, X.; Tran, T.N.; Yoo, S.S.; Dai, G.; Karande, P. Design and fabrication of human skin by three-dimensional bioprinting. *Tissue Eng. C* **2014**, *20*, 473–484. [CrossRef] [PubMed]
206. Skardal, A.; Atala, A. Biomaterials for integration with 3-D bioprinting. *Ann. Biomed. Eng.* **2015**, *43*, 730–746. [CrossRef] [PubMed]
207. De la Puente, P.; Ludeña, D. Cell culture in autologous fibrin scaffolds for applications in tissue engineering. *Exp. Cell. Res.* **2014**, *322*, 1–11. [CrossRef] [PubMed]
208. Xu, T.; Gregory, C.A.; Molnar, P.; Cui, X.; Jalota, S.; Bhaduri, S.B.; Boland, T. Viability and electrophysiology of neural cell structures generated by the inkjet printing method. *Biomaterials* **2006**, *27*, 3580–3588. [CrossRef] [PubMed]
209. Lee, Y.B.; Polio, S.; Lee, W.; Dai, G.; Menon, L.; Carroll, R.S.; Yoo, S.S. Bio-printing of collagen and VEGF-releasing fibrin gel scaffolds for neural stem cell culture. *Exp. Neurol.* **2010**, *223*, 645–652. [CrossRef] [PubMed]

4

Immobilization of Colloidal Monolayers at Fluid–Fluid Interfaces

Peter T. Bähler [1], Michele Zanini [1], Giulia Morgese [2], Edmondo M. Benetti [2] and Lucio Isa [1,*]

1 Laboratory for Interfaces, Soft matter and Assembly, Department of Materials, ETH Zurich, Vladimir-Prleog-Weg 5, 8093 Zurich, Switzerland; baehler.p@gmx.ch (P.T.B.); michele.zanini@mat.ethz.ch (M.Z.)
2 Laboratory for Surface Science and Technology, Department of Materials, ETH Zurich, Vladimir-Prleog-Weg 5, 8093 Zurich, Switzerland; giulia.morgese@mat.ethz.ch (G.M.); edmondo.benetti@mat.ethz.ch (E.M.B.)
* Correspondence: lucio.isa@mat.ethz.ch

Academic Editor: Clemens K. Weiss

Abstract: Monolayers of colloidal particles trapped at an interface between two immiscible fluids play a pivotal role in many applications and act as essential models in fundamental studies. One of the main advantages of these systems is that non-close packed monolayers with tunable inter-particle spacing can be formed, as required, for instance, in surface patterning and sensing applications. At the same time, the immobilization of particles locked into desired structures to be transferred to solid substrates remains challenging. Here, we describe three different strategies to immobilize monolayers of polystyrene microparticles at water–decane interfaces. The first route is based on the leaking of polystyrene oligomers from the particles themselves, which leads to the formation of a rigid interfacial film. The other two rely on in situ interfacial polymerization routes that embed the particles into a polymer membrane. By tracking the motion of the colloids at the interface, we can follow in real-time the formation of the polymer membranes and we interestingly find that the onset of the polymerization reaction is accompanied by an increase in particle mobility determined by Marangoni flows at the interface. These results pave the way for future developments in the realization of thin tailored composite polymer-particle membranes.

Keywords: colloids; fluid interfaces; immobilization

---

1. Introduction

Monolayers of colloidal particles are widely used both as model systems [1–3] as well as in a large range of applications, spanning from the fabrication of lithography masks [4,5] and anti-reflective coatings [6] to emulsion stabilizers [7] and patterning elements for ultra-thin polymeric membranes [8]. Standard protocols to prepare monolayers of colloidal particles adsorbed onto a solid substrate involve direct adsorption from solution (including sedimentation [3] and electrostatic adsorption [9,10]), spin-coating [11], controlled drying [12], convective assembly [13] and electric-field-assisted deposition [14,15]. In all of these cases, dense particle monolayers are produced, mostly with particles sharing contacts in closely packed arrays. Applications, e.g., in biosensing [16] and patterning [5,17,18], and fundamental studies, e.g., in tailoring surface adhesion [19] and friction [20] by particle adsorption, often also require non-close-packed monolayers with control on both the particle size and the spacing between them. This can be achieved by first depositing a close-packed monolayer and then by reducing the particle size by reactive etching [21], or alternatively by depositing the particles onto stretchable elastomers [22]. Albeit successful and easy to implement, these methods present some limitations. The former process does not afford independent control on size and spacing, i.e., the two are linked by the initial particle diameter. Moreover, sizes can only be reduced up to a given point before the particles lose their shape integrity. The latter strategy involves high-temperature steps, which restrict its applicability to inorganic particles. Another alternative exploits the self-assembly of colloids at fluid–fluid interfaces [4,23]. In this case, monolayers of particles are first formed at the interface between two fluids, i.e., oil-water or air-water, and are then deposited onto a solid substrate. The presence of long-range electrostatic forces between particles of suitable surface charge and wettability enables the possibility to obtain regular, non-close-packed 2D lattices [24,25]. These forces are dipolar in origin and can either be generated by charges on the particle–water surface or on the particle–nonpolar fluid (air or oil) surface, depending on the materials of interest. In particular, in the case of dominant charges on the particle-nonpolar fluid surface, it has been shown that adsorption of ions from air in atmospheric conditions, or migration of ions from the water side, can greatly reduce the effective surface charge density and thus lead to weaker repulsion [26,27]. These contributions are less important in the case of highly water-insoluble oils, such as purified alkanes, which tend to maintain higher unscreened surface charges leading to stronger electrostatic repulsion. For the purpose of structural tailoring, spacings up to ten particle diameters can be achieved and smoothly tuned by controlling the number of particles per unit area injected at the interface [4] or by monolayer compression [26,28]. The most complicated part of the process is the transfer of the particles from the interface to the substrate without destroying the monolayer's integrity. Different approaches have been devised, either using solvent exchange [23] or by choosing oil phases with suitable viscosities and volatilities [4]. The process remains nevertheless challenging due to limited adhesion between the particles and the substrate. Additional routes have been pursued by using soft deformable particles that show an increased area of contact and thus higher adhesion with the template, but, in this case, the range of available inter-particle spacing is limited and it is dictated by the range of soft steric interactions at the interface [29].

In this paper, we describe three different approaches that can be used to immobilize particle monolayers directly at a fluid–fluid interface before deposition, which could therefore be used in the future to circumvent the issues described above. The paper is organized as follows. We initially present a first route, where immobile monolayers of polystyrene (PS) particles are created spontaneously at a water–decane interface without the addition of any external ingredient. The kinetics of the immobilization process is followed by in situ microscopy and particle tracking. We then move to another approach, where two different interfacial polymerization routes are followed to immobilize the particles at the water–decane interface into thin polymer membranes. The presentation and discussion of the experimental results is followed by conclusions and a perspective for future work. An experimental section detailing the materials and methods used in our work closes the manuscript.

2. Results and Discussion

Figure 1 illustrates schematically the three different approaches that we have investigated to study PS particle monolayer immobilization at the water–decane interface. Results and discussion follow for each route.

2.1. Spontaneous Immobilization

Experiments aimed at studying particle dynamics within a monolayer at a water–decane interface revealed that particle mobility was a strong function of the residence time of the particles at the interface. Systematic investigations showed that, given a sufficient amount of time, the particles became completely immobilized without distorting the interface microstructure. The experiments were carried out as follows (additional details in the Experimental Section). A given number of fluorescent polystyrene particles of either 2.8 or 1.08 μm diameter was injected using a micropipette at a macroscopically flat oil-water interface created in a customized sample cell. A solution of 50:50 ultra-pure water and isopropanol was used to aid the spreading of the particles at the interface [30]. The latter was obtained by filling a metallic ring to its rim with water and carefully covering it with decane, after which the sample cell was sealed at the top with a glass slide. The cell was then placed under a microscope, where videos of the particles at the interface were captured by fluorescence microscopy as a function of waiting/residence time at the interface. The videos were successively processed to extract particle trajectories and to compute the particle mean squared displacements (MSD) as a function of time. The MSD is a standard quantity to describe the modality of the motion of colloidal particles driven by thermal fluctuations. In the case of freely diffusing particles, the MSD plotted as a function of time shows a slope of 1 in a log-log plot. Slopes below 1 indicate sub-diffusive behavior characteristic of crowded or super-cooled systems, and flat lines indicate complete local caging of the particles that cannot move beyond a specific distance, as in the case of completely immobilized particles.

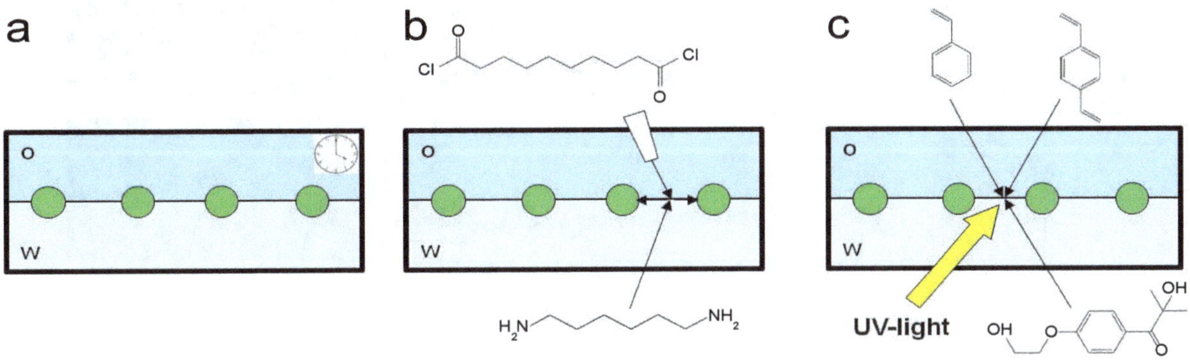

Figure 1. Schematics of the three different immobilization routes. (**a**) spontaneous immobilization at a water–decane interface. Particles within the monolayer simply become immobile with time; (**b**) nylon interfacial polymerization. After the spontaneous interfacial adsorption of 1,6-diaminohexane from the water phase, sebacoyl dichloride is injected into the organic phase to start the polymerization; (**c**) polystyrene interfacial polymerization. After the spontaneous interfacial adsorption of the monomer (styrene) and crosslinker (*p*-divinylbenzene) from the oil phase and of the initiator (Irgacure 2959) from the water phase, the system is illuminated by UV light to initiate the free radical polymerization of styrene at the interface.

Figures 2 and 3 show the MSDs of the 2.8 and 1.08 μm diameter particles, respectively, for different residence times t_w at the interface, where $t_w = 0$ corresponds to the time at which the particles were injected at the interface. The data are accompanied by a series of representative microscopy images showing the interface microstructure at the various waiting times. Starting from the data in Figure 2, we describe the mobility and the arrangement of the PS particles at the interface. The first observation to be made is that the particles self-assembled into non-close-packed crystals due to electrostatic repulsion. The corresponding MSD at short waiting times displayed the expected behavior. At short times, the MSD increased sub-linearly with time, indicating sub-diffusive motion of the particles interacting with their neighbors. After a few seconds, the MSDs plateaued and became flat, indicating that the particles became completely caged by their neighbors. This situation corresponds to particles rattling around their lattice positions. Remarkably, the rattling happened over distances much smaller than the typical particle size or inter-particle distance, indicating strong local trapping. After a waiting time of approximately 16 h, the situation was radically different. Here, the local rattling of the particles in the lattice had completely disappeared and the MSD was flat at all times. The value of the MSD corresponded to displacements of the order of one tenth of the particle diameter, which basically constitutes the resolution limit of the particle tracking. We can then conclude that, after a few hours, the particles stopped moving completely and a fully immobile monolayer was formed. Very interestingly, no appreciable changes in the interface microstructure were found. The particle monolayer maintained its non-close-packed crystalline structure, as can be seen by the micrographs. Analogous results are seen in Figure 3 for the smaller PS particles.

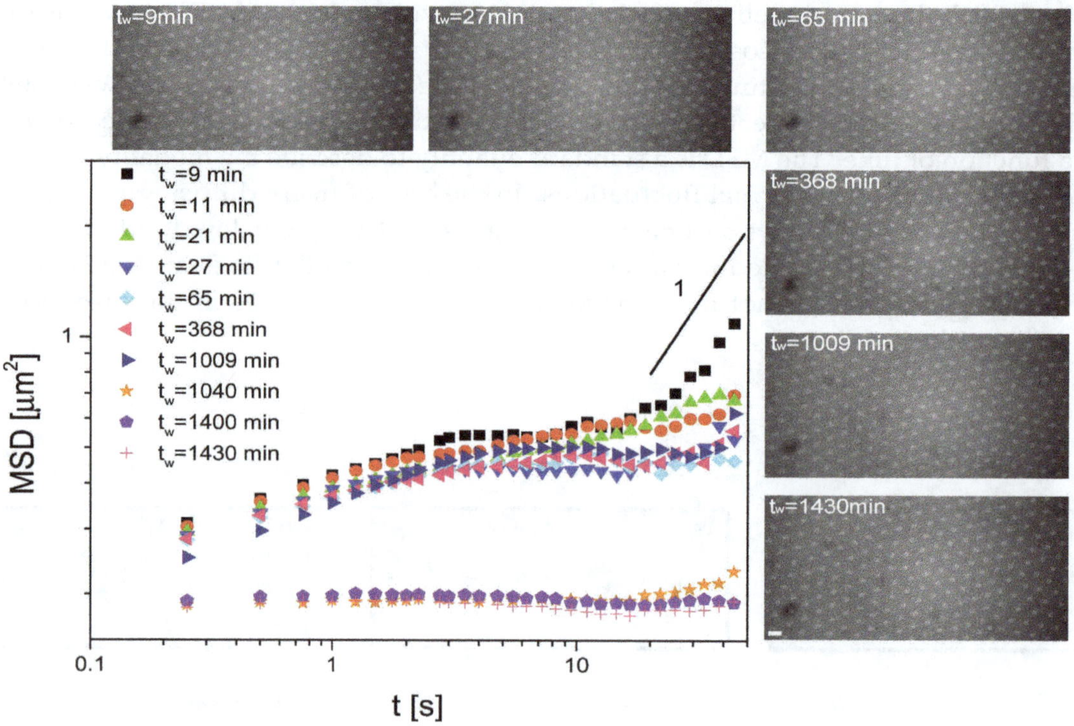

Figure 2. Mean squared displacements (MSD) versus time for 2.8 μm polystyrene (PS) particles at the water–decane interface as a function of residence time at the interface t_w. A solid line with slope 1 representative of freely diffusive motion is included for comparison. The micrographs show the interface microstructure at various waiting times. The scale bar is 10 μm.

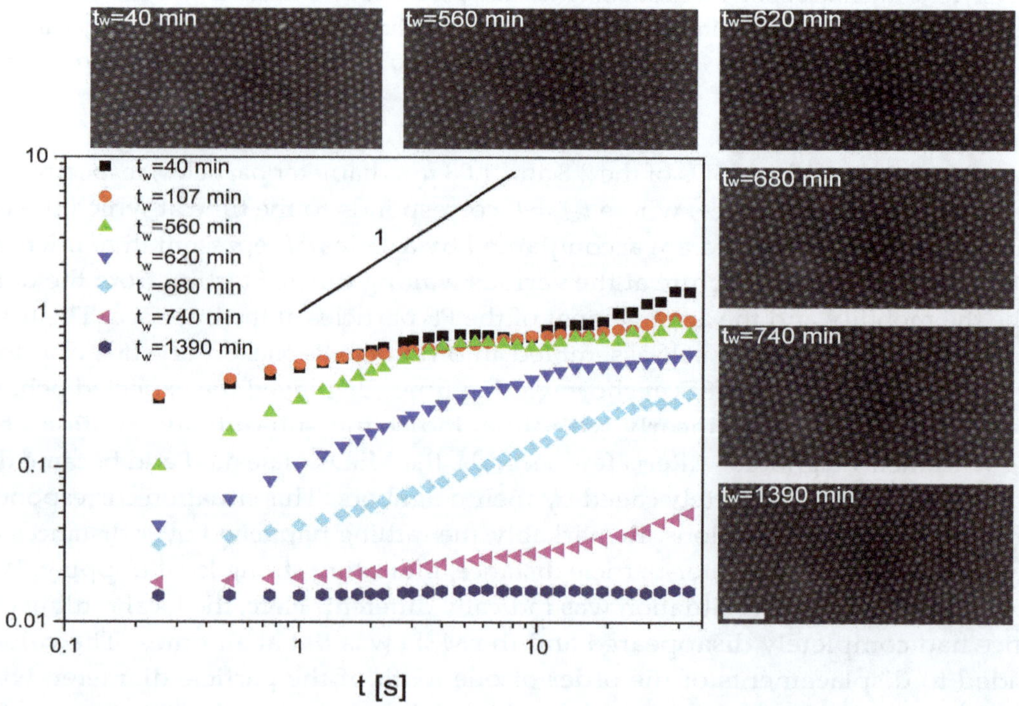

Figure 3. MSD versus time for 1.08 μm PS particles at the water–decane interface as a function of residence time at the interface t_w. A solid line with slope one representative of freely diffusive motion is included for comparison. The micrographs show the interface microstructure at various waiting times. The scale bar is 10 μm.

The complete immobilization of the particles suggests that a strong elastic film was formed at the interface as a function of the particle residence time. This result was initially surprising, given that no other component was added to the system during the experiments. We therefore proceeded to unravel the reason for the formation of this very stiff membrane. After testing that this was not due to the presence of surface-active contaminations coming from the sample cells or the liquids used, the only conclusion left was that the particles themselves were responsible for their immobilization. This was presumably due to the appearance of surface-active impurities leaking from the particles into decane and then adsorbing at the interface, as we later confirmed by the following experiment. After drying 0.1 mL of the PS particles aqueous stock suspension at 40 °C overnight in a vacuum oven in an Eppendorf tube, they were subsequently redispersed in 1.5 mL of purified decane, sonicated for at least 10 min and allowed to sediment overnight. We then took the supernatant and measured its interfacial tension against ultra-pure water. We measured a strong reduction of the interfacial tension from the pure water–decane value of 53 mN/m down to 32 mN/m (1.08 μm) and 40 mN/m (2.8 μm) after 3 h (significantly less than the timescales for interfacial arrest reported in Figures 2 and 3), which confirmed the leaking of surface-active species from the PS particles into decane. We then proceeded to identify these surface-active species by running UV-Vis spectroscopy on the same supernatants. The results are shown in Figure 4. The spectra for the supernatants of the 1.08 and 2.8 μm particles (red and green curves) showed distinctive bands in two regions. A first very intense band in the 250 nm range and a second less intense band in the 400–550 nm region. In order to identify and confirm the origin of these bands, we also ran UV-Vis spectroscopy on decane solutions of the same dye molecules used to render our PS particles fluorescent and on decane solutions of PS oligomers having two molecular weights, 300 and 2200 Da respectively. The first band for the particle supernatants was compatible with the peak of the PS oligomers, as shown by comparing the green and red curves in Figure 4 with the blue and purple ones obtained for the pure polystyrene oligomers. The fact that the peak position of the green and red curves was identical and lay in between the ones for PS300 and PS2200 indicates that oligomers of comparable molecular weight, between 300 and 2200 Da, were released into decane from both particles. These were most likely un-reacted and non-crosslinked PS chains left behind after particle synthesis. By comparing the peaks in the longer-wavelength region, we confirmed that dye molecules were also leaking out into decane. Identical spectral features were in fact found in the supernatant and in the pure dye solutions, as evidenced in the inset to Figure 4. In order to assess which one of the two substances leaking from the particles was responsible for the creation of the strong elastic film at the interface, we measured the surface tension reduction of a pure PS solution and of the dye solutions at the water–decane interface and compared it to the one measured for the particle supernatants. Pendant drop tensiometry showed that, albeit surface-active, the dye molecules showed a saturation of the water–decane interfacial tension reduction at a few mN/m below the pure water–decane level over several hours, and therefore cannot be responsible for the stronger interfacial tension reduction measured for the supernatants. Measurements on a PS300 solution (and we expect similar results for PS2200) instead showed a dramatic reduction of interfacial tension, which saturated at 20 mN/m for high PS concentrations already after a few minutes. We can therefore conclude with certainty that the formation of the strong interfacial film immobilizing the particles at the interface was due to interfacial adsorption of non-crosslinked PS oligomers leaking from the particles when exposed to decane.

2.2. Nylon Interfacial Polycondensation

The results presented in the previous section showed that it was possible to immobilize particles into stable regular lattices at the water–decane interface by solely waiting for a sufficient amount of time for the PS oligomers to build up an elastic monolayer. This approach, spontaneously occurring for our PS beads trapped at a water–decane interface, is very simple but time-consuming and only applicable to systems releasing surface-active species capable of creating highly elastic films. Despite being "frozen" in place, the particles are only surrounded by a very thin polymer layer, whose composition and thickness cannot be externally controlled. This section and the next one describe two different possible routes to embed these regular particle arrangements into thicker and more controlled polymeric membranes.

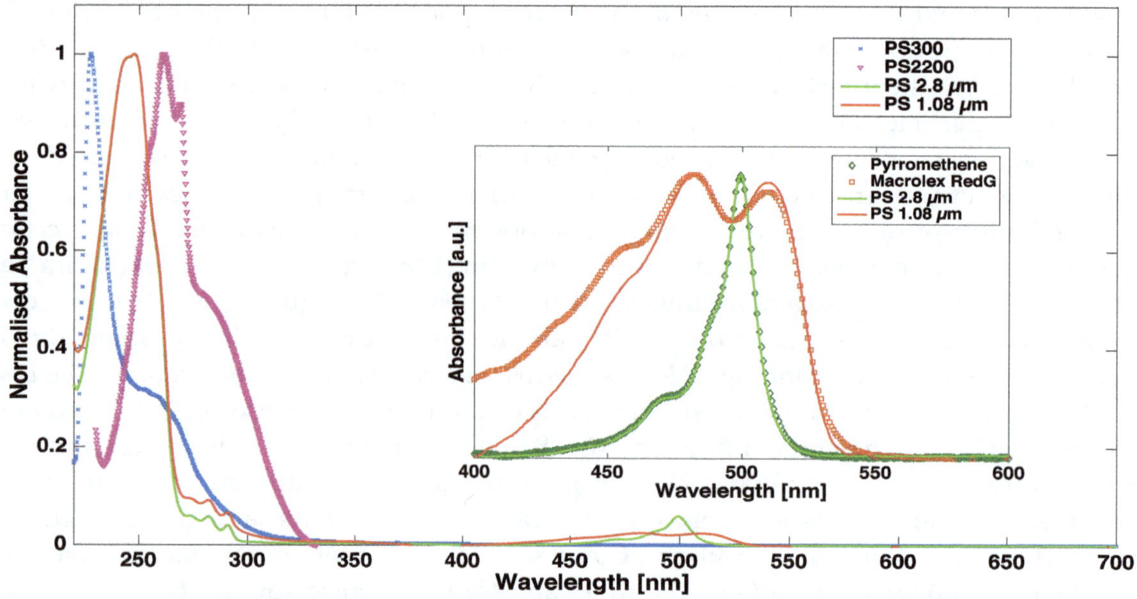

Figure 4. UV-Vis spectra of supernatants after overnight exposure to decane for the 2.8 (**green** solid line) and 1.08 μm (**red** solid line) diameter PS particles. The spectra are combined with the UV-Vis spectra of reference decane solutions of pure PS (molecular weight 300 Da (**blue** crosses) and 2200 Da (**purple** triangles)) and of the red (Macrolex RedG (**red** squares)) and green (Pyrromethene (**green** diamonds)) dyes, zoomed in the inset for the relevant wavelength region.

The first route, schematically shown in Figure 1b, is based on the interfacial polycondensation reaction commonly known as the "nylon rope trick". In this case, one monomer, sebacoyldichloride, reacts with a second one, 1.6-diaminohexane, under the elimination of hydrogen chloride [31]. This reaction is particularly suited for our purpose since the two monomers can be dissolved into immiscible solvents, e.g., water and decane, and the reaction takes place only at the interface between these two fluids, where the monomers meet. In the presence of particles at the interface, these are entrapped by the nylon membrane as it grows.

We carried out these experiments by using a customized sample cell, as schematically depicted in the inset to Figure 5a. The cell consisted of a metallic ring with a sharpened edge, used to contain the water phase and pin the liquid interface, glued onto a microscope slide and enclosed by a cut-out centrifuge tube, whose cap was also glued onto the slide. The metallic ring was filled with a 6 mM water solution of 1.6-diaminohexane to achieve a flat water–air interface and then carefully covered by 0.5 mL of pure decane to create the water-oil interface. At these concentrations, the 1.6-diaminohexane rapidly formed an interfacial layer, as monitored separately by pendant drop tensiometry. A pipette tip, filled with the particles dispersed in the spreading solution previously described, was then brought in contact with the interface to produce a particle monolayer surrounded by 1.6-diaminohexane. At this stage, 100 μL of a 34 mM sebacoyldichloride solution in decane were added to the oil phase and the dynamics of the particles at the interface was monitored as discussed above. Alternatively, a round piece of filter paper (MN 615, prewetted by decane) could be mounted above the interface and blocked by the screw-on cap of the centrifuge tube and additional oil (1–4 mL) could be inserted. The filter paper acted as a physical barrier to avoid flow at the interface caused by the injection of the additional decane containing the sebacoyldichloride. After a given amount of time, the second monomer reached the interface and the polymerization reaction started. The evolution of particle mobility was monitored as a function of time after the injection of the second monomer t_w and it is shown in Figure 5a by plotting the MSD of the 2.8 μm PS particles at ten seconds (MSD_{10}) vs. t_w. The black square symbols refer to the case with no filter paper and the red circles to the case with filter paper. In both cases, the same qualitative behavior was observed. After an initial time where the MSD remained fairly constant, a sudden very large increase of particle mobility was seen (as highlighted by the arrows). Visual inspection showed that this corresponded to very violent convective flows at the interface, which then stopped fairly rapidly. After that, the particles became completely immobilized, with values of the MSD comparable to the ones reported in Figure 2. The strong convective flows were not due to mechanical disturbance of the interface upon injecting the second monomer, since they occurred after minutes and they also appeared in the presence of the filter paper. They were instead due to Marangoni stresses caused by gradients of surface tension at the interface when it was reached by the second monomer. In fact, due to diffusion, different amounts of sebacoyldichloride reached the interface in different places and at different times, causing local gradients of concentration and therefore of interfacial tension leading in turn to Marangoni flows [32]. The flows eventually stopped, when the polymerization reaction had formed a sufficiently stiff membrane at the interface. This phenomenon was unavoidable. Changing both the porosity of the filter paper, as well as the concentration and amount of injected sebacoyldichloride, did not lead to any qualitative change and the only differences were in the waiting time before the onset of the flow. Very interestingly, the occurrence of the Marangoni flows and the polymerization did not disrupt significantly the microstructure of the interface. Figure 5b–e show fluorescence micrographs of the interface before and after polymerization, with and without the filter paper, for two monolayer densities. No qualitative differences were visible.

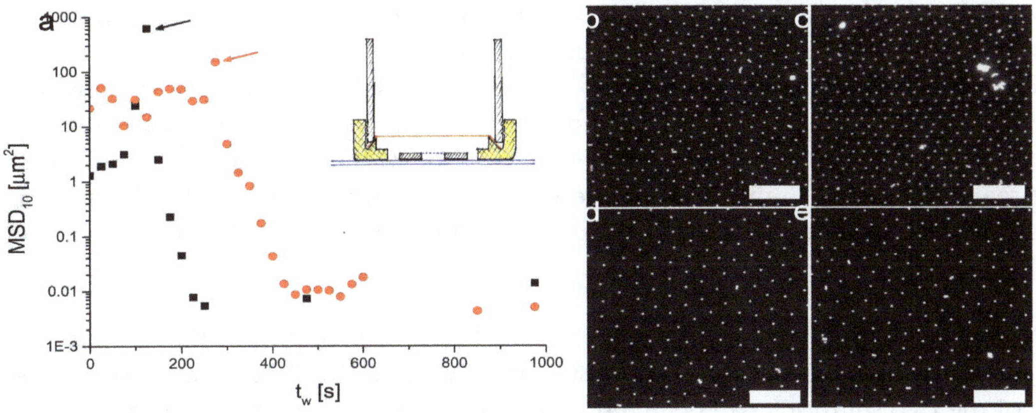

Figure 5. (**a**) MSD calculated after 10 seconds for 2.8 μm PS particles at a water–decane interface during the interfacial polymerization of nylon plotted as a function of waiting time after injection of sebacoyldichloride, with (**red**) and without (**black**) filter paper. The arrows mark the points of highest particle mobility due to convective Marangoni flows at the interface. Inset: schematic of the sample cell used for the measurements. The horizontal dashed line represents the interface and the red line depicts the filter paper membrane; (**b**–**e**) fluorescence images of the particles at the interface before (**b** and **d**) and after polymerization (**c** and **e**), showing no qualitative difference in the microstructure. Images (**b**) and (**c**) are taken without the filter paper membrane; (**d**) and (**e**) with the filter paper membrane. The scale bars are 50 μm.

This led us to believe that this route yielded a very fast and efficient immobilization mechanism of particle monolayers into polymeric membranes, but additional inspection of the nylon membranes by bright-field microscopy, as reported in Figure 6, showed instead that the completion of the polymerization reaction required significantly longer times. Visible droplets were in fact formed onto the membrane a few minutes after particle immobilization and disappeared over the course of several hours. These were most likely sebacoyldichloride pockets that accumulated on the initially formed membrane and that were then slowly consumed by the diaminohexane diffusing through the continuously growing membrane.

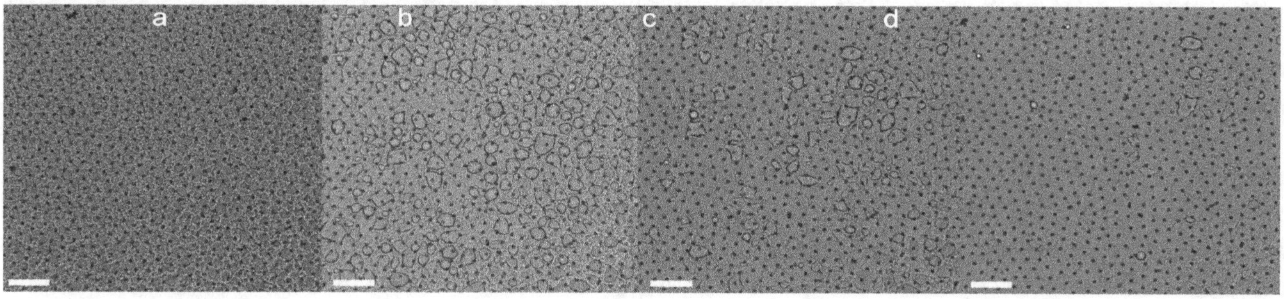

Figure 6. Sequence of bright field microscopy images of the nylon membrane forming at the water–decane interface taken 30 (**a**); 120 (**b**); 180 (**c**) and 240 (**d**) minutes after sebacoyldichloride injection. Scale bars are 50 μm.

2.3. UV Interfacial Polymerization of Styrene

Despite its simplicity, the interfacial polymerization of nylon presents some limitations in terms of the control over the final properties of the membrane since the reaction basically proceeds until all the monomers are consumed. We therefore explored a second route where the polymerization reaction can be externally triggered by UV illumination. The reaction scheme we used is depicted schematically in Figure 1c and it is based on the free radical polymerization of styrene. Radical polymerization reactions belong to the family of chain-growth polymerizations and require an initiator to get started. The polymerization continues as long as the active chain end is not terminated by recombination with an other radical or when all the monomer is consumed [33]. The initiation can be triggered by external stimuli, such as UV light as in our case [33]. In order to improve the mechanical stability of the forming polymer film, we also added *p*-divinylbenzene as crosslinker. The different polarities of all the species make these reactions very suitable to be carried out at fluid–fluid interfaces, where the initiator and monomer can be dispersed into immiscible fluids and meet only at the interface, such as in the case of emulsions [34,35].

In the specific case of our experiments, we had a water-soluble initiator (Irgacure 2959) and decane-soluble monomer and crosslinker (styrene and *p*-divinylbenzene). While details of the preparation of the two solutions will be given in the Experimental Section, we report here the essential steps for the particle immobilization experiments. We used a slightly modified version of the sample cell used for the nylon polymerization as sketched in the top inset to Figure 7a. Here, a thicker metal ring was used to form the interface between 112 μL of a 4.46 mM water solution of the initiator and 2 mL of decane containing 1.56 mM of styrene and 0.14 mM of *p*-divinylbenzene. The particles were injected at the interface as previously described. After particle injection, the top part of the cell was completely filled with the decane solution and the cell was carefully sealed with a glass slide. Observation of the particles at the interface was carried out with a custom-built microscope as shown in Figure 8, which allowed simultaneous imaging in reflection and sample illumination with a UV LED lamp emitting at 365 nm. Pendant drop tensiometry investigations showed that, at these concentrations, the water–decane interface was readily covered by a stable layer containing monomer, crosslinker and initiator, which are all surface-active species. After equilibration, the UV lamp was turned on and particle mobility was followed by tracking.

Figure 7a shows the evolution of MSD at ten seconds for 2.8 μm PS particles as a function of time after illumination start-up. Representative MSD curves are shown in Figure 7b for various values of t_w. From these graphs, we can make similar observations as in the case of the nylon polymerization. We observe that, before illumination, the particles showed the standard expected behavior for particles in the interfacial crystals, as reported earlier in Figure 2. When the photopolymerization began, we observed a significant increase of the particle mobility, again due to the presence of concentration gradients of the reacted species at the interface. Upon continuing UV illumination, the polymerization was completed and the particles were fully immobilized in a polystyrene membarne. We point out that membranes could also be formed in the absence of particles, confirming that their presence does not affect the polymerization reaction significantly.

Despite the fact that all the components necessary for the polymerization were already at the interface before the reaction started, as opposed to the nylon polymerization, the UV photopolymerization disrupted entirely the crystalline arrangement of the particles at the interface, as can be seen in the fluorescence micrographs of Figure 7c–e. This was probably due to the fact that, as the polystyrene membrane grew at the interface, it pushed the particles around in a random fashion, distorting completely the pre-existing arrangement of the particles. Partial swelling in styrene, albeit not detectable by optical microscopy, could also be partly responsible for altering the electrostatic interactions between particles during membrane formation. Similar results were observed when using the 1.08 μm PS particles.

Finally, we attempted the deposition of the photo-polymerized composite membrane onto a solid support to investigate its structural and mechanical properties. The bottom inset to Figure 7a shows a flake of such a membrane deposited on a pitted silicon wafer with 5×5 μm^2 holes. The membrane was scooped up from the interface onto the substrate. Unfortunately, the deposited membrane was still too fragile to perform atomic-force microscopy (AFM) investigations and tended to rupture over the cavities on the silicon wafer.

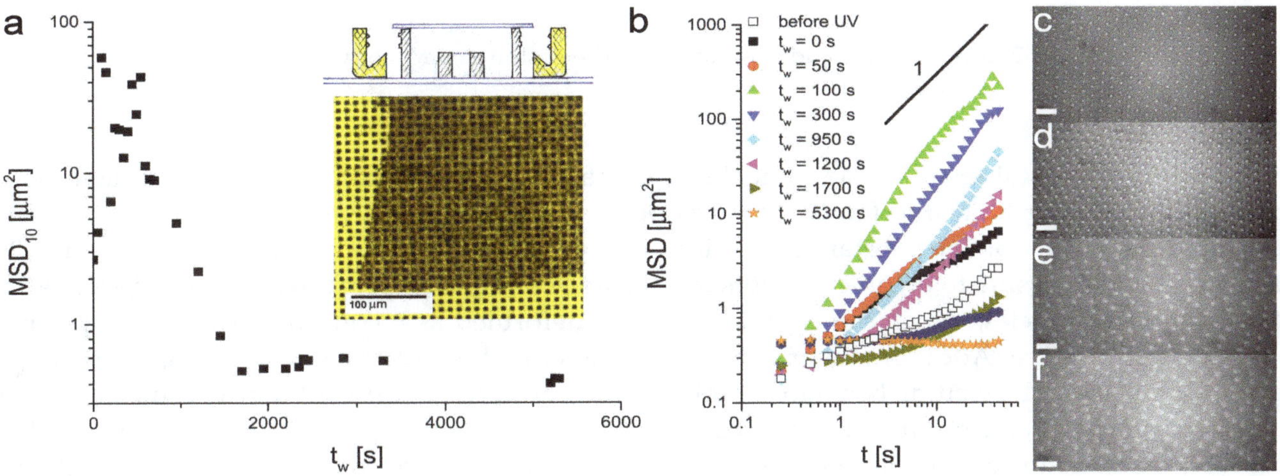

Figure 7. (**a**) MSD calculated after 10 s for 2.8 μm PS particles at a water–decane interface during the interfacial photopolymerization of polystyrene as a function of time after starting the UV illumination t_w. Top inset: schematics of the sample cell. Bottom inset: micrograph of a composite PS particle-polystyrene membrane deposited onto a pitted silicon wafer with 5×5 μm^2 cavities; (**b**) MSD versus time for different representative times during the photopolymerization; (**c–f**) fluorescence images of the 2.8 μm PS particles at the interface before UV illumination (**c**) and after 50 s (**d**), 40 (**e**) and 90 (**f**) minutes. Scale bar: 25 μm.

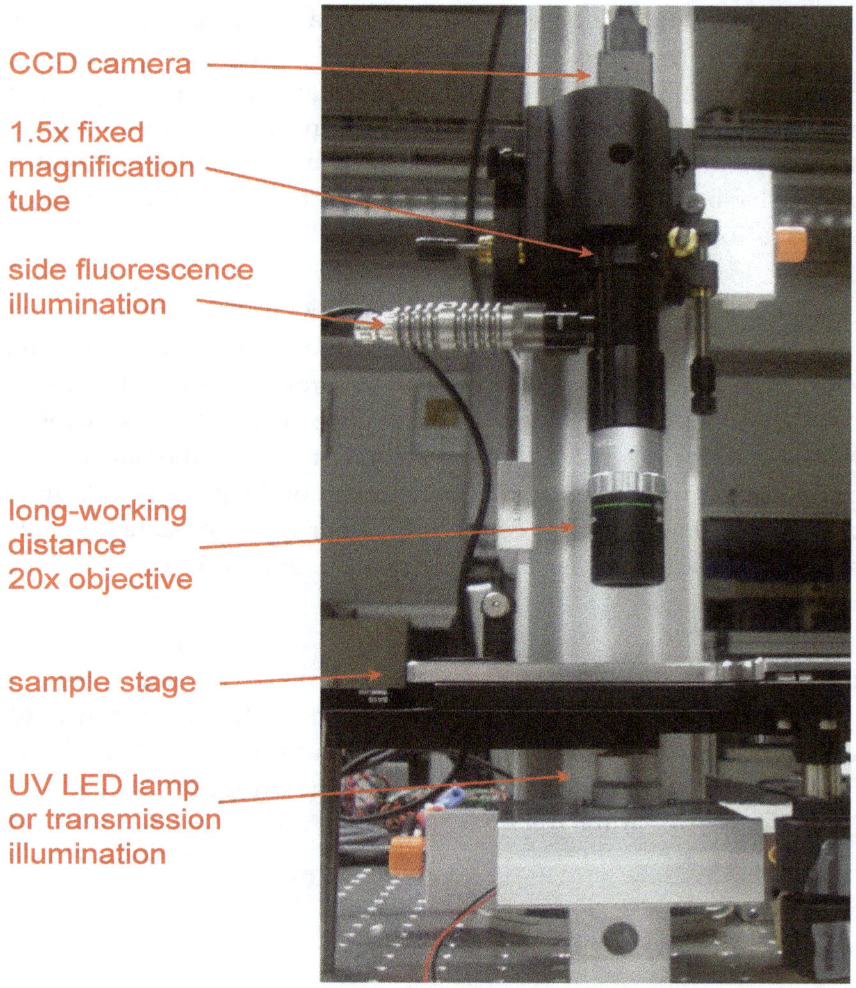

Figure 8. Image and description of the custom microscopy setup.

3. Conclusions

Our experimental results show that there are several open routes for the immobilization of colloidal particles at fluid interfaces. The first option involves the formation of thin elastic films generated by surface-active substances leaking from the particles themselves. Previous work [36] has shown that exposure to organic solvents (and relatively high temperatures) causes plasticization of polystyrene particles, which can even be strongly deformed as a consequence of adsorption at an oil–water interface. Additional work has also shown that the presence of soluble impurities can very strongly affect the wetting behavior of PS ellipsoids at water–decane interfaces, which had to be subjected to rather harsh cleaning conditions in order to eliminate any undesired contamination [37]. Our work instead showed that this process can be used to our advantage to create very stable crystalline arrays of colloids at the interface. The other two approaches that we investigated went one step further and immobilized the particle monolayer within a polymer membrane.

Immobilizing particle monolayers at an interface has proved to be a very successful way to measure *a posteriori* the contact angle of particles at the interface. A wide range of strategies have been proposed to this end [38], including freezing [39] or gelling the water phase [40], growing metallic caps [41] or swelling the particles [42]. A recent method also involved the growth of a thin cyanoacrylate glue layer at the water–air interface, embedding particles that could be later imaged in an scanning electron microscope (SEM) [43]. Our strategies could provide an extension to the latter technique to oil–water interfaces, even though additional studies elucidating the role of membrane growth on the particle position relative to the interface (e.g., by comparison with the methods above) would be required.

On the other hand, composite particle membranes obtained from interfacial assembly have been previously demonstrated [44,45], and we proposed here two conceptually very simple alternative approaches. Interestingly, both approaches showed that the onset of the polymerization reactions coincided with a a strong increase of particle mobility at the interface caused by Marangoni flows. The interplay between polymerization kinetics and the evolution of the microstructure of the interfacial composite membrane becomes then a very interesting direction for future research, where new strategies to reduce or harness Marangoni flows need to be proposed. The next steps to be carried out after this work are to devise suitable strategies to transfer the interfacial membranes onto solid supports to investigate their structure, e.g., thickness or roughness, and mechanical properties as a function of polymerization conditions in a systematic way.

4. Experimental Section

4.1. Materials

The particles used in the experiments were polystyrene colloids purchased from microParticles GmbH (Berlin, Germany). We used 2.8 ± 0.04 μm diameter green-fluorescent particles and 1.08 ± 0.04 μm diameter red-fluorescent particles. The particles were received as 2.5 *w/v*% water stock suspensions and diluted to 0.75 and 0.5 *w/v*% in 50:50 ultra-pure water:isopropanol mixtures, for the green and red particles, respectively. Isopropanol (99.95%, Fisher Chemicals, Leicestershire, UK), acetone (99.9%, Sigma-Aldrich, St. Louis, MO, USA), 1,6-Diaminohexane (Sigma-Aldrich), sebacoyldichloride (Merck, Darmstadt, Germany) and 2-Hydroxy-4′-(2-Hydroxyethoxy)-2-Methylpropiophenone (Irgacure 2959) (98%, Aldrich Chemistry, St. Louis, MO, USA) were used as received. Decane (99%, ABCR, Karlsruhe, Germany) was purified to remove polar impurities by five consecutive basic alumina column filtrations, using fresh powder for each one. The purified decane showed a stable interfacial tension of 53 ± 0.5 mN/m against water for a minimum of 2 h. Styrene (>99%, TCI, Portland, OR, USA) and *p*-divinylbenzene (80%, Aldrich Chemistry) were each filtered through a basic alumina column to remove the inhibitor. The activated liquids were kept in the fridge until use. Borosilicate glass slides and the pitted silicon wafers (MakroPore-12-70, Smartmembranes, Germany) were cleaned by consecutive ultra-sonication for 10 min in acetone, isopropanol and then ultra-pure water. This was followed by rinsing with ultra-pure water and drying with a nitrogen stream. Shortly before use they were placed in a UV/ozone cleaner for 2 min (UV/Ozone Procleaner Plus, Bioforce Nanosciences, Ames, IA, USA). Finally, monodisperse polystyrene 300 Da (Polyscience Inc., Warrington, PA, US), monodisperse polystyrene 2200 Da (Alfa Aeser, Heysham, UK), pyrromethene (BODIPY © 493/503, Invitrogen, Carlsbad, CA, US) and Macrolex RedG (Lanxess, Leverkusen, Germany) were used to prepare the test solutions for UV-visible spectroscopy and pendant drop tensiometry.

The chemicals for the UV photo-polymerization of styrene were prepared as follows. The water phase and the oil phase were prepared separately inside two round-bottom flasks. The initiator Irgacure 2959 was dissolved in ultra-pure water at a concentration of 4.46 mM and stirred for approximately half an hour. Afterwards, remaining particulate was removed by pressing the solution through a syringe filter. The oil-phase was prepared by adding styrene and *p*-divinylbenzene to decane at concentrations of 1.56 mM and of 0.14 mM, respectively. Before use, oxygen was removed by bubbling the two separate solutions with nitrogen for at least one hour. Fresh solutions were prepared before each experiment.

4.2. Sample Cells

Sample cells for observation under the microscope were custom-made by cutting 50 mL TPP centrifuge tubes (Sigma-Aldrich) and gluing them onto borosilicate glass slides using the UV-curable adhesive Norland NOA 61 Norland Products, Cranbury, NJ, US. The interface was created and pinned at polished aluminium or galvanized steel rings of thicknesses and diameters in the range of few millimeters. MN 615 (MACHEREY-NAGEL, Duren, Germany) filter paper was used as a physical barrier for the nylon interfacial polymerization and mounted above the interface by screwing the sides of a TPP centrifuge tube onto its cap glued onto the glass slide.

4.3. Microscopy Experiments

Microscopy experiments were carried out on the custom-built optical line shown in Figure 8. The setup allowed for simultaneous bright-field or fluorescence imaging and UV illumination by means of a UV LED lamp with a wavelength of 365 nm. The interface was viewed through a 20× infinity-corrected long-working-distance objective with a 20 mm working distance and an additional fixed in-line magnification of 1.5×. Images were captured with a xiQ USB3 CCD camera (Ximea, Munster, Germany) and recorded using the freeware μManager (Micro-Manager, US). The sample cell was placed on a stage which allowed for illumination of the interface both in reflection (fluorescence) and transmission. Image series of the particles at the interface were acquired at four frames per second at regular intervals to follow the time evolution of the particle dynamics. The images were analyzed using freely available Matlab code [46] based on the standard code by Cocker and Grier [47].

4.4. Pendant Drop Tensiometry and UV-Visible Spectroscopy Experiments

The interfacial activity of the various substances described in the Results section was measured by a pendant drop device (DSA100, Krüss GmbH, Hamburg, Germany). Depending on the substances of interest, the interfacial tension of either aqueous droplets in decane or decane droplets in water (with an inverted needle) was measured as a function of time. The accuracy of the measurements is of ±0.5 mN/m. Spectroscopic analysis to prove the presence of polystyrene oligomers and fluorescent dyes released by the PS in decane was carried out with a UV-Visible spectrometer (Jasco V660, Hachioji, Japan) calibrated against pure decane. Each sample was diluted appropriately before the measurements, since only the nature of the released impurities was of interest and no quantitative analysis was carried out.

Acknowledgments: Lucio Isa and Michele Zanini acknowledge financial support from the Swiss National Science Foundation (grant *PP00P2_144646/1*).

Author Contributions: All authors have carried out the experiments, analyzed and discussed the data and contributed to the writing of the manuscript.

References

1. Ebert, F.; Keim, P.; Maret, G. Local crystalline order in a 2D colloidal glass former. *Euro. Phys. J. E* **2008**, *26*, 161–168.
2. Bohlein, T.; Mikhael, J.; Bechinger, C. Observation of kinks and antikinks in colloidal monolayers driven across ordered surfaces. *Nat. Mater.* **2012**, *11*, 126–130.
3. Skinner, T.O.E.; Aarts, D.G.A.L.; Dullens, R.P.A. Grain-Boundary Fluctuations in Two-Dimensional Colloidal Crystals. *Phys. Rev. Lett.* **2010**, *105*, 168301.
4. Isa, L.; Kumar, K.; Müller, M.; Grolig, J.; Textor, M.; Reimhult, E. Particle Lithography from Colloidal Self-Assembly at Liquid-Liquid Interfaces. *ACS Nano* **2010**, *4*, 5665–5670.
5. Vogel, N.; Weiss, C.K.; Landfester, K. From soft to hard: the generation of functional and complex colloidal monolayers for nanolithography. *Soft Matter* **2012**, *8*, 4044–4061.
6. Ray, M.A.; Shewmon, N.; Bhawalkar, S.; Jia, L.; Yang, Y.Z.; Daniels, E.S. Submicrometer Surface Patterning Using Interfacial Colloidal Particle Self-Assembly. *Langmuir* **2009**, *25*, 7265–7270.
7. Binks, B.S.; Horozov, T.S. *Colloidal Particles at Liquid Interfaces*; Cambridge University Press: Cambridge, UK, 2006.

8. Kang, C.; Ramakrishna, S.N.; Nelson, A.; Cremmel, C.V.M.; vom Stein, H.; Spencer, N.D.; Isa, L.; Benetti, E.M. Ultrathin, freestanding, stimuli-responsive, porous membranes from polymer hydrogel-brushes. *Nanoscale* **2015**, *7*, 13017–13025.
9. Hanarp, P.; Sutherland, D.; Gold, J.; Kasemo, B. Nanostructured model biomaterial surfaces prepared by colloidal lithography. *Nanostruct. Mater.* **1999**, *12*, 429–432.
10. Hanarp, P.; Sutherland, D.S.; Gold, J.; Kasemo, B. Control of nanoparticle film structure for colloidal lithography. *Coll. Surf. A Physicochem. Eng. Asp.* **2003**, *214*, 23–36.
11. Min, W.L.; Jiang, P.; Jiang, B. Large-scale assembly of colloidal nanoparticles and fabrication of periodic subwavelength structures. *Nanotechnology* **2008**, *19*, 7.
12. Denkov, N.D.; Velev, O.D.; Kralchevsky, P.A.; Ivanov, I.B.; Yoshimura, H.; Nagayama, K. Mechanism of formation of 2-dimensional crystals from latex-particles on substrates. *Langmuir* **1992**, *8*, 3183–3190.
13. Gu, Z.Z.; Fujishima, A.; Sato, O. Fabrication of High-Quality Opal Films with Controllable Thickness. *Chem. Mater.* **2002**, *14*, 760–765.
14. Zhang, K.Q.; Liu, X.Y. In situ observation of colloidal monolayer nucleation driven by an alternating electric field. *Nature* **2004**, *429*, 739–743.
15. Velev, O.; Bhatt, K.H. On-chip micromanipulation and assembly of colloidal particles by electric fields. *Soft Matter* **2006**, *2*, 738–750.
16. Reimhult, E.; Kumar, K. Membrane biosensor platforms using nano- and microporous supports. *Trends Biotechnol.* **2008**, *26*, 82–89.
17. Elnathan, R.; Isa, L.; Brodoceanu, D.; Nelson, A.; Harding, F.J.; Delalat, B.; Kraus, T.; Voelcker, N.H. Versatile Particle-Based Route to Engineer Vertically Aligned Silicon Nanowire Arrays and Nanoscale Pores. *ACS Appl. Mater. Interfaces* **2015**, *7*, 23717–23724.
18. Rey, B.M.; Elnathan, R.; Ditcovski, R.; Geisel, K.; Zanini, M.; Fernandez-Rodriguez, M.A.; Naik, V.V.; Frutiger, A.; Richtering, W.; Ellenbogen, T.; et al. Fully Tunable Silicon Nanowire Arrays Fabricated by Soft Nanoparticle Templating. *Nano Lett.* **2016**, *16*, 157–163.
19. Ramakrishna, S.N.; Clasohm, L.Y.; Rao, A.; Spencer, N.D. Controlling Adhesion Force by Means of Nanoscale Surface Roughness. *Langmuir* **2011**, *27*, 9972–9978.
20. Ramakrishna, S.N.; Nalam, P.C.; Clasohm, L.Y.; Spencer, N.D. Study of Adhesion and Friction Properties on a Nanoparticle Gradient Surface: Transition from JKR to DMT Contact Mechanics. *Langmuir* **2013**, *29*, 175–182.
21. Vogel, N.; Goerres, S.; Landfester, K.; Weiss, C.K. A Convenient Method to Produce Close- and Non-close-Packed Monolayers using Direct Assembly at the Air–Water Interface and Subsequent Plasma-Induced Size Reduction. *Macromol. Chem. Phys.* **2011**, *212*, 1719–1734.
22. Li, X.; Wang, T.; Zhang, J.; Yan, X.; Zhang, X.; Zhu, D.; Li, W.; Zhang, X.; Yang, B. Modulating Two-Dimensional Non-Close-Packed Colloidal Crystal Arrays by Deformable Soft Lithography. *Langmuir* **2010**, *26*, 2930–2936.
23. Ray, M.A.; Jia, L. Micropatterning by non-densely packed interfacial colloidal crystals. *Adv. Mater.* **2007**, *19*, 2020–2022.
24. Pieranski, P. Two-dimesional interfacial colloidal crystals. *Phys. Rev. Lett.* **1980**, *45*, 569–572.
25. Horozov, T.S.; Aveyard, R.; Clint, J.H.; Binks, B.P. Order-Disorder Transition in Monolayers of Modified Monodisperse Silica Particles at the Octane-Water Interface. *Langmuir* **2003**, *19*, 2822–2829.
26. Petkov, P.V.; Danov, K.D.; Kralchesvky, P.A. Surface Pressure Isotherm for a Monolayer of Charged Colloidal Particles at a Water/Nonpolar–Fluid Interface: Experiment and Theoretical Model. *Langmuir*, **2014**, *30*, 2768–2778.
27. Petkov, P.V.; Danov, K.D.; Kralchesvky, P.A. Monolayers of charged particles in a Langmuir trough: Could particle aggregation increase the surface pressure? *J. Coll. Interface Sci.* **2016**, *462*, 223–234.
28. Bonales, L.J.; Rubio, J.E.F.; Ritacco, H.; Vega, C.; Rubio, R.G.; Ortega, F. Freezing Transition and Interaction Potential in Monolayers of Microparticles at Fluid Interfaces. *Langmuir* **2011**, *27*, 3391–3400.
29. Rey, M.; Fernandez-Rodriguez, M.A.; Steinacher, M.; Scheidegger, L.; Geisel, K.; Richtering, W.; Squires, T.M.; Isa, L. Isostructural solid-solid phase transition in monolayers of soft core-shell particles at fluid interfaces: Structure and mechanics. *Soft Matter* **2016**, *12*, 3545–3557.
30. Reynaert, S.; Moldenaers, P.; Vermant, J. Control over colloidal aggregation in monolayers of latex particles at the oil-water interface. *Langmuir* **2006**, *22*, 4936–4945.

31. Morgan, P.W.; Kwolek, S.L. The nylon rope trick: Demonstration of condensation polymerization. *J. Chem. Educ.* **1959**, *36*, 182.
32. Barnes, G.T.; Gentle, I.R. *Interfacial Science: An Introduction*, 2nd ed.; Oxford University Press: Oxford, UK, 2011.
33. Kaiser, W. *Kunststoffchemie Für Ingenieure*, 3rd ed.; Hanser: Munich, Germany, 2011.
34. Hunkeler, D.; Candau, F.; Pichot, C.; Hemielec, A.E.; Xie, T.Y.; Barton, J.; Vaskova, V.; Guillot, J.; Dimonie, M.V.; Reichert, K.H. Heterophase polymerizations: A physical and kinetic comparison and categorization. In *Theories and Mechanism of Phase Transitions, Heterophase Polymerizations, Homopolymerization, Addition Polymerization*; Springer-Verlag Berlin Heidelberg: Berlin, Germany, 1994; Volume 112 of the series Advances in Polymer Science, pp. 115–133.
35. Cunningham, M.F. Living/controlled radical polymerizations in dispersed phase systems. *Prog. Polym. Sci.* **2002**, *27*, 1039–1067.
36. Park, B.J.; Furst, E.M. Fabrication of Unusual Asymmetric Colloids at an Oil–Water Interface. *Langmuir* **2010**, *26*, 10406–10410.
37. Coertjens, S.; Moldenaers, P.; Vermant, J.; Isa, L. Contact Angles of Microellipsoids at Fluid Interfaces. *Langmuir* **2014**, *30*, 4289–4300.
38. Zanini, M.; Isa, L. Particle contact angles at fluid interfaces: Pushing the boundary beyond hard uniform spherical colloids. *J. Phys. Condens. Matter* **2016**, *28*, 313002.
39. Isa, L.; Lucas, F.; Wepf, R.; Reimhult, E. Measuring single-nanoparticle wetting properties by freeze-fracture shadow-casting cryo-scanning electron microscopy. *Nat. Commun.* **2011**, *2*, 438.
40. Paunov, V.N. Novel Method for Determining the Three-Phase Contact Angle of Colloid Particles Adsorbed at Air–Water and Oil–Water Interfaces. *Langmuir* **2003**, *19*, 7970–7976.
41. Sabapathy, M.; Kollabattula, V.; Basavaraj, M.G.; Mani, E. Visualization of the equilibrium position of colloidal particles at fluid–water interfaces by deposition of nanoparticles. *Nanoscale* **2015**, *7*, 13868–13876.
42. Lu, Z.; Zhou, M. Fabrication of large scale two-dimensional colloidal crystal of polystyrene particles by an interfacial self-ordering process. *J. Coll. Interface Sci.* **2011**, *361*, 429–435.
43. Vogel, N.; Ally, J.; Bley, K.; Kappl, M.; Landfester, K.; Weiss, C.K. Direct visualization of the interfacial position of colloidal particles and their assemblies. *Nanoscale* **2014**, *6*, 6879–6885.
44. Kiesow, I.; Marczewski, D.; Reinhardt, L.; Mühlmann, M.; Possiwan, M.; Goedel, W.A. Bicontinuous Zeolite Polymer Composite Membranes Prepared via Float Casting. *J. Am. Chem. Soc.* **2013**, *135*, 4380–4388.
45. Xu, H.; Goedel, W.A. From Particle-Assisted Wetting to Thin Free-Standing Porous Membranes. *Angew. Chem. Int. Ed.* **2003**, *42*, 4694–4696.
46. MATLAB. Available online: http://people.umass.edu/kilfoil/downloads.html (accessed on 1 April 2015).
47. Crocker, J.C.; Grier, D.G. Methods of digital video microscopy for colloidal studies. *J. Coll. Interface Sci.* **1996**, *179*, 298–310.

5

Self-Healing Supramolecular Hydrogels based on Reversible Physical Interactions

Satu Strandman and X.X. Zhu *

Département de Chimie, Université de Montréal, C.P. 6128, Succursale Centre-ville, Montreal, QC H3C 3J7, Canada; satu.strandman@umontreal.ca

* Correspondence: julian.zhu@umontreal.ca

Academic Editor: Clemens K. Weiss

Abstract: Dynamic and reversible polymer networks capable of self-healing, *i.e.*, restoring their mechanical properties after deformation and failure, are gaining increasing research interest, as there is a continuous need towards extending the lifetime and improving the safety and performance of materials particularly in biomedical applications. Hydrogels are versatile materials that may allow self-healing through a variety of covalent and non-covalent bonding strategies. The structural recovery of physical gels has long been a topic of interest in soft materials physics and various supramolecular interactions can induce this kind of recovery. This review highlights the non-covalent strategies of building self-repairing hydrogels and the characterization of their mechanical properties. Potential applications and future prospects of these materials are also discussed.

Keywords: hydrogels; self-healing; supramolecular materials; non-covalent interactions; dynamic cross-links; physical gels; transient networks; mechanical failure and recovery; self-assembly; host-guest chemistry

1. Introduction

Hydrogels are soft solids or solid-like materials that immobilize a large amount of water in a three-dimensional (3D) network held together by covalent bonds, non-covalent or topological interactions [1–4]. These materials are appealing to various applications especially in biology and medicine owing to their compositional and structural versatility that allows introducing responsiveness to external stimuli and adapting them to biological interfaces, high water content and physical properties that are similar to soft tissues or at nanoscale, similar to extracellular matrix [5–13]. Some of the applications of hydrogels in (bio)medical fields include sensors [14–17], actuators [18,19], device coatings [20–22], wound dressings and adhesives [23–25], liquid-absorbing hygiene products [26], delivery vehicles for active compounds or cells [27–30], soft contact lenses [31–33], and matrices and implants in tissue engineering and regenerative medicine [34–37].

There is a continuous need towards extending the lifetime, improving the safety, and enhancing the performance of both soft and hard materials. Mechanical deformation may lead to micro- or macroscale cracks, leading to gradual deterioration or sudden loss of mechanical properties [38,39]. One strategy to improve the performance and extend the lifetime of gels is to introduce self-healing ability, that is, a capacity to restore the initial properties after material failure, using dynamic and reversible linkages [39–45]. A visual demonstration of the self-healing of a hydrogel is shown in Figure 1. The dynamic linkages can be based on reversible covalent chemical bonds, formed for example by Diels-Alder reaction [46–50], disulfide [51–53], imine [54,55], oxime [56], or acylhydrazone [50,52] formation, photocrosslinking [57], radical reactions [58,59], or phenylboronate complexation [60–62], and these have been highlighted in several reviews. Dynamic linkages can also be based on supramolecular non-covalent (physical) interactions, such as crystallization, hydrogen bonds, host-guest, hydrophobic or polymer-nanocomposite interactions, or multiple combined interactions. Some of these non-covalent networks can be built from pre-formed colloidal systems, such as liposomes or mixed micelles.

Figure 1. Visual evidence of self-healing of a supramolecular hydrogel through non-covalent host-guest interactions between β-cyclodextrin- and bile acid-bearing polymers: **(I)** initial hydrogel sample; **(II)** sliced sample; **(III)** sliced sample rejoined for healing; and **(IV)** self-healed sample. Reprinted from [63].

Physical transient gels are considered as a model system in soft materials physics with their well-defined structural properties, ease of preparation, and relatively simple linear rheological behavior, although their non-linear rheological response can be highly complex [38]. Recent reviews have focused on hard and/or chemically crosslinked self-healing materials [41,43,45,48,64] or on certain classes of supramolecular polymers, such as hydrogen-bonded or metal-ligand-coordinated systems [40,42,44,65–67]. Wang and Heilshorn [68] highlighted the potential of adaptable hydrogels in cell encapsulation. As non-covalently crosslinked gels often show rapid structure recovery, they are gaining increasing interest in injectable and extrudable systems used in therapeutic delivery or additive manufacturing (3D printing). These applications require the knowledge of the precise recovery rate and mechanical properties of the gels. For example, the tailored elasticity of the extracellular matrix-mimicking hydrogels as bio-inks is crucial for the cell differentiation and survival, while injectable gels should have sufficient strength after structure recovery to hold them in place. These reviews have emphasized the chemistry for the making of the physical gels rather than their resulting physical properties. The self-healing phenomenon is mostly based on visual demonstrations. Further understanding of the relation of the design, structure, and properties is in the call for new developments in this area. We attempt to fill this gap by presenting the preparation strategies in relation to the mechanical properties of the self-healing gels through examples along with a review of the methods used for mechanical testing of self-healing characteristics. We will then discuss the potential applications where gel strength and fast recovery play an important role and present the future prospects of reversible physical gels.

2. Characterization of the Self-Healing Behavior

The term "self-healing" has been used to describe various structure recovery processes of hydrogels, where the gels revert to their original state and recover the mechanical strength completely or partially after the deformation or fracture. The gelation and structure recovery can take place via different interactions and self-assembling processes. The viscoelastic properties of physical networks are determined by the number and lifetime of crosslinks, and these further affect the kinetics of the self-repair [69,70]. In addition to macroscopic healing, transient chain association causes temporal evolution of microscopic topology, leading to changes in mechanical properties [71]. Therefore, both macroscopic rheology and microscopic fluorescence results have been used as a basis for theoretical models that aim at quantitative prediction of chain association dynamics in physical networks [69–71].

Common experimental methods to characterize the self-healing behavior of hydrogels include:

- *Visual observation.* Freshly cut or damaged pieces of a hydrogel are brought in contact and their rejoining to form a uniform gel is followed both visually and using a simple qualitative mechanical deformation, that is, stretching or bending the sample. Examples are shown in Figure 1 and in several figures below. Visual observation can be accompanied by microscopic imaging techniques (confocal, scanning electron microscopy SEM) to reveal the micro- or nanoscale structure recovery.

- *Oscillatory rheology: Step strain measurement.* After determining the yield stress/strain and the recovery time of the gel sample, the gel will be exposed to strain that alternates periodically between low (structure conservation or recovery) and high values (structure breaking) at constant oscillation frequency. This test allows for the quantitative determination of equilibrium moduli and the extent of structure recovery (recovery rate). Several examples of step strain tests will be shown in Section 3.

- *Cyclic compression/tensile testing*. Self-healing of hydrogels can be tested by cyclic compression or tensile tests, each cycle followed by a recovery period. The changes in stress/strain curves and fracture point or in initial compressive modulus give information on the recovery of ruptured crosslinks. Tensile tests can be used to quantify the self-healing efficiency after joining the fractured surfaces of a ruptured hydrogel by comparing the elongation at break for an intact and re-joint hydrogel. Figure 2 shows an example of a compression test of a physically crosslinked hydrogel and a hybrid gel with combined chemical and physical crosslinking, as well as the compression and elongation curves for a physically crosslinked gel (no chemical crosslinker). Here, the nominal stress σ_{nom} is the force per cross-sectional area of the un-deformed gel specimen and the strain is represented by λ, the deformation ratio (deformed length/initial length). The hysteresis in the compression and elongation curves indicates the decreased number of crosslinks and thus, reduced structure recovery over the deformation cycles [72].

One of the challenges of physically crosslinked gels is optimizing the strength and self-healing properties, as the rigidity of gels limits the chain diffusion to the damaged site. Despite the versatility of supramolecular hydrogels, they suffer from relatively low storage moduli (G' in Pa–kPa range) and so far, combining them with reversible or irreversible covalent networks or harder domains has provided the best mechanical strength. For example, a single component PVA hydrogel with H-bonded crystalline domains could reach a fracture stress of ~280 kPa initially and ~200 kPa after 48 h healing (72% of original strength) [73], and nanocomposite gels of hydrophobically modified polyacrylamide and graphene oxide (GO) could achieve tensile stress of 243 kPa, but the recovery rate was low, 44%. For comparison, the same gel without the reinforcing GO showed a tensile strength of 52 kPa [74].

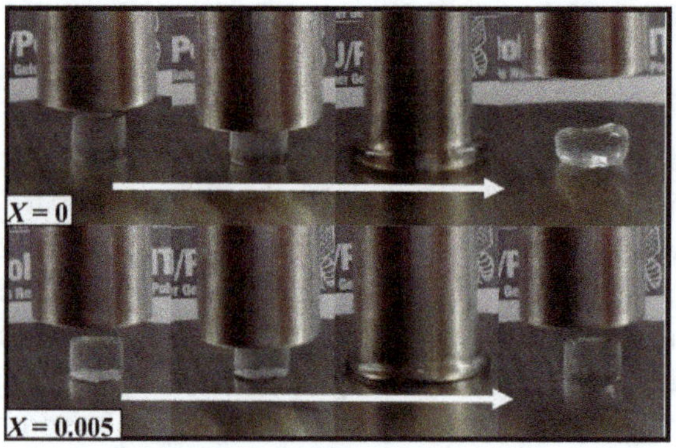

(A)

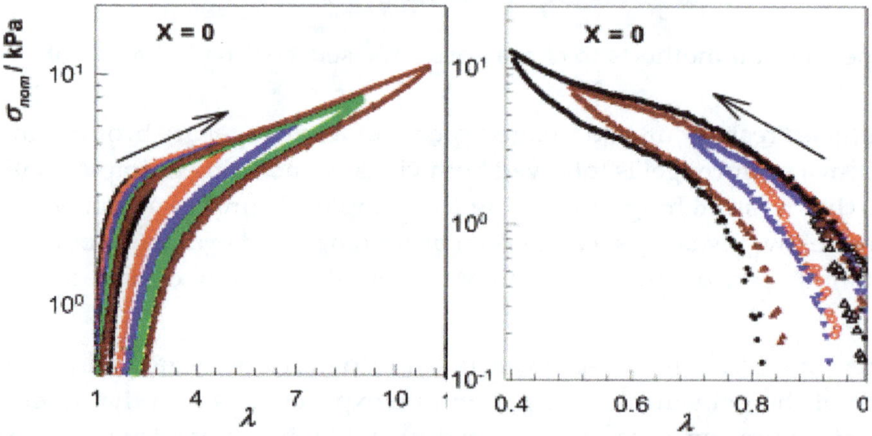

(B)

Figure 2. (**A**) Photographs of the physical (molar ratio of chemical crosslinker *vs.* monomer $X = 0$) and hybrid gels with combined chemical and physical crosslinking ($X = 0.005$) during the compression tests. The physical gel remains deformed after removal of the stress (**top-right** image); (**B**) Nominal stress σ_{nom} *vs.* deformation ratio λ curves from cyclic elongation (**left**) and compression tests (**right**) for a physically crosslinked gel ($X = 0$). Reprinted from [72].

3. Preparation Strategies

3.1. *Hydrophobic Interactions*

Hydrophobicity gives rise to unusual properties of aqueous solutions of nonpolar compounds and plays a key role in a variety of chemical and biophysical phenomena, such as protein folding or self-assembly of amphiphiles into micelles and membranes [75]. Hydrophobic interactions are different from other noncovalent interactions as they do not depend on direct intermolecular attraction between interacting species but are rather driven by the tendency of water molecules to retain their H-bonded network intact around a nonpolar solute, altered by increased temperature, the presence of cosolutes, and size and curvature of nonpolar species [76]. Resulting molecular rearrangement can lead to complex colloidal behavior of amphiphilic molecules in aqueous solutions. Polymer-based hydrogels formed through hydrophobic interactions can be made by introducing hydrophobic sequences within or in the ends of hydrophilic polymer chains [77]. The transient network formation through interchain interactions depends on polymer concentration, fraction of hydrophobic moieties, and polymer architecture [77–79]. Such non-covalent hydrogels may exhibit self-healing capacity owing to the dynamic and reversible nature of the junctions [80].

Self-healing associating networks of hydrophobically modified water-soluble polymers have been made by micellar copolymerization of (1) hydrophilic comonomers, such as acrylic acid (AAc), acrylamide (AAm) or *N*-alkylacrylamides (*N,N*-dimethylacrylamide, *N*-isopropylacrylamide); with (2) large hydrophobic monomers, such as stearyl methacrylate (C18) [80–84], dococyl acrylate (C22) [81];

octylphenyl polyethoxyether acrylate [85] in the presence of (3) a surfactant (sodium dodecyl sulfate SDS, cetyltrimethyl ammonium bromide CTAB) and depending on the hydrophobe; (4) salt (NaCl, NaBr) or a cosurfactant. Micellar radical copolymerization in aqueous solution is a common technique for the synthesis of associative copolymers. However, unlike smaller hydrophobic *N*-alkylacrylamides or *N*-alkyl(meth)acrylates (C4–C12), long-chain alkyl(meth)acrylates have very low water-solubility. The addition of sufficient amount of salt or cosurfactant induces the growth and/or morphological transition of surfactant micelles. Larger micelles can then solubilize a high amount of hydrophobes, whereupon they can grow further (catanionic CTAB-SDS system) or adopt a different morphology (SDS-NaCl system) for thermodynamic feasibility [81,82]. C18- and C22-acrylamide copolymer gels with a SDS-NaCl system reached final elastic modulus G' of around 1 kPa at 1 Hz and tan δ of 0.5–0.9, monitored with a rotational rheometer during the polymerization.

Okay group found that after the copolymerization, the hydrogels became mechanically stronger upon swelling and extraction of surfactant micelles, as the hydrophobic interactions became enhanced without surfactant [81]. The tensile strength of a C18-acrylamide copolymer gel increased from 12 ± 1 kPa to 78 ± 6 kPa after the removal of SDS micelles by equilibrium swelling in water, but at the same time the elongation at break decreased from 2200% $\pm$ 350% to 650% $\pm$ 80%, indicating increased stiffness of the gel. At the same time, the gels lost their ability to self-heal because of longer lifetimes of hydrophobic associations. The self-healing property was tested by a tensile strength test on a series of hydrogel samples before and after joining the fractured surfaces of ruptured gels. The results indicate that the self-healing capacity of hydrophobically modified PAAm gels requires the presence of hydrophobe-solubilizing surfactant micelles [81] and possibly a charged comonomer for electrostatic trapping of oppositely charged micelles, such as AAc for trapping mixed micelles of C18 and CTAB (Figure 3) [80,84]. The self-healing efficiency increases with decreasing lifetime of dynamic crosslinks due to favorable chain diffusion across the fractured surfaces. Therefore, improving the gel strength by enchancing the hydrophobic interactions or even combining physical and covalent crosslinking (hybrid gels) [72], takes a toll on the self-healing capacity and balancing these two characteristics is important in optimizing the gel performance.

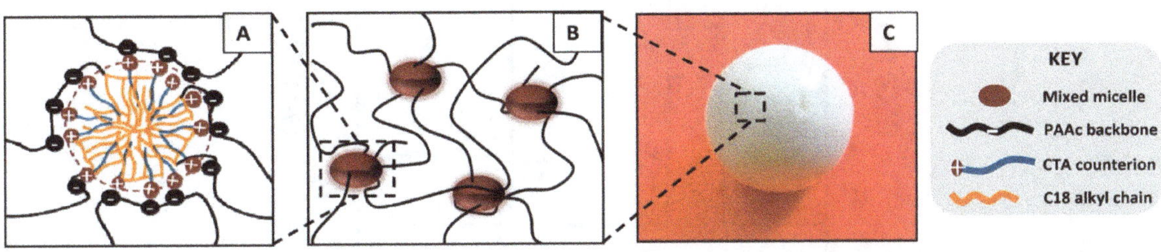

Figure 3. (**A**,**B**) Cartoon showing the cross-link in self-healing hydrophobically modified poly(acrylic acid) (PAAc) hydrogels; (**C**) Image of a PAAc hydrogel sample in the form of a sphere in equilibrium with water. The mixed micelles consist of stearyl methacrylate (C18) and CTAB in aqueous NaBr solution. Reprinted with permission from [80].

Another method for preparing a hydrophobically associated self-healing hydrogel is introducing reversible network junctions through liposomes that can anchor the hydrophobic moieties of (co)polymers into their bilayer. As liposomes have found clinical applications as drug carriers [86], their delivery to a specific site in the body could be achieved via injectable, rapidly self-healing hydrogels. Mixing a telechelic cholesterol (Chol) end group-bearing poly(ethylene glycol) (PEG) with dimethyldioctadecylammonium bromide (DODAB) liposomes resulted in an elastic self-healing gel where Chol-PEG-Chol acted as a dynamic crosslinker between the liposomes (Figure 4) [87]. High liposome concentration and low temperature favored the network formation through higher probability of bridging and lower mobility of Chol groups into and out of liposome bilayers, respectively. The plateau modulus G_0 at high oscillatory frequencies in rheological experiments showed a crosslinker concentration (C) dependence of $G_0{\sim}C^{4.3}$ (*vs.* ${\sim}C^{2.25}$ of Cates model for living polymers and flexible wormlike micelles), which suggests that the interactions between the liposome and crosslinker are more complex than those in a simple polymer network. The highest Chol-PEG-Chol concentration (6% *w*/*v*) gave G_0 of 5.5 kPa. The gels showed instantaneous self-healing after subjecting them to low stress (yield stress 59 Pa at C = 3% *w*/*v*), which is assigned to the re-insertion of Chol groups into liposomes after stress-induced breaking of Chol-PEG-Chol bridges. This gives promise to applications in injectable drug delivery systems and 3D tissue engineering scaffolds [87]. However, the observed body temperature-induced softening of the gel may limit its medical applications and further tuning of the gel strength may be required.

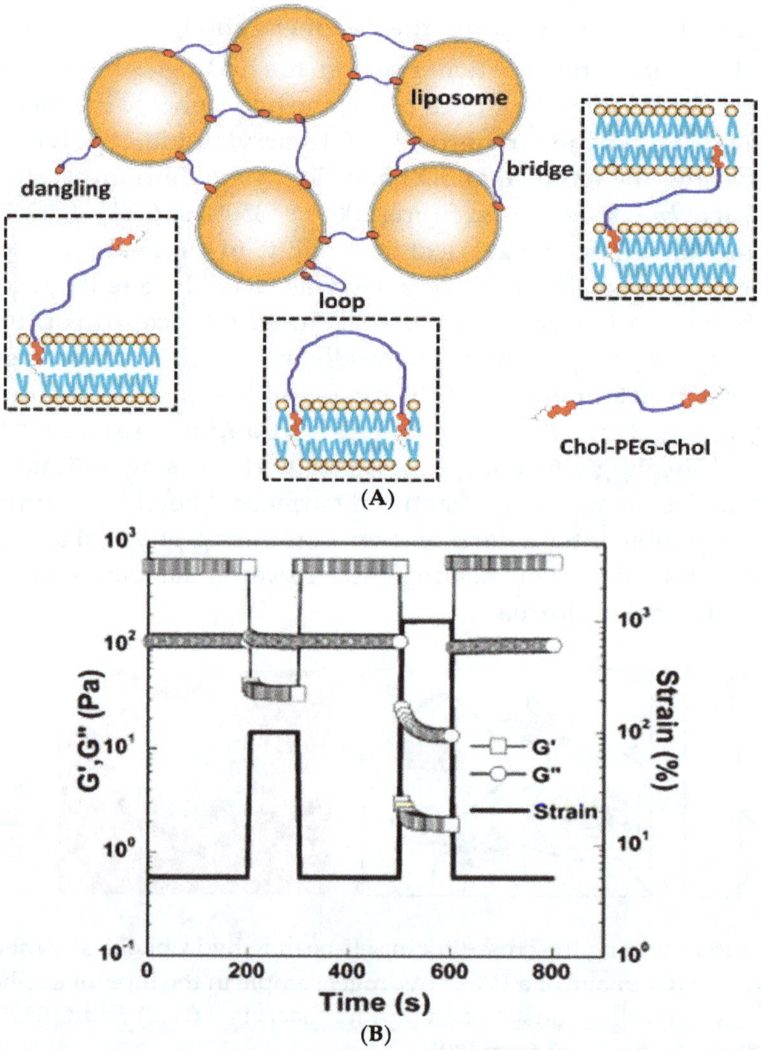

Figure 4. (**A**) Schematic illustration of liposome gel. Cholesterol end groups of Chol-PEG-Chol were embedded in the liposome bilayers, forming bridge, loop or dangling; (**B**) Evolution of storage G′ (□) and loss modulus G″ (○) of liposome gel with time following two successive pulses of high deformation (solid line).

3.2. *Host-Guest Interactions*

Host-guest interactions represent specific non-covalent interactions that are based on selective inclusion complexation between macrocyclic hosts, such as cucurbit[n]urils, cyclodextrins, crown ethers, calix[n]arenes, resorcinarenes, and pillar[n]arenes, and smaller guest molecules [42,88–90]. Although the size of the cavity and portal of a host molecule are good predictors of binding, the selectivity of a host towards a guest goes beyond a simple hole-fitting concept, as solvent effects, multiple binding sites or secondary interactions can be involved [90,91]. Self-healing gels exploiting inclusion complexation can be built by (1) mixing host- and guest-equipped polymers or (2) copolymerizing vinyl group-bearing pre-formed host-guest inclusion complexes with comonomers [42,89]. Although all the macrocycles above show potential for building self-healing gels, mainly cyclodextrins and cucurbit[n]urils have so far been used in fabricating hydrogels and will be discussed more in detail below. Crown ether derivatives have been used to fabricate linear and crosslinked supramolecular polymers owing to their affinity towards cationic species [88,92,93], but self-healing has only been reported for organogels formed upon the complexation of polymer-bound cationic guests by a bis(crown ether) [94] or crown-ether-bearing polymers by bis-ammonium compounds [95].

3.2.1. Cyclodextrins

Cyclodextrins (CDs) are popular hosts in supramolecular chemistry owing their availability and selectivity according to the cavity size (γ-CD > β-CD > α-CD), where hydrophobic and van der Waals interactions between the inner surface of CD ring and hydrophobic guests of suitable size are responsible for inclusion complexation [96,97]. Guest molecules can range from polar compounds such as alcohols, acids, amines, and amino acids to less polar compounds such as linear and branched alkyls, cycloalkanes, aromatic molecules, and steroidal compounds or even to polymeric molecules, such as in the case of poly(pseudo)rotaxanes [98].

Following the first strategy of mixing two functionalized polymers, self-healing gels have been prepared from CD-bearing polymers, such as poly(meth)acrylates, poly(*N*-alkyl)acrylamides, poly(ethylene glycol) or poly(ethylene imine), and guests such as ferrocene [99–102], imidazole [103], bromonaphthalene [104], and azobenzene [104]. Even larger guests can be bound, such as bile acids, which are physiologically important steroidal compounds and thus, ideal building blocks for polymeric biomaterials [105–108]. They can form inclusion complexes with β-CD and the complexation can be inversed upon the addition of competing guest, potassium 1-adamantylcarboxylate [109]. Host-guest complexation between poly(*N*,*N*-dimethylacrylamide) chains bearing either β-CD (8 mol%) or cholic acid moieties (2 mol%) yielded self-healing hydrogels, which showed highest storage modulus G' at 1:1 host guest ratio and high polymer concentrations (12.5 wt%). The gel elasticity was recovered rapidly (in 30 s) after the shear-induced rupture of supramolecular crosslinks (Figure 5). The plateau modulus G_0 of 1:1 gels was ~10–1000 Pa at 6.5–12.5 wt% [63]. Another steroidal molecule, cholesterol (Chol), has been used as a guest in self-healing gels based on the inclusion complexation between poly(L-glutamic acid) (PLGA)-bound β-CD (48 mol%) and PLGA-*b*-PEG-*b*-PLGA triblock copolymer furnished with pendent Chol groups (20 mol%) [110]. The healing time of these gels was 60 s, determined by oscillatory step strain experiments, and high compressive modulus (46 kPa) was obtained with a high-molar-mass linker with long PLGA blocks (~30 kDa). The gels showed good biocompatibility and slow degradation of up to 72 days *in vitro* at 37 °C [110].

Self-healing can also be achieved through reversible polypseudorotaxane formation, which involves binding of a polymer chain inside several polymer-bound macrocyclic hosts. Complexation between β-CD-grafted alginate (Alg-*g*-CD) and Pluronic® 108 yielded self-healing biocompatible and degradable hydrogels (Figure 6), where multiple cyclodextrin hosts bound to the middle block of Pluronic PEG-*b*-PPG-*b*-PEG triblock copolymer (PPG = poly(propylene glycol)) [111]. Rapid recovery from shear (in 10 s) was observed and the gels remained stable over long periods of time (up to 5 days). As Pluronic® 108 is thermosensitive, forming micelles at body temperature, the association of the β-CD-bound block copolymer upon increasing temperature led to significant stiffening of gels and the strongest gels showed elastic shear modulus G' of 10–12 kPa at f = 10 Hz. The gels exhibited gradual release of a globular protein of 66.5 kDa (bovine serum albumin, BSA) *in vitro* and thus, applications as injectable delivery vehicles were proposed [111].

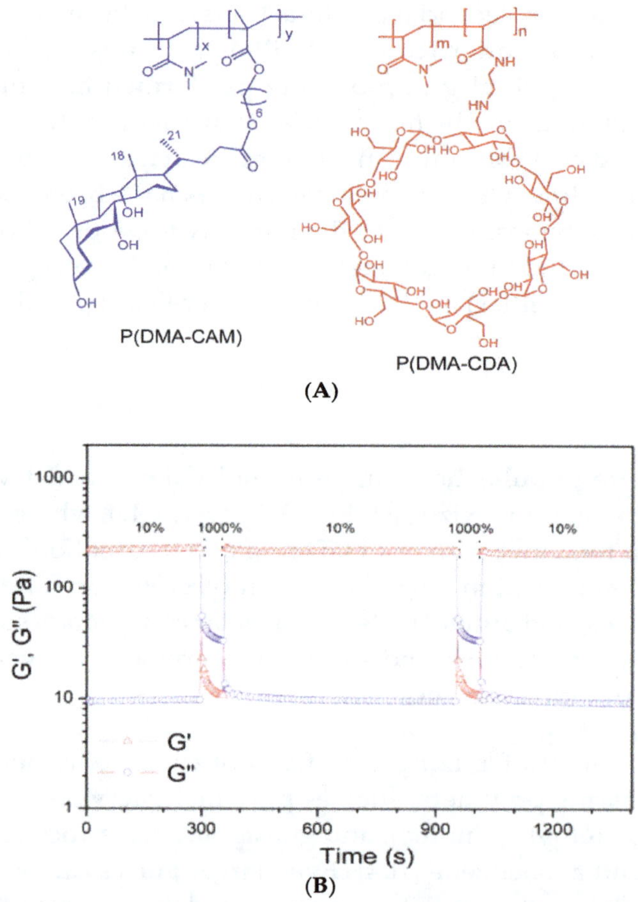

Figure 5. (**A**) Structures of cholic acid- (**left**) and β-CD-bearing (**right**) poly(*N*,*N*-dimethylacrylamides), P(DMA-CAM) and P(DMA-CDA), respectively; (**B**) G′ (squares) and G″ (circles) values of the P(DMA-CAM-2%)/P(DMA-CDA) hydrogel (8.3 wt%) in continuous step strain measurements (25 °C). Large strain (1000%) inverted the values G′ and G″ to give the sol state. G′ was recovered under a small strain (10%) within 30 s.

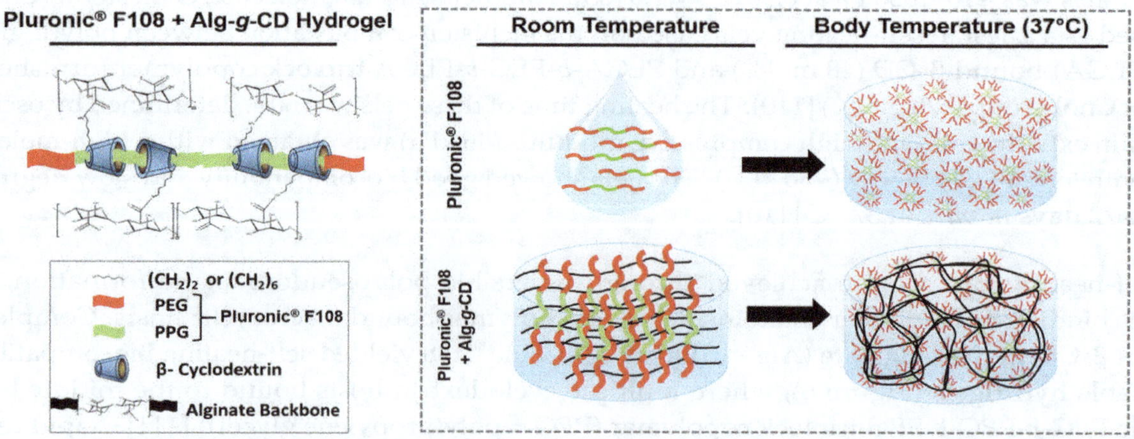

Figure 6. Physical cross-linking between Alg-*g*-CD macromolecules and Pluronic® F108 (**left**) and the thermo-response of the hydrogel network (**right**). Pluronic® F108 forms micelles and self-cross-links at body temperature. β-CD (host) conjugated onto the alginate backbone (Alg-*g*-CD) formed a physically cross-linked supramolecular inclusion complex with the guest, the PPG block (green) of Pluronic® F108.

The host and guest moieties can also coexist in the same polymer. Harada group prepared a series of tough and flexible self-healing gels from a variety of polyacrylamides and polyacrylates, where β-CD and adamantane were side groups of the same polymer chain, leading to interpolymer interactions and thus crosslinking [112]. High host and guest content in the same chain led to poorer mechanical properties through the heterogeneity of the resulting network. The polymers showed potential as self-healing coatings in the dry state, where healing through supramolecular interactions could be induced upon exposing the damaged area to water that acts as a plasticizer and increases the mobility of the polymer chains. The synthetic approach was taken even further in redox-sensitive self-healing gels based on polyacrylamides furnished with the abovementioned β-CD and Ad units as well as a ferrocene side group, which can be bound to β-CD in its reduced state but expelled in oxidized form (Figure 7) [113]. The strongest gels based on β-CD-Ad-Fc-PNIPAAm achieved Young's modulus E of up to 20 kPa, obtained by tensile testing. Self-healing occurred in both oxidized and native states although the healing efficiency was lower for the oxidized gel. Here, the self-healing took at least 2 h and 70% recovery ratio was achieved for native gel in >70 h. Oxidization also led to gel swelling due to reduced crosslinking density. Interestingly, the gels showed also a shape memory effect based on the release of Fc upon oxidation and rebinding to a new β-CD site upon reduction [113].

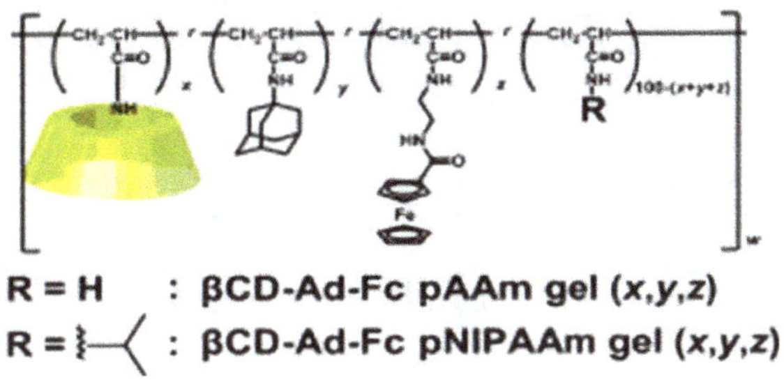

(A)

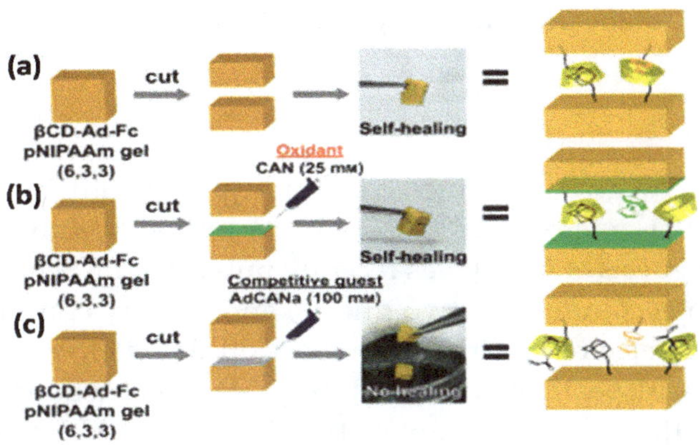

(B)

Figure 7. (**A**) Structures of copolymers bearing β-CD, adamantane (Ad) and ferrocene (Fc) in the same chain; x, y and z indicate mol% of respective units in the chain; (**B**) Self-healing properties of a poly(*N*-isopropylacrylamide) copolymer (**a**) with Fc in reduced form; (**b**) Fc in oxidized form; and (**c**) in the presence of competitive guest sodium adamantanecarboxylate (AdCANa).

Figure 8 presents an example of the second type of preparation strategy for self-healing supramolecular gels, where host-guest complexes are formed prior to the copolymerization of host α-CD or β-CD) and guest (adamantane, Ad or n-butyl, n-Bu) molecules that both bear polymerizable vinyl groups [114].The aim was to improve the self-healing capacity upon preorganization of interacting domains. Greater degree of self-repair was obtained with Ad guest (99% after 24 h, bound to β-CD) than with n-Bu (74% after 24 h, bound to α-CD) and the elastic modulus G′ of β-CD-Ad gel was 100–1000 times higher than that of α-CD-nBu gel (680 kPa *vs.* 0.33 kPa). Self-healing of the former gel could be prevented upon adding a competitive guest, sodium adamantanecarboxylate (AdCANa), or free host, β-CD [114].

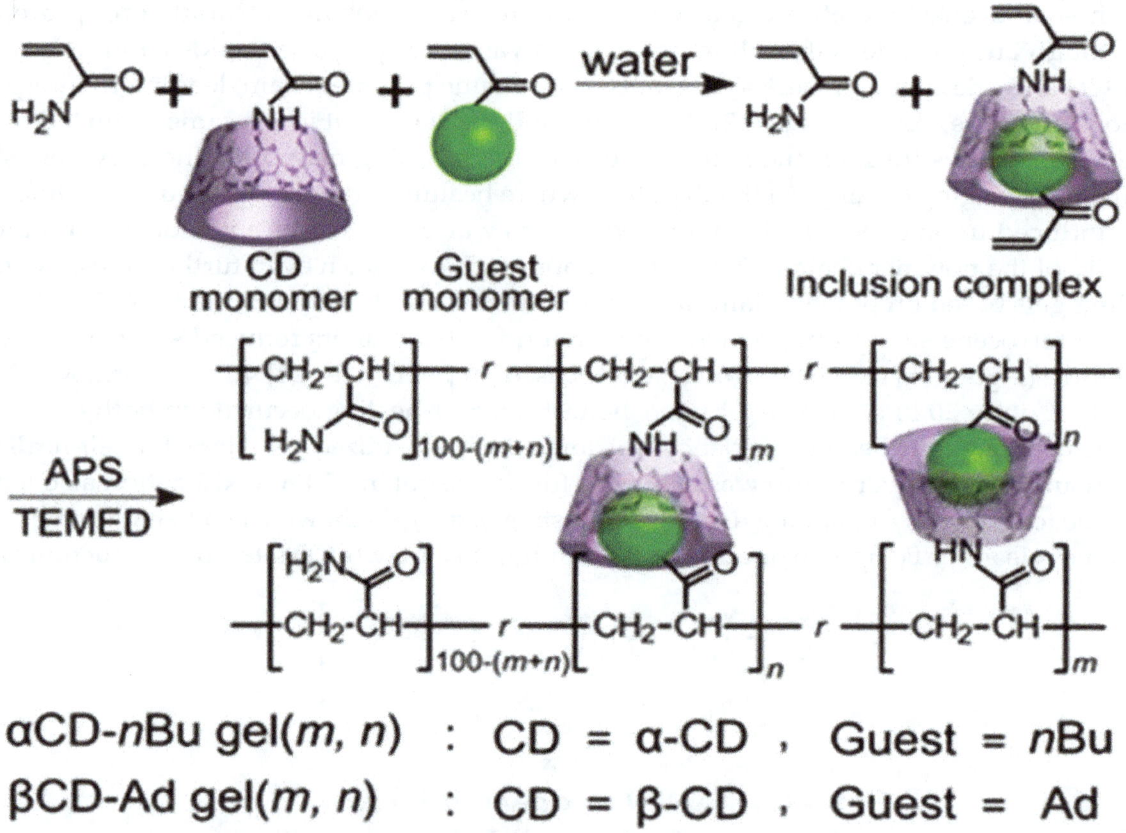

Figure 8. Preparation of host-guest supramolecular αCD-nBu gels (*m*, *n*) and βCD-Ad gels (*m*, *n*); *m* and *n* denote the mol% of the host unit and guest unit, respectively.

3.2.2. Cucurbit[n]urils

Macrocycles composed on glycouril units, cucurbit[n]urils (CB[n], where $n = 5$–8, 10; $n = 6$ most abundant), can form binary 1:1 or ternary 1:1:1 host-guest complexes with a variety of guest molecules. A comprehensive review of guests of CB[n] was recently published by Scherman *et al.* [115]. The cavity of CB[8] is large enough to accommodate two guests, and ternary complexes can be formed upon simultaneous binding of an electron-rich and an electron-deficient guests, such as naphthyl and methylviologen derivatives, respectively [116,117]. Because the functionalization of cucurbit[n]urils can be challenging [118,119], inclusion complexation-induced crosslinking and gelation is rather achieved upon ternary binding of polymer-bound guests by free CB[n] [120–122].

Self-healing hydrogels were prepared from naphthyl-functionalized hydroxyethyl cellulose (HEC-Np) and a viologen-functionalized poly(vinyl alcohol) (PVA-MV) [121]. The latter gels were reinforced by adding colloidal nanofibrillated cellulose (NFC) hydrogel to a fixed composition of supramolecular HEC-Np/PVA-MV/CB[8] hydrogel. Figure 9 depicts the resulting interpenetrating network (IPN). Supramolecular hydrogel alone showed rapid and complete self-healing after repeated shear, owing to rapid association kinetics of the ternary complex of CB[8] (association constant k_a ~10^8 $M^{-1} \cdot s^{-1}$) [121]. The strongest hybrid hydrogel has elastic modulus G′ of 2 kPa at 0.1% strain and it was stronger (up to 50-fold increase in G′), had higher yield strain (up to >4-fold enhancement) and showed improved and faster structure recovery than NFC gel alone. Supramolecular HEC-Np/PVA-MV/CB[8] hydrogel bridged the denser floc-like domains of colloidal NFC gel and mediated the healing of NFC network [122]. The recent research on ternary CB[n] complexes [115,123–125] gives promise to further developments in self-healing host-guest materials.

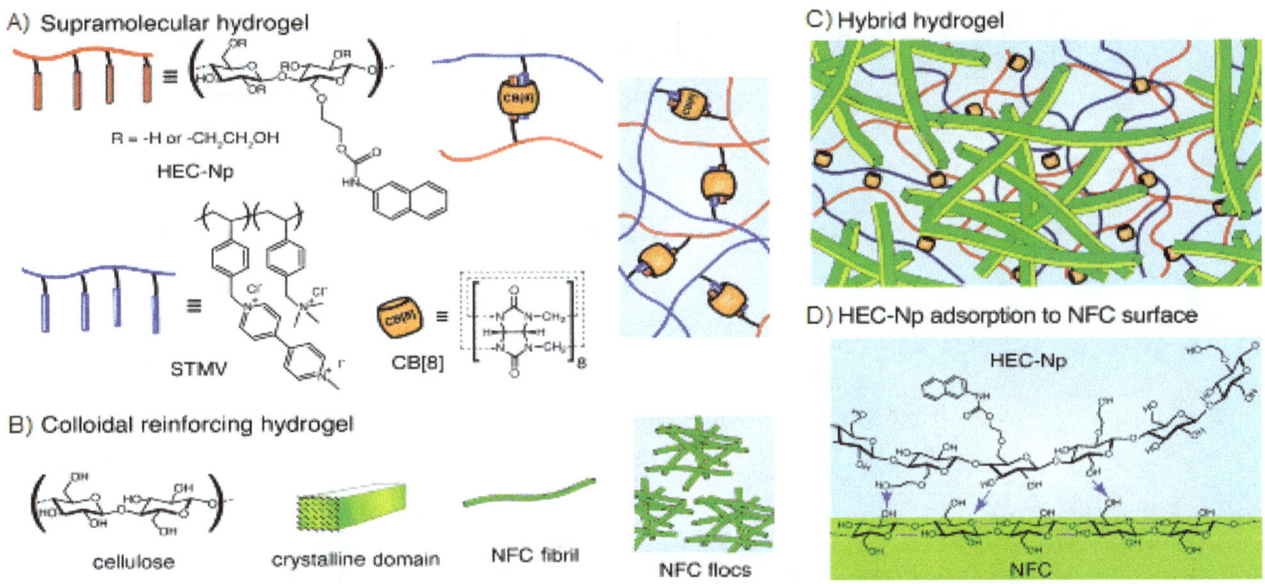

Figure 9. (**A**) Supramolecular hydrogel consisting of HEC-Np, STMV, and the CB[8] host motif capable of binding the first guest naphthyl and the second guest viologen highly dynamically; (**B**) Colloidal reinforcing nanofibrillated cellulose, also showing the denser and less dense network regimes; (**C**) Interpenetrating hybrid hydrogel consisting of the molecular-level supramolecular and colloidal-level NFC hydrogel; and (**D**) Surface adsorption of HEC-Np onto the NFC surface. Possible hydrogen bonding is schematically shown.

3.3. Hydrogen Bonding

Dynamic supramolecular polymers are typically synthesized by introducing complementary H-bonding donor and acceptor motifs into the building blocks [65,66,126], which is also an elegant strategy for building reversible networks with self-healing capacity. In aqueous medium, the H-bonding occurs in competition with water molecules, whose contribution can be diminished by using multiple H-bonding motifs with high dimerization affinity, such as 2-ureido-4[1*H*]pyrimidinone (UPy) units. The dimerization can be further enhanced by shielding the motifs from water with hydrophobic substituents, such as adamantyl (Ad) or alkyl groups [127]. Such motifs have been used to induce reversible physical network formation of a copolymer of *N,N′*-dimethylacrylamide (DMA) and a methacroyl monomer bearing a Ad-functionalized UPy unit [127] or a PEG chain bearing UPy moieties shielded from water by apolar isophorone [128] or alkyl spacers [129,130] (Figure 10a,b). The transient network formation of telechelic UPy-bearing polymers arose from the entanglement of supramolecular self-assembled fibrils at elevated concentrations (Figure 10c), which was further dependent on the length of the alkyl spacer and temperature [129]. These hydrogelators showed a reversible transition from viscous liquids to solid-like gels upon cooling, and the gel strength and self-healing properties could be modulated by the addition of short-chain monofunctional UPy-PEGs [130]. The strongest gel showed elastic modulus G' of 18 kPa at 20 °C at angular frequency of 10 rad/s. Despite the multivalency of interactions, the structure recovery of the gels could take from hours to days, similar to many low-molecular-weight hydrogelators (LMWHs) that also form networks of bundled self-assembled fibers, studied extensively by the groups of Weiss, van Esch, and Zhu, among others [131–134]. The slow recovery raises a question of whether such systems can be strictly considered as self-healing materials, as rapid healing and fast macromolecular dynamics are often required for potential applications. Nevertheless, the transient network formation of telechelic UPy-alkyl-PEGs was fast enough for the successful encapsulation and injection delivery of an anti-fibrotic growth factor protein BMP7 *in vivo* in the kidney capsule of rats [129].

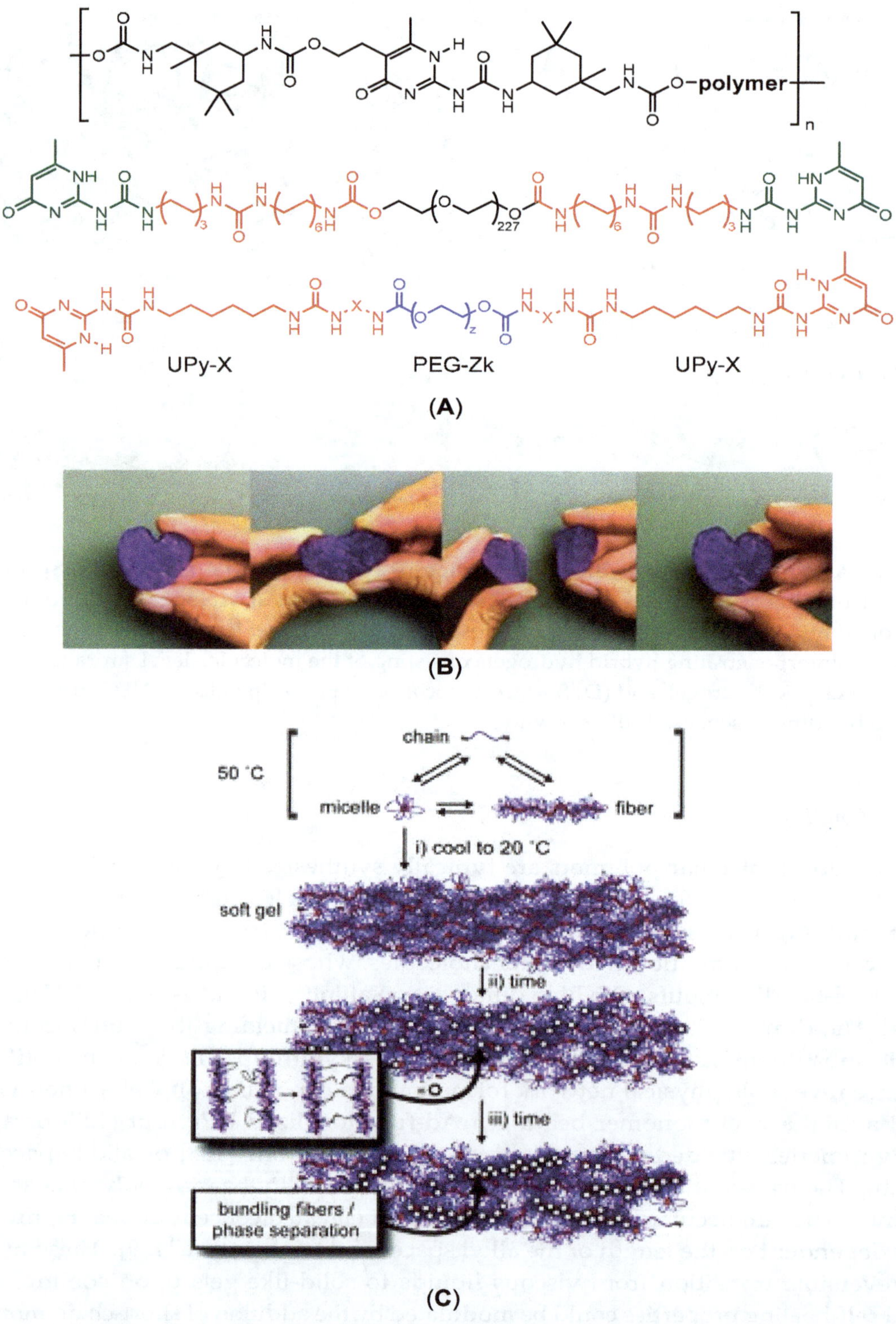

Figure 10. (**A**) Examples of the structures of PEG-based H-bonding polymers that contain UPy units as part of main chain [128] or as end groups [129,130]; (**B**) Visual demonstration of the self-healing of a 15 wt% hydrogel of the main-chain-functionalized polymer of (a). Purple dye was added for demonstration purposes [128]; (**C**) At high temperatures (above 50 °C) and in dilute solution, the hydrogelators are present as three species: single chain, spheric micelle, and fiber: (**i**) Upon cooling, or increase in concentration, a soft hydrogel is formed; (**ii**) After a time span of approximately 16–24 h, the strength of the gel is increased due to the formation of supramolecular cross-links; and (**iii**) The fibers may bundle and phase separate forming ordered domains in the hydrogel network.

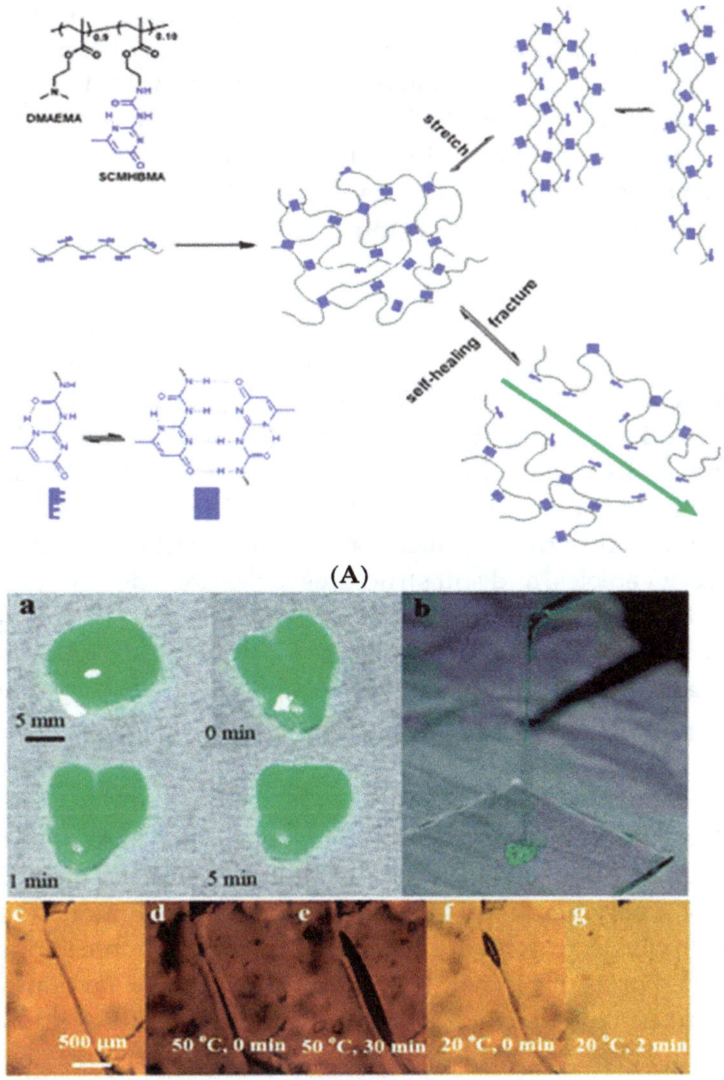

(B)

Figure 11. (**A**) Chemical structure of the copolymer containing DMAEMA and SCMHBMA and the schematic model of self-healing and stretching of the hydrogel formed by the copolymer; (**B**) Demonstration of the self-healing (**a**,**c**–**g**) and the stretching (**b**) properties of the DMAEMA-SCMHBMA hydrogel at pH 8. The gel in (**a**,**b**) was colored with methyl blue for better imaging. Optical microscopy images (**c**–**g**) were obtained from a hydrogel film with an incision (**c**) after annealing at 50 °C (**d**,**e**) and subsequent cooling to 20 °C (**f**,**g**).

UPy units can also be introduced as side groups that form interchain crosslinks upon H-bonded dimerization. A copolymer composed of 2-(dimethylamino)ethyl methacrylate (DMAEMA) and a UPy-unit-bearing monomer 2-(3-(6-methyl-4-oxo-1,4-dihydropyrimidin-2-yl)ureido)ethyl methacrylate (SCMHBMA) (Figure 11A) was thermo-responsive with lower critical solution temperature (LCST) of 40 °C at pH 8 [135]. The copolymer yielded viscous solutions at acidic pH where DMAEMA units were protonated, whereas self-healing hydrogels were obtained at neutral to basic pH (pH 7–8) at 20 °C (Figure 11B). Here, the self-healing properties were only observed visually and using polarized optical microscopy (POM). Increasing the temperature above the LCST prevented self-healing, because the collapse of polymer chains restricted their diffusion to the damaged site and H bonds could not be restored. Self-healing was also observed for covalently crosslinked gels with the same composition, as well as for other copolymers of SCMHBMA with hydrophilic or thermo-sensitive comonomers, such as 2-hydroxyethyl methacrylate (HEMA), 2-(2-methoxyethoxy)ethyl methacrylate (MEO2MA), *N*-isopropylacrylamide (NIPAAm), and *N,N′*-dimethylacrylamide (DMA) [135]. Later, this monomer was copolymerized into crosslinked polyurethane-PEG-methacrylate (PU-PEGMA) networks to yield highly deformable (elongation at break up to 2000%) self-healing hydrogels, which were capable of recovering up to 87% of their original tensile strength of 382 kPa in 10 min after bringing the cut pieces into contact. The gels also showed high compression strength of up to 4.5 MPa and did not rupture even when exposed to 90% compressive strain [136].

Varghese and coworkers [137] showed that poly(6-acryloyl-6-aminocaproic acid)-based hydrogels (PA6ACA) underwent rapid (<2 s) and repeatable H-bonding-induced self-repair in acidic conditions (pH $\leqslant$3) where carboxylic acid groups were protonated. A6ACA monomer contains an alkyl spacer between the acrylamide and –COOH group. The hydrogels healed for 10 s could withstand higher than 2 kPa tensile stresses and the rupture occurred within the bulk region of the gel, not at the welded interface. The self-healing ability depended strongly on the availability of amide bond for H-bonding, that is, the length of the spacer between –COOH and amide group. Based on proof-of-concept testing, the gels are expected to find their applications as acid-resistant sealants, muco-adhesive tissue adhesives and drug delivery vehicles, and soft structures for various devices [137]. Self-healing of these gels could also be induced by divalent metal complexation (Cu^{2+}) by A6ACA units [138]. Recently, the copolymers of an amino acid-based monomer, *N*-acryloyl glycinamide (NAGA) and acrylamide (AAm) showed thermoplasticity and temperature-driven H-bonding between the amino acid moieties, which yielded strong self-healing hydrogels (G′ up to 1 MPa) with the healing efficiency of up to 84% [139].

3.4. Ionic Interactions

Ionic interactions have long been used to reinforce elastomeric materials, as a relatively small concentration of acid or ionic groups can substantially alter the physical, mechanical, optical, dielectric, and dynamic properties of a polymer [140,141]. Ionomers, *i.e.*, copolymers with typically less than 15 mol% of ionic groups, and polyelectrolytes are widely used to create ionic supramolecular systems and crosslinked networks [141]. For example, self-healing materials based on non-aqueous ionic networks have been made from poly(acrylic acid) (PAA) and a phosphonium ionic liquid linker, where the length of the alkyl substituents and number of phosphonium groups (di- or monofunctional) determined the mechanical properties of the material [142]. Ionic crosslinks can also be used to dissipate energy and induce healing of covalently crosslinked hydrogels. For instance, tough stretchable hydrogels have been synthesized by mixing ionically crosslinked alginate (crosslinking by Ca^{2+} ions) into a photocrosslinked polyacrylamide (PAAm) network. These gels were stronger than individual PAAm or alginate gels and stretchable to nearly 20 times of their length prior to rupture. Of the network strength, 74% was recovered after 1 day of healing at elevated temperature (80 °C) [143].

The interacting charges may also exist in the same monomer. Zwitterions are dipolar species, in which the cation and the anion are separate in the same monomer unit and can be completely dissociated, thus maintaining the overall electroneutrality. Zwitterionic polymers have been shown to form reversible self-healing hydrogels through ionic interactions. Examples of such systems are carboxy- or sulfobetaines and their copolymers with neutral comonomers. A carboxybetaine acrylate (AAZ) homopolymer yielded biocompatible rapidly healing ($\leqslant$100 s) hydrogels, where the strongest physically crosslinked gels exhibited elastic modulus G′ of ~800 Pa, tensile strength of around 60 kPa and Young's modulus E of ~25 Pa. The self-healing of gels (Figure 12a) was induced by the electrostatic attraction ("zwitterionic fusion") between the quaternary ammonium group and carboxylate group of the same monomer. The self-healing could take place even after a long separation time of the cut fragments both in physically and chemically crosslinked gel samples [144], while most examples of self-healing of gels have been tested on fresh surfaces. Moreover, the fusion of cell-laden pieces of hydrogel did not cause a significant loss in cell viability, which suggests potential applications in tissue engineering and regenerative medicine.

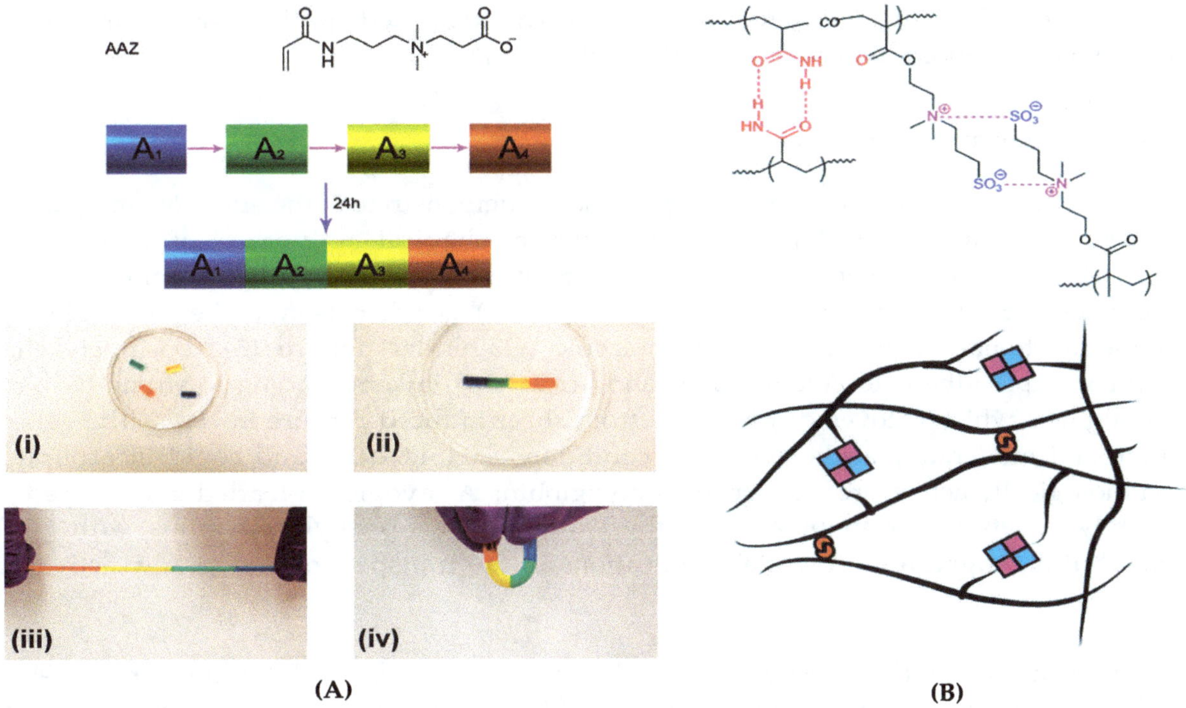

Figure 12. (**A**) Carboxybetaine acrylate (AAZ) monomer and the fusion of its homopolymer hydrogels from separate blocks (A_1–A_4, (**i**)) into a uniform gel (**ii**) that can be stretched (**iii**) and bent (**iv**) without breaking. A_1 (**blue**), A_2 (**green**), A_3 (**yellow**) and A_4 (**red**) have the same composition [144]; (**B**) Reversible hydrogen bonding and electrostatic interactions of a P(AAm-*co*-DMAPS) copolymer, resulting in crosslinked hydrogel formation.

In another carboxybetaine-based system, nanocomposite hydrogels were prepared by the copolymerization of a carboxybetaine methacrylamide CBMAA-3 ((3-methacryloylaminopropyl)-(2-carboxyethyl)dimethylammonium-(carboxybetaine methacrylamide)) and 2-hydroxyethyl methacrylate (HEMA) in the presence of an inorganic clay, Laponite XLG [146]. Laponites are hydrophilic disk-shaped inorganic clay platelets that can be uniformly dispersed in water and act as physical crosslinkers. The presence of Laponite enhanced the mechanical properties of composite hydrogels (Young's modulus up to 80 kPa, elongation at break up to 1800%) and reduced the water uptake of gels due to higher crosslinking density, while the higher zwitterion (CBMAA-3) content was associated with faster structure recovery [146]. These materials were proposed to act as non-fouling protein-resistant biomaterials. Interestingly, the nanocomposite hydrogels of Laponite with the copolymers of DMA and a sulfobetaine acrylamide (*N*,*N*-dimethyl(acrylamidopropyl) ammonium propane sulfonate) (DMAAPS) demonstrated upper critical solution temperature (UCST) behavior with UCST of $\leqslant$9.3 °C when DMA content was $\leqslant$10 mol% [147]. The hydrogels showed self-healing behavior above but not below the UCST, similar to the H-bonded LCST network described in Section 3.3 [135], because the diffusion of dangling polymers chains across the damaged region and interactions with clay platelets were disrupted upon the collapse of chains below the UCST. Here, the strongest gel showed elastic modulus G' of ~1.6 kPa at 0 °C and $f = 1$ Hz, the tensile strength below 5 °C was ~65 kPa and elongation at break was ~1800%. However, only 20% recovery in tensile strength was achieved after 24 h self-healing above the UCST [147].

Combined electrostatic and H-bonding interactions were responsible for the self-healing of a zwitterionic hydrogel based on P(AAm-*co*-DMAPS) polymers, where AAm = acrylamide and DMAPS = 3-dimethyl(methacryloyloxyethyl) ammonium propane sulfonate (Figure 12b) [145]. High healing efficiency (up to 80%) and fast recovery (~35 s) were observed for gels with AAm/DMAPS = 1 and the strongest gel showed elastic modulus G′ of ~5.5 kPa. The suggested applications of these hydrogels are in the field of enhanced oil recovery (EOR) [145].

3.5. Polymer-Nanocomposite Interactions

Some of the above-mentioned examples have demonstrated the use of inorganic clay nanoplatelets (Laponite) [146,147] to enhance the mechanical properties of supramolecular hydrogels. Aida and coworkers [148] prepared rapidly self-healing hydrogels from aqueous clay nanosheet (CNS) solution with sodium polyacrylate (ASAP) as a dispersant and exfoliator by adding a guanidine-functionalized dendritic macromolecule (G3-binder, Figure 13). Positively charged guanidium groups adhere to anionic CNSs and act as crosslinkers, leading to strong (G′ >5 kPa) free-standing gels which recovered repeatedly from shear-induced rupture in 600 s. The gels could resist brine and moderately acidic and basic conditions (pH 4.0–10.0) and could incorporate and maintain biologically active proteins, such as myoglobin. A myoglobin-loaded gel retained ~71% of the catalytic activity of myoglobin in the oxidation of *o*-phenylenediamine with H_2O_2 in phosphate buffer, suggesting potential applications in the transport of biologically active agents [148].

Guanidinium-based (PG_n) blocks have also been used to build luminescent self-healing hydrogels via physical crosslinking of PG_n-*b*-PEO_{230}-*b*-PG_n block copolymers with anionic nanosized inorganic transition-metal oxide clusters, polyoxometalates (POMs). The resulting self-healable hydrogels showed enhanced luminescence properties, elastic modulus G′ of up to 30 kPa and rapid structure recovery in 20 s [149]. In another example of nanoclay composites, the interaction of a cationic $PDMAEMA_6$-*b*-PEO_{109}-*b*-$PDMAEMA_6$ block copolymer with Laponite XLG clay platelets resulted in a self-healing CO_2-responsive hydrogel, where protonated PDMAEMA blocks were bound to the anionic surface of nanoclay and formed interplatelet bridges. The protonation of PDMAEMA was induced by bubbling CO_2 through the clay-polymer solution and it could be reversed by N_2 flow, while the self-healing of the gel arose from the electrostatic interactions. The strongest gel showed elastic modulus G′ of 5 kPa at $f = 1$ Hz [150]. The interaction with nanoclay could also be based on H-bonding, such as in the case of self-healing Laponite nanoclay composites with water-soluble poly(*N,N*′-dimethylacrylamide)(PDMA) or thermo-responsive poly(*N*-isopropylacrylamide) (PNIPAAm) [151].

Self-healing hydrogels based on hydrophobic interactions between a hydrophobically modified water-soluble polymer and graphene oxide (GO) sheets were prepared by the copolymerization of stearyl methacrylate and acrylamide in the presence of GO and sodium dodecyl benzene sulfonate (SDBS) [74]. As discussed in Section 3.1, hydrophobically modified copolymers alone can show self-healing properties. Here, the GO sheets acted as physical crosslinking junctions resulting in a dramatic increase in mechanical strength but decreased self-healing efficiency (53% and 66% with 5 wt% GO and without GO, respectively).

The strongest gels exhibited tensile strength of 243 kPa with elongation at break of 1700% and elastic modulus of ~65 kPa. The nanocomposite gels could efficiently remove hydrophobic compounds from water, tested by the absorption and release of methylene blue (MB) and congo red (CR) dyes, and could be recycled cyclically for dye absorption without compromising the mechanical properties. Hence, applications in water purification have been suggested [74]. Other self-healing GO nanocomposites have been made for example from chitosan via ionic interactions between cationic chitosan and the anionic surface of GO sheets [152] or poly(6-acryloyl-6-aminocaproic acid) (PA6ACA) in the presence of Ca^{2+} via combined coordination and H-bonding interactions [153].

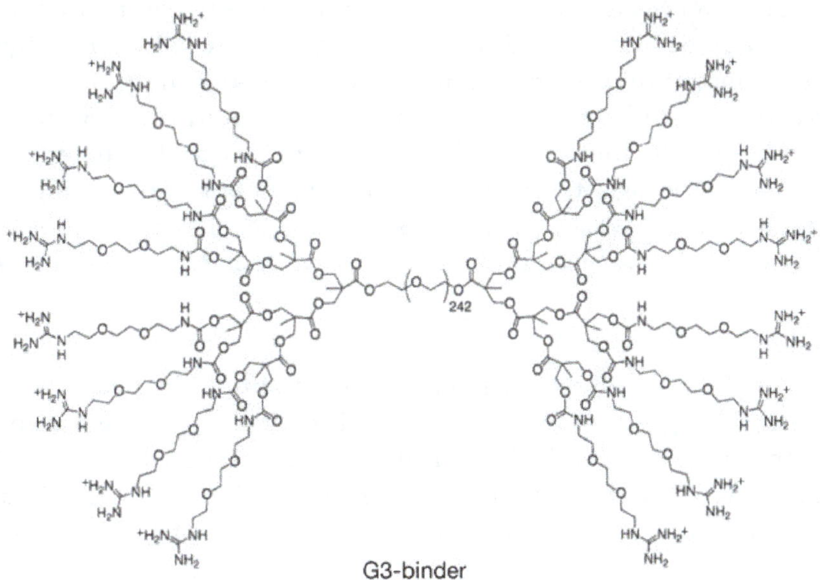

(A)

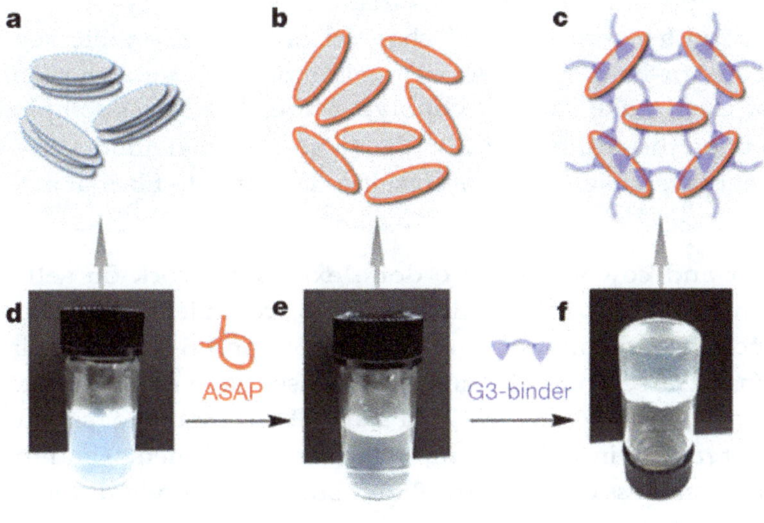

(B)

Figure 13. (**A**) Structure of guanidine-functionalized G3-binder; (**B**) Non-covalent preparation of hydrogels. (**a–c**) Proposed mechanism for hydrogelation. CNSs, entangled with one another (**a**), are dispersed homogeneously by interaction of their positive-charged edge parts with anionic ASAP (**b**). Upon addition of Gn-binder, exfoliated CNSs are crosslinked to develop a 3D network (**c**). (**d–f**) Optical images of an aqueous suspension of CNSs (**d**), an aqueous dispersion of CNSs and ASAP (**e**) and a physical gel upon the addition of G3-binder to the dispersion (**f**).

3.6. Other Crosslinking Mechanisms: Crystallization, Transient Protein Interactions

Crystallization of macromolecules can lead to gelation in aqueous solution through locking of the polymer chains into regular structures that act as physical crosslinks. Such molecular assembly is typical for protein- or peptide-based systems [68,154,155], but is not uncommon even with synthetic macromolecules. For example, poly(vinyl alcohol) (PVA) formed physically crosslinked self-healing hydrogels when exposed to repeated freeze-thaw cycles, and the crystallinity as well as the mechanical strength of gels increased with polymer concentration and the number of cycles [73]. PVA gels of 35 wt% showed repeatable rapid healing and evolution of fracture stress: after 10 s recovery, the gels could withstand a stress of around 10 kPa and after 1 h, 40% of initial tensile strength was regained. The crystallites act as physical crosslinks in a hydrogel, while self-healing is favored by concentration-dependent interchain H-bonding and the diffusion of PVA chains over the fractured surface. Hence, self-healing could not occur below a limiting concentration (20 wt%) [73]. The mechanical properties of PVA hydrogels were further enhanced by the addition of H-bonding agent, melamine, that acts as both H-bonding acceptor and donor [156].For example, a gel with 1.5 wt%melamine displayed a tensile strength of 3.11 MPa and 525% elongation at break, while the values for a gel without melamine were 0.74 MPa and 325%, respectively.

However, adding melamine reduced the self-healing ability PVA hydrogels (from 86% to 52%) through increasing the gel rigidity, which further led to reduced chain interdiffusion and H-bonding over the fractured surfaces. [157]. Similarly, an interpenetrating hydrogel network of chemically crosslinked PEG and physically crosslinked PVA showed reduced self-healing capacity due to lower PVA chain mobility and lower H-bond forming–OH group density [158].

Examples of proteins and polypeptides that have been shown to form self-healing hydrogels include silk-collagen-like block copolypeptides that showed pH-dependent fibrillar self-assembly [159], diblock copolypeptides composed of charged and hydrophobic segments that exhibited conformation-dependent (α-helix *vs.* β-sheet) gelation and high thermal stability up to 90 °C [160], micelles of elastin-like diblock copolypeptide that self-assembled into rapidly shear-recovering hydrogels upon the addition of Zn^{2+} [161], and bis-maleimide-PEG-linked globular actin (G-actin) that underwent reversible salt-induced self-assembly into fibrillar actin (F-actin) [162].

Pochan, Schneider and coworkers have done extensive work on self-assembling β-hairpin peptides with shear-thinning and self-healing behavior for injectable therapeutic delivery [163]. Recently, engineered peptides containing aggregation-prone and β-sheet-forming C-terminal amyloid β-protein sequences (associated with Alzheimer's disease) gave rise to amyloid fibril network and weak self-healing hydrogel formation (G' 40–250 Pa at 10 Hz) [164]. The resulting gels recovered from deformation in 15 min of rest (or in 100 s during step strain experiments) and the recovery mechanism was associated with the stickiness of the hydrophobic surface of amyloid fibrils.

Heilshorn and coworkers [165] prepared an elegant injectable and biocompatible hydrogel that combined synthetic and peptide-based systems and underwent two different physical crosslinking mechanisms. This double network system was based on two components: (1) 8-arm star-like PEG conjugated with one thermo-responsive PNIPAAm chain and seven proline-rich peptide domains (P1); and (2) linear protein copolymer consisting of P1-recognizing units and cell adhesion-favoring RGD arginine-glycine-aspartic acid domains connected by hydrophilic spacers (Figure 14).

The first crosslinking occurred *ex vivo* upon mixing the two components, which led to peptide-based recognition and very weak hydrogels with elastic modulus of 13 Pa. The second crosslinking occurred at body temperature upon the collapse of PNIPAAm above its LCST and led to increase in gel strength ($G' \sim 100$ Pa). The gels showed reversible rapid self-healing (<2 s) due to reformation of physical network junctions and the process was faster than with gels based on protein-ligand interactions, passive re-entanglement of polymer chains or stepwise reassembly into nanofibers. The gels enhanced stem cell retention at the desired site *in vivo*, reducing the number of injections and cells required for transplantation [165].

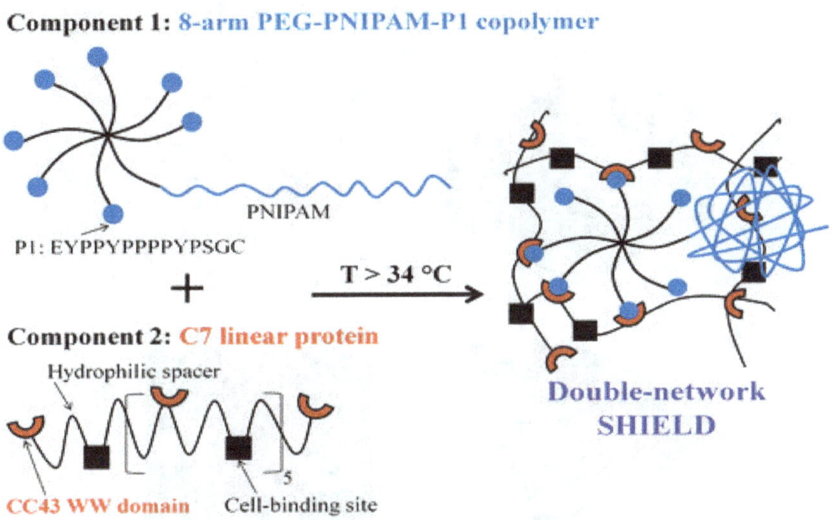

Figure 14. Schematic of a two-component double network "SHIELD", where **component 1** is an 8-arm PEG with 1 arm conjugated with PNIPAAm and the other 7 arms conjugated with proline-rich peptide (denoted as P1) domains. **Component 2** is a recombinant C7 linear protein copolymer bearing CC43 WW (denoted as C) domains and RGD (arginine-glycine-aspartic acid) cell-binding domains connected by hydrophilic spacers. The formation of double network occurs via peptide recognition and collapse of thermo-responsibe PNIPAAm chains.

4. Potential Applications

Rapid shear thinning or solid-liquid transition and fast structure recovery are key requirements both for injectable hydrogels and hydrogel inks for additive manufacturing. Additive manufacturing technologies, such as 3D printing, enable the fabrication of complex micrometer- and millimeter-scale structures, where hydrogels can be used as "bio-inks" to form constructs that replace, augment or model tissues [166–169]. The strategies and molecular design criteria of 3D-printable hydrogel inks have been highlighted in a recent review, where non-covalent gelators were categorized into supramolecular polymers, low molecular weight gelators, and macromolecular and colloidal solid particle-based systems [169]. Among these, the naturally occurring, slightly modified biopolymers have so far shown the highest potential for biofabrication owing to their favorable biocompatibility and capability to support cell survival and differentiation in culture. For example, Burdick and coworkers [170] used host-guest complexation-based self-healing gels to construct patterned multicellular structures (Figure 15), where mesenchymal stem cells (MSCs) and 3T3 fibroblasts loaded within a biocompatible network of adamantane- and β-CD-functionalized biopolysaccharide, hyaluronic acid (HA), were successfully printed to their own compartments. Here, a strong 4 wt% HA gel with 40 mol% of β-CD (host) and Ad (guest) units was used as a support gel ($G' \sim 1$ kPa at $f = 1$ Hz) and softer 5 wt% shear-thinning gel with 25 mol% of host and guest units was used as an ink ($G'' > G'$ at $f = 1$ Hz indicating liquid-like behavior; G' ~300 Pa and $G' > G''$ below 1% strain indicating solid-like behavior). Cyclodextrins have widely been used in numerous injectable hydrogels owing to their excellent biocompatibility, weak immunogenicity, and selectivity as supramolecular hosts [44,171]. Secondary covalent crosslinking was introduced via photopolymerization of pendent methacrylate groups (Me-HA) to stabilize the printed gel against chemical or physical perturbations, such as perfusion for the convection of nutrients and removal of cell metabolic waste. As a result, the gels showed increased elastic moduli (G' >10 kPa) and reduced frequency dependence [170]. Spatially controlled bioprinting of multiple cell types enables the development of organ or tissue constructs that do not require substantial vascularization as well as mini-tissue models for pharmaceutical or cosmetic testing and disease studies [172].

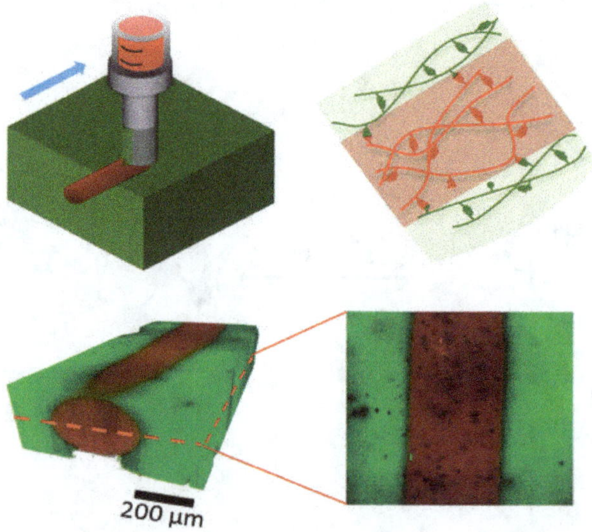

Figure 15. (**Above**): Schematic illustration of 3D printing of a supramolecular gel based on adamantane (Ad, guest) and β-cyclodextrin (β-CD, host) conjugated to hyaluronic acid (HA). Green represents a supramolecular support gel and red the printed ink; (**Below**): 3D reconstruction of a confocal Z-stack of an ink filament (rhodamine-labeled, **red**) printed into a support gel (fluorescein-labeled, **green**). Scalebar: 200 μm.

The use of injectable hydrogels provides a minimally invasive strategy for the administration of bioactive molecules and cells in the body [8,173,174]. Simple injection instead of a complicated surgical procedure reduces the infection risks and recovery time in patients and injections can be done to sites that are hard to reach by surgical techniques. Hydrogels can protect protein-based drugs from enzymatic degradation and provide targeted delivery and prolonged release of the drug to the target site [174]. They also provide a specific biological and mechanical three-dimensional environment for cells that guides the regulation of cellular functions during the formation of regenerated tissue. Injection delivery can be done via a syringe or via a catheter and recently, an H-bonded UPy-unit containing hydrogels loaded with a small-molecule drug pirfenidone or a fluorescent model protein mRuby2 were injected into a pig heart via an intramyocardial catheter [175]. The injectable system was a Newtonian fluid with low viscosity at high pH and formed a stiff gel with $G' = 10–20$ kPa (at angular frequency $\omega = 0.1–100$ rad/s) after neutralization. Here the liquid-like behavior was necessary for the delivery via a catheter. The gradual degradation of the supramolecular gel and release of the model protein shows promise of these self-healing gels as delivery systems for growth factors and exosomes. In cellular therapy, rapid self-healing of the hydrogel localizes the cells to the targeted area and increases the probability of cell retention after transplantation [165]. The above-mentioned supramolecular HA/β-CD/Ad hydrogel provided improved delivery and retention of endothelial progenitor cells (EPCs) *in vivo* in ischemic myocardium, aiming at the revascularization of myocardium after heart failure. Significant increase in the vascularization, improvement of ventricular function, and reduction of scar formation were observed [176]. Here, both examples of therapies target at the regeneration of ischemic myocardium after infarction.

One potential application of self-healing hydrogels is the coating of medical implants. Biocompatible hydrogel coatings with good structural stability can be used to protect orthopedic implants or implanted sensors from biofouling [21,177,178] or to deposit growth factors, cells or peptides on the implant surface [178,179]. The limitation in many bio-applications is the gel strength, which can be enhanced by secondary crosslinking or mixing with other polymers. The latter approach has been used in the preparation of an enteric elastomer (EE, Figure 16) from a mixture of H-bonding PA6ACA polymer, which is structurally similar to traditional enteric polymers, and poly(methacrylic acid-*co*-ethyl acrylate) (EUDRAGIT L 100-55) [180]. The stiffness of the resulting elastomer could be tuned by the ratio of these two polymers and the strongest gel reached the tensile strength of ~3.3 MPa and >900% elongation at break. The elastomer was used to construct a gastric-retentive device with alternating pieces of harder polycaprolactone (PCL) linked together by EE gels. *In vivo* evaluation of the gastric device demonstrated that the device was intact in the acidic conditions of stomach, but dissociated into smaller pieces upon passing to small and large intestine with neutral pH, where H-bonds start breaking upon the deprotonation of acid groups [180]. Hence, these supramolecular materials and their elastomers have strong potential in new gastric-resistant devices for weight control, ingestible electronics, and prolonged drug delivery. Recently, Gong and coworkers demonstrated the reprocessability of tough polyelectrolyte complex-based self-healing hydrogels into capsules, films, and fibers at ambient temperature [181].

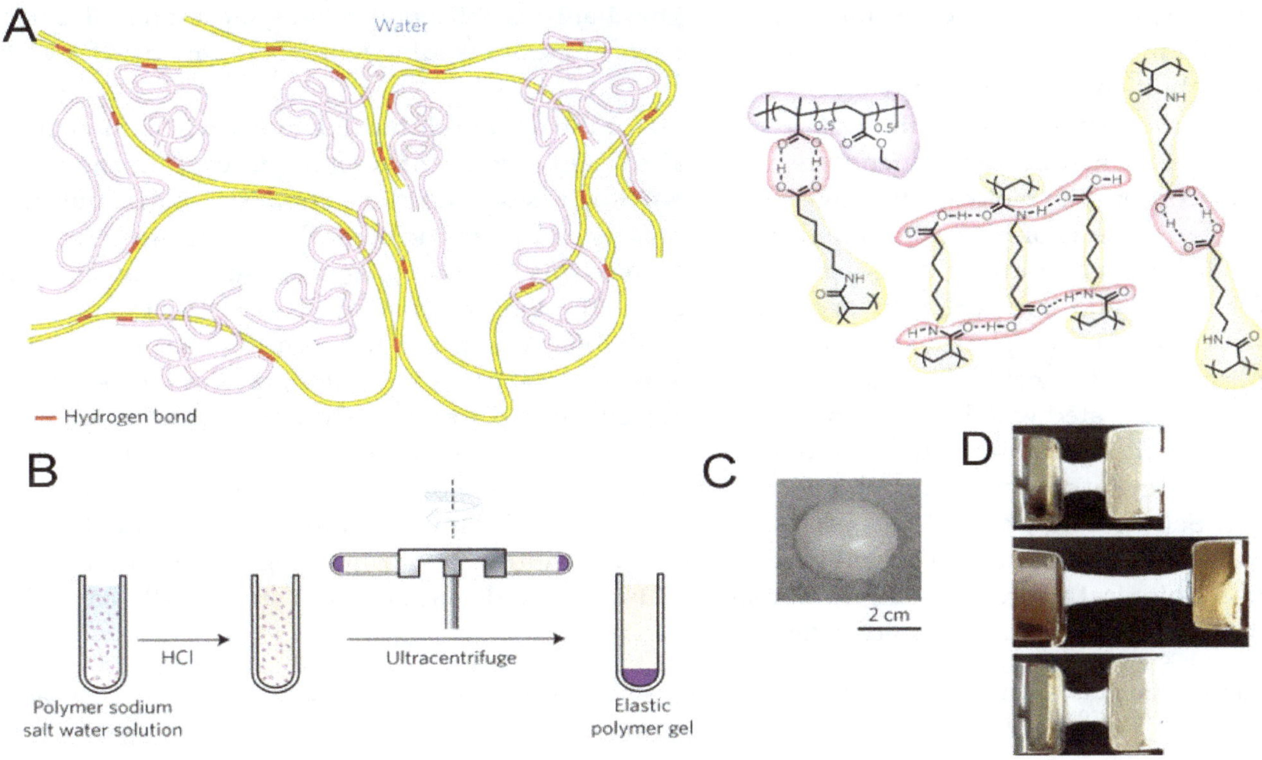

Figure 16. (**A**) Proposed supramolecular polymer gel network. Structures in yellow, synthesized poly(acryloyl 6-aminocaproic acid) (PA6ACA); structures in purple, linear poly(methacrylic acid-co-ethyl acrylate) (EUDRAGIT L 100-55); **red** part, inter-polymer hydrogen bonds; (**B**) Manufacturing process of the polymer gel. (**Left**), the homogeneous solution of PA6ACA sodium salt solution and L 100-55 sodium salt solution with varying polymer weight ratios. (**Middle**), the addition of HCl solution resulting in precipitation; (**Right**), formation of the elastic polymer gel after ultracentrifugation; (**C**) Photo of a piece of the polymer gel obtained after ultracentrifugation; and (**D**) Images of stretch and recovery testing of a polymer gel with PA6ACA:L 100-55 = 1:2. (**Top**), 1.5 cm piece of polymer gel held between two clamps. (**Middle**), stretching of the polymer gel to three times its initial length; (**Bottom**), recovery of the polymer gel 5 min after the external force was removed. Scale bar is 2 cm for (**C**) and (**D**).

5. Future Prospects and Conclusions

Self-healing of hydrogels is relatively easier to achieve than in harder materials due to the higher mobility of the chains and the dynamic bonding interactions of the functional groups. Although supramolecular materials have been under extensive research for decades, the focus has only recently shifted on investigating how the reversibility of their physical interactions can be exploited at macroscopic scale for instance, in medical devices or delivery systems. The rapid structure recovery physical gels arising from the diffusion of polymer chains and interacting sites to the damaged area is beneficial in injection- and extrusion-based applications, but achieving a balance between the mechanical properties and recovery rate is challenging. The experimental data of mechanical strength and recovery of the various hydrogels reported are rather scattered and it is difficult to compare them directly due to the differences in experimental conditions and characterization methods used. It is thus difficult to indicate a clear trend that would guide in the selection of the most suitable preparation strategies. Often the choice of the preparation methods depends on the final use of the hydrogels in addition to the required physical properties. Among the physical interactions, hydrogen bonding seems to provide hydrogels of higher strengths with tensile and compression strengths of several MPa. Various techniques to reinforce the hydrogels are highlighted in this review and the most efficient ones combine several crosslinking mechanisms or involve mixing the supramolecularly crosslinking polymer with nano-fibers, -sheets or -particles, or blending with other polymers for better processability and enhanced lifetime.

These approaches show the highest potential for the use of self-healing hydrogels in therapeutic systems and medical devices and are expected to gain significant research interest in the future. Also, incorporating biologically active compounds, such as bile acids or peptides, as supramolecular building blocks is of increasing interest. The presence of pharmacologically active agents or cells may alter the crosslinking and degradation mechanisms of hydrogels, leading to changes in gel deformation and structure recovery and therefore, the long-term stability of each new hydrogel system would need to be investigated under physiological or cell culture conditions for biomedical applications. Physical hydrogels often exhibit time-dependent deformation, referred to as "creep", which is associated with high dissipative loss modulus in relation to elastic modulus and may affect the stem cell fate choice and tissue development in cell-seeded hydrogels [182]. In addition, the supramolecular carrier system should not interfere with the biological activity of the therapeutic agent. Self-healing physical gels provide a promising platform for building dynamic devices and systems, especially upon the recent innovations in 3D printing, and combining supramolecular and biotechnological approaches shows a great promise for the development of structurally controlled materials with tailored mechanical properties.

Acknowledgments: Financial support for research from NSERC of Canada and FRQNT of Quebec is gratefully acknowledged.

Author Contributions: Satu Strandman gathered the literature material, wrote the review and assembled the figure panels. X. X. Zhu organized the topics, contributed to the discussion of the contents and corrected the manuscript.

Abbreviations

The following abbreviations are used in this manuscript:

Ad	adamantane
AdCANa	sodium adamantanecarboxylate
AAm	acrylamide
AAZ	carboxybetaine acrylate
Alg	alginate
ASAP	sodium polyacrylate
CB[n]	cucurbit[n]uril, n = number of glycouril units
CBMAA-3	(3-methacryloylaminopropyl)-(2-carboxyethyl)dimethylammonium-(carboxybetaine methacrylamide)
CD	cyclodextrin
Chol	cholesterol

CNS	clay nanosheets
DMAEMA	2-(dimethylamino)ethyl methacrylate
DMAPS	3-dimethyl(methacryloyloxyethyl) ammonium propane sulfonate
CTAB	cetyl trimethyl ammonium bromide
DMA	*N,N′*-dimethylacrylamide
DODAB	dimethyldioctadecylammonium bromide
EE	enteric elastomer
EPC	endothelial progenitor cell
Fc	ferrocene
G′	storage modulus
G″	loss modulus
GO	graphene oxide
HA	hyaluronic acid
HEC	hydroxyethyl cellulose
HEMA	2-hydroxyethyl methacrylate
MEO2MA	2-(2-methoxyethoxy)ethyl methacrylate
MSC	mesenchymal stem cell
MV	methylviologen
NFC	nanofibrillar cellulose
NIPAAm	*N*-isopropylacrylamide
Np	naphthyl
PA6ACA	poly(6-acryloyl-6-aminocaproic acid)
PAAc	poly(acrylic acid)
PEG	poly(ethylene glycol)
PLGA	poly(L-glutamic acid)
PPG	poly(propylene glycol)
PVA	poly(vinyl alcohol)
RGD	arginine-glycine-aspartic acid
SCMHBMA	2-(3-(6-methyl-4-oxo-1,4-dihydropyrimidin-2-yl)ureido)ethyl methacrylate
SDS	sodium dodecyl sulfate
SEM	scanning electron microscopy
STMV	methyl viologen-functionalized cationic polystyrene
UPy	2-ureido-4[1*H*]pyrimidinone

References

1. Almdal, K.; Dyre, J.; Hvidt, S.; Kramer, O. Towards a phenomenological definition of the term "Gel". *Polym. Gels Netw.* **1993**, *1*, 5–17. [CrossRef]
2. Flory, P.J. Introductory lecture. *Faraday Discuss. Chem. Soc.* **1974**, *57*, 7–18. [CrossRef]
3. Raghavan, S.R.; Douglas, J.F. The conundrum of gel formation by molecular nanofibers, wormlike micelles, and filamentous proteins: Gelation without cross-links? *Soft Matter* **2012**, *8*, 8539–8546. [CrossRef]
4. Ito, K. Slide-ring materials using topological supramolecular architecture. *Curr. Opin. Solid State Mater. Sci.* **2010**, *14*, 28–34. [CrossRef]
5. Peppas, N.A.; Hilt, J.Z.; Khademhosseini, A.; Langer, R. Hydrogels in biology and medicine: From molecular principles to bionanotechnology. *Adv. Mater.* **2006**, *18*, 1345–1360. [CrossRef]
6. Billiet, T.; Vandenhaute, M.; Schelfhout, J.; Van Vlierberghe, S.; Dubruel, P. A review of trends and limitations in hydrogel-rapid prototyping for tissue engineering. *Biomaterials* **2012**, *33*, 6020–6041. [CrossRef] [PubMed]
7. Hoffmann, A.S. Hydrogels for biomedical applications. *Adv. Drug Deliv. Rev.* **2012**, *64*, 18–23. [CrossRef]

8. Nguyen, Q.V.; Huynh, D.P.; Park, J.H.; Lee, D.S. Injectable polymeric hydrogels for the delivery of therapeutic agents: A review. *Eur. Polym. J.* **2015**, *72*, 602–619. [CrossRef]
9. Calo, E.; Khutoryanskiy, V.V. Biomedical applications of hydrogels: A review of patents and commercial products. *Eur. Polym. J.* **2015**, *65*, 252–267. [CrossRef]
10. Buwalda, S.J.; Boere, K.W.M.; Dijkstra, P.J.; Feijen, J.; Vermonden, T.; Hennink, W.E. Hydrogels in a historical perspective: From simple networks to smart materials. *J. Control. Release* **2014**, *190*, 254–273. [CrossRef] [PubMed]
11. Kopecek, J. Hydrogels: From soft contact lenses and implants to self-assembled nanomaterials. *J. Polym. Sci. A Polym. Chem.* **2009**, *47*, 5929–5946. [CrossRef] [PubMed]
12. Bartnikowski, M.; Wellard, R.M.; Woodruff, M.; Klein, T. Tailoring hydrogel viscoelasticity with physical and chemical crosslinking. *Polymers* **2015**, *7*, 2650–2669. [CrossRef]
13. Lim, H.L.; Hwang, Y.; Kar, M.; Varghese, S. Smart hydrogels as functional biomimetic systems. *Biomater. Sci.* **2014**, *2*, 603–618. [CrossRef]
14. Richter, A.; Paschew, G.; Klatt, S.; Lienig, J.; Arndt, K.F.; Adler, H.J.P. Review on Hydrogel-based pH sensors and microsensors. *Sensors* **2008**, *8*, 561–581. [CrossRef]
15. Deligkaris, K.; Tadele, T.S.; Olthuis, W.; van den Berg, A. Hydrogel-based devices for biomedical applications. *Sens. Actuator B Chem.* **2010**, *147*, 765–774. [CrossRef]
16. Mateescu, A.; Wang, Y.; Dostalek, J.; Jonas, U. Thin hydrogel films for optical biosensor applications. *Membranes* **2012**, *2*, 40–69. [CrossRef] [PubMed]
17. Yetisen, A.K.; Butt, H.; Volpatti, L.R.; Pavlichenko, I.; Humar, M.; Kwok, S.J.J.; Koo, H.; Kim, K.S.; Naydenova, I.; Khademhosseini, A.; *et al.* Photonic hydrogel sensors. *Biotechnol. Adv.* **2015**. [CrossRef] [PubMed]
18. Ionov, L. Hydrogel-based actuators: Possibilities and limitations. *Mater. Today* **2014**, *17*, 494–503. [CrossRef]
19. Carpi, F.; Smela, E. *Biomedical Applications of Electroactive Polymer Actuators*; John Wiley & Sons: Chichester, UK, 2009; pp. 5–100.
20. Morais, J.; Papadimitrakopoulos, F.; Burgess, D. Biomaterials/Tissue interactions: Possible solutions to overcome foreign body response. *AAPS J.* **2010**, *12*, 188–196. [CrossRef] [PubMed]
21. Campoccia, D.; Montanaro, L.; Arciola, C.R. A review of the biomaterials technologies for infection-resistant surfaces. *Biomaterials* **2013**, *34*, 8533–8554. [CrossRef] [PubMed]
22. Cavallaro, A.; Taheri, S.; Vasilev, K. Responsive and "smart" antibacterial surfaces: Common approaches and new developments. *Biointerphases* **2014**, *9*, 029005. [CrossRef] [PubMed]
23. Boateng, J.S.; Matthews, K.H.; Stevens, H.N.E.; Eccleston, G.M. Wound healing dressings and drug delivery systems: A review. *J. Pharm. Sci.* **2008**, *97*, 2892–2923. [CrossRef] [PubMed]
24. Madaghiele, M.; Demitri, C.; Sannino, A.; Ambrosio, L. Polymeric hydrogels for burn wound care: Advanced skin wound dressings and regenerative templates. *Burns Trauma* **2014**, *2*, 153–161. [CrossRef]
25. Ghobril, C.; Grinstaff, M.W. The chemistry and engineering of polymeric hydrogel adhesives for wound closure: A tutorial. *Chem. Soc. Rev.* **2015**, *44*, 1820–1835. [CrossRef] [PubMed]
26. Kabiri, K.; Omidian, H.; Zohuriaan-Mehr, M.J.; Doroudiani, S. Superabsorbent hydrogel composites and nanocomposites: A review. *Polym. Comp.* **2011**, *32*, 277–289. [CrossRef]
27. Wan, J. Microfluidic-based synthesis of hydrogel particles for cell microencapsulation and cell-based drug delivery. *Polymers* **2012**, *4*, 1084–1108. [CrossRef]
28. Olabisi, R.M. Cell microencapsulation with synthetic polymers. *J. Biomed. Mater. Res. A* **2015**, *103*, 846–859. [CrossRef] [PubMed]
29. Wang, C.; Varshney, R.; Wang, D.A. Therapeutic cell delivery and fate control in hydrogels and hydrogel hybrids. *Adv. Drug Deliv. Rev.* **2010**, *62*, 699–710. [CrossRef] [PubMed]
30. Kearney, C.J.; Mooney, D.J. Macroscale delivery systems for molecular and cellular payloads. *Nat. Mater.* **2013**, *12*, 1004–1017. [CrossRef] [PubMed]
31. Nicolson, P.C.; Vogt, J. Soft contact lens polymers: An evolution. *Biomaterials* **2001**, *22*, 3273–3283. [CrossRef]
32. Rubinstein, M.P. Applications of contact lens devices in the management of corneal disease. *Eye* **2003**, *17*, 872–876. [CrossRef] [PubMed]
33. Xinming, L.; Yingde, C.; Lloyd, A.W.; Mikhalovsky, S.V.; Sandeman, S.R.; Howel, C.A.; Liewen, L. Polymeric hydrogels for novel contact lens-based ophthalmic drug delivery systems: A review. *Contact Lens Anterior Eye* **2008**, *31*, 57–64. [CrossRef] [PubMed]

34. Kharkar, P.M.; Kiick, K.L.; Kloxin, A.M. Designing degradable hydrogels for orthogonal control of cell microenvironments. *Chem. Soc. Rev.* **2013**, *42*, 7335–7372. [CrossRef] [PubMed]
35. Gibbs, D.M.R.; Black, C.R.M.; Dawson, J.I.; Oreffo, R.O.C. A review of hydrogel use in fracture healing and bone regeneration. *J. Tissue Eng. Regen. Med.* **2014**. [CrossRef] [PubMed]
36. Hastings, C.L.; Roche, E.T.; Ruiz-Hernandez, E.; Schenke-Layland, K.; Walsh, C.J.; Duffy, G.P. Drug and cell delivery for cardiac regeneration. *Adv. Drug Deliv. Rev.* **2015**, *84*, 85–106. [CrossRef] [PubMed]
37. Muehleder, S.; Ovsianikov, A.; Zipperle, J.; Redl, H.; Holnthoner, W. Connections matter: Channeled hydrogels to improve vascularization. *Front. Bioeng. Biotechnol.* **2014**, *2*, 1–7.
38. Ligoure, C.; Mora, S. Fractures in complex fluids: The case of transient networks. *Rheol. Acta* **2013**, *52*, 91–114. [CrossRef]
39. Garcia, S.J. Effect of polymer architecture on the intrinsic self-healing character of polymers. *Eur. Polym. J.* **2014**, *53*, 118–125. [CrossRef]
40. Wei, Z.; Yang, J.H.; Zhou, J.; Xu, F.; Zrinyi, M.; Dussault, P.H.; Osada, Y.; Chen, Y.M. Self-healing gels based on constitutional dynamic chemistry and their potential applications. *Chem. Soc. Rev.* **2014**, *43*, 8114–8131. [CrossRef] [PubMed]
41. Yang, Y.; Ding, X.; Urban, M.W. Chemical and physical aspects of self-healing materials. *Prog. Polym. Sci.* **2015**, *49–50*, 34–59. [CrossRef]
42. Yang, X.; Yu, H.; Wang, L.; Tong, R.; Akram, M.; Chen, Y.; Zhai, X. Self-healing polymer materials constructed by macrocycle-based host-guest interactions. *Soft Matter* **2015**, *11*, 1242–1252. [CrossRef] [PubMed]
43. Herbst, F.; Döhler, D.; Michael, P.; Binder, W.H. Self-healing polymers via supramolecular forces. *Macromol. Rapid Commun.* **2013**, *34*, 203–220. [CrossRef] [PubMed]
44. Dong, R.; Pang, Y.; Su, Y.; Zhu, X. Supramolecular hydrogels: Synthesis, properties and their biomedical applications. *Biomater. Sci.* **2015**, *3*, 937–954. [CrossRef] [PubMed]
45. An, S.Y.; Arunbabu, D.; Noh, S.M.; Song, Y.K.; Oh, J.K. Recent strategies to develop self-healable crosslinked polymeric networks. *Chem. Commun.* **2015**, *51*, 13058–13070. [CrossRef] [PubMed]
46. Sanyal, A. Diels-Alder cycloaddition-cycloreversion: A powerful combo in materials design. *Macromol. Chem. Phys.* **2010**, *211*, 1417–1425. [CrossRef]
47. Gandini, A. The furan/maleimide Diels-Alder reaction: A versatile click-unclick tool in macromolecular synthesis. *Prog. Polym. Sci.* **2013**, *38*, 1–29. [CrossRef]
48. Liu, Y.L.; Chuo, T.W. Self-healing polymers based on thermally reversible Diels-Alder chemistry. *Polym. Chem.* **2013**, *4*, 2194–2205. [CrossRef]
49. Wei, Z.; Yang, J.H.; Du, X.J.; Xu, F.; Zrinyi, M.; Osada, Y.; Li, F.; Chen, Y.M. Dextran-based self-healing hydrogels formed by reversible Diels-Alder reaction under physiological conditions. *Macromol. Rapid Commun.* **2013**, *34*, 1464–1470. [CrossRef] [PubMed]
50. Yu, F.; Cao, X.; Du, J.; Wang, G.; Chen, X. Multifunctional hydrogel with good structure integrity, self-healing, and tissue-adhesive property formed by combining Diels-Alder click reaction and acylhydrazone bond. *ACS Appl. Mater. Interfaces* **2015**, *7*, 24023–24031. [CrossRef] [PubMed]
51. Gyarmati, B.; Némethy, Á.; Szilágyi, A. Reversible disulphide formation in polymer networks: A versatile functional group from synthesis to applications. *Eur. Polym. J.* **2013**, *49*, 1268–1286. [CrossRef]
52. Deng, G.; Li, F.; Yu, H.; Liu, F.; Liu, C.; Sun, W.; Jiang, H.; Chen, Y. Dynamic hydrogels with an environmental adaptive self-healing ability and dual responsive sol–gel transitions. *ACS Macro Lett.* **2012**, *1*, 275–279. [CrossRef]
53. Casuso, P.; Odriozola, I.; Pérez-San Vicente, A.; Loinaz, I.; Cabañero, G.; Grande, H.J.; Dupin, D. Injectable and self-healing dynamic hydrogels based on metal(I)-thiolate/disulfide exchange as biomaterials with tunable mechanical properties. *Biomacromolecules* **2015**, *16*, 3552–3561. [CrossRef] [PubMed]
54. Haldar, U.; Bauri, K.; Li, R.; Faust, R.; De, P. Polyisobutylene-based pH-responsive self-healing polymeric gels. *ACS Appl. Mater. Interfaces* **2015**, *7*, 8779–8788. [CrossRef] [PubMed]
55. Zhang, Y.; Tao, L.; Li, S.; Wei, Y. Synthesis of multiresponsive and dynamic chitosan-based hydrogels for controlled release of bioactive molecules. *Biomacromolecules* **2011**, *12*, 2894–2901. [CrossRef] [PubMed]
56. Mukherjee, S.; Hill, M.R.; Sumerlin, B.S. Self-healing hydrogels containing reversible oxime crosslinks. *Soft Matter* **2015**, *11*, 6152–6161. [CrossRef] [PubMed]
57. Fiore, G.L.; Rowan, S.J.; Weder, C. Optically healable polymers. *Chem. Soc. Rev.* **2013**, *42*, 7278–7288. [CrossRef] [PubMed]

58. Amamoto, Y.; Kamada, J.; Otsuka, H.; Takahara, A.; Matyjaszewski, K. Repeatable photoinduced self-healing of covalently cross-linked polymers through reshuffling of trithiocarbonate units. *Angew. Chem.* **2011**, *123*, 1698–1701. [CrossRef]
59. Imato, K.; Nishihara, M.; Kanehara, T.; Amamoto, Y.; Takahara, A.; Otsuka, H. Self-healing of chemical gels cross-linked by diarylbibenzofuranone-based trigger-free dynamic covalent bonds at room temperature. *Angew. Chem. Int. Ed.* **2012**, *51*, 1138–1142. [CrossRef] [PubMed]
60. Roberts, M.C.; Hanson, M.C.; Massey, A.P.; Karren, E.A.; Kiser, P.F. Dynamically restructuring hydrogel networks formed with reversible covalent crosslinks. *Adv. Mater.* **2007**, *19*, 2503–2507. [CrossRef]
61. He, L.; Fullenkamp, D.E.; Rivera, J.G.; Messersmith, P.B. pH responsive self-healing hydrogels formed by boronate-catechol complexation. *Chem. Commun.* **2011**, *47*, 7497–7499. [CrossRef] [PubMed]
62. Deng, C.C.; Brooks, W.L.A.; Abboud, K.A.; Sumerlin, B.S. Boronic acid-based hydrogels undergo self-healing at neutral and acidic pH. *ACS Macro Lett.* **2015**, *4*, 220–224. [CrossRef]
63. Jia, Y.-G.; Zhu, X.X. Self-healing supramolecular hydrogel made of polymers bearing cholic acid and β-cyclodextrin pendants. *Chem. Mater.* **2015**, *27*, 387–393. [CrossRef]
64. Hillewaere, X.K.D.; Du Prez, F.E. Fifteen chemistries for autonomous external self-healing polymers and composites. *Prog. Polym. Sci.* **2015**, *49–50*, 121–153. [CrossRef]
65. Roy, N.; Bruchmann, B.; Lehn, J.-M. DYNAMERS: Dynamic polymers as self-healing materials. *Chem. Soc. Rev.* **2015**, *44*, 3786–3807. [CrossRef] [PubMed]
66. Zhang, Y.; Barboiu, M. Constitutional dynamic materials—Toward natural selection of function. *Chem. Rev.* **2015**. [CrossRef] [PubMed]
67. Krogsgaard, M.; Nue, V.; Birkedal, H. Mussel-inspired materials: Self-healing through coordination chemistry. *Chem. Eur. J.* **2016**, *22*, 844–857. [CrossRef] [PubMed]
68. Wang, H.; Heilshorn, S.C. Adaptable hydrogel networks with reversible linkages for tissue engineering. *Adv. Mater.* **2015**, *27*, 3717–3736. [CrossRef] [PubMed]
69. Stukalin, E.B.; Cai, L.-H.; Kumar, N.A.; Leibler, L.; Rubinstein, M. Self-healing of unentangled polymer networks with reversible bonds. *Macromolecules* **2013**, *46*, 7525–7541. [CrossRef] [PubMed]
70. Ahmadi, M.; Hawke, L.G.D.; Goldansaz, H.; van Ruymbeke, E. Dynamics of entangled linear supramolecular chains with sticky side groups: Influence of hindered fluctuations. *Macromolecules* **2015**, *48*, 7300–7310. [CrossRef]
71. Hackelbusch, S.; Rossow, T.; van Assenbergh, P.; Seiffert, S. Chain dynamics in supramolecular polymer networks. *Macromolecules* **2013**, *46*, 6273–6286. [CrossRef]
72. Tuncaboylu, D.C.; Argun, A.; Algi, M.P.; Okay, O. Autonomic self-healing in covalently crosslinked hydrogels containing hydrophobic domains. *Polymer* **2013**, *54*, 6381–6388. [CrossRef]
73. Zhang, H.; Xia, H.; Zhao, Y. Poly(vinyl alcohol) hydrogel can autonomously self-heal. *ACS Macro Lett.* **2012**, *1*, 1233–1236. [CrossRef]
74. Cui, W.; Ji, J.; Cai, Y.-F.; Li, H.; Ran, R. Robust, anti-fatigue, and self-healing graphene oxide/hydrophobically associated composite hydrogels and their use as recyclable adsorbents for dye wastewater treatment. *J. Mater. Chem. A* **2015**, *3*, 17445–17458. [CrossRef]
75. Berne, B.J.; Weeks, J.D.; Zhou, R. Dewetting and hydrophobic interaction in physical and biological systems. *Ann. Rev. Phys. Chem.* **2009**, *60*, 85–103. [CrossRef]
76. Otto, S.; Engberts, J.B.F.N. Hydrophobic interactions and chemical reactivity. *Org. Biomol. Chem.* **2003**, *1*, 2809–2820. [CrossRef] [PubMed]
77. Winnik, M.A.; Yekta, A. Associative polymers in aqueous solution. *Curr. Opin. Colloid Interface Sci.* **1997**, *2*, 424–436. [CrossRef]
78. Chassenieux, C.; Nicolai, T.; Benyahia, L. Rheology of associative polymer solutions. *Curr. Opin. Colloid Interface Sci.* **2011**, *16*, 18–26. [CrossRef]
79. Hietala, S.; Strandman, S.; Järvi, P.; Torkkeli, M.; Jankova, K.; Hvilsted, S.; Tenhu, H. Rheological properties of associative star polymers in aqueous solutions: Effect of hydrophobe length and polymer topology. *Macromolecules* **2009**, *42*, 1726–1732. [CrossRef]
80. Gulyuz, U.; Okay, O. Self-healing poly(acrylic acid) hydrogels with shape memory behavior of high mechanical strength. *Macromolecules* **2014**, *47*, 6889–6899. [CrossRef]
81. Tuncaboylu, D.C.; Sari, M.; Oppermann, W.; Okay, O. Tough and Self-Healing Hydrogels Formed via Hydrophobic Interactions. *Macromolecules* **2011**, *44*, 4997–5005. [CrossRef]

82. Akay, G.; Hassan-Raeisi, A.; Tuncaboylu, D.C.; Orakdogen, N.; Abdurrahmanoglu, S.; Oppermann, W.; Okay, O. Self-healing hydrogels formed in catanionic surfactant solutions. *Soft Matter* **2013**, *9*, 2254–2261. [CrossRef]
83. Algi, M.P.; Okay, O. Highly stretchable self-healing poly(*N*,*N*-dimethylacrylamide) hydrogels. *Eur. Polym. J.* **2014**, *59*, 113–121. [CrossRef]
84. Gulyuz, U.; Okay, O. Self-healing poly(*N*-isopropylacrylamide) hydrogels. *Eur. Polym. J.* **2015**, *72*, 12–22. [CrossRef]
85. Jiang, G.; Liu, C.; Liu, X.; Zhang, G.; Yang, M.; Chen, Q.; Liu, F. Self-healing Mechanism and Mechanical Behavior of Hydrophobic Association Hydrogels with High Mechanical Strength. *J. Macromol. Sci. A* **2010**, *47*, 335–342. [CrossRef]
86. Chang, H.-I.; Yeh, M.-K. Clinical development of liposome-based drugs: Formulation, characterization, and therapeutic efficacy. *Int. J. Nanomed.* **2012**, *7*, 49–60.
87. Rao, Z.; Inoue, M.; Matsuda, M.; Taguchi, T. Quick self-healing and thermo-reversible liposome gel. *Colloids Surf. B Biointerfaces* **2011**, *82*, 196–202. [CrossRef] [PubMed]
88. Wei, P.; Yan, X.; Huang, F. Supramolecular polymers constructed by orthogonal self-assembly based on host-guest and metal-ligand interactions. *Chem. Soc. Rev.* **2015**, *44*, 815–832. [CrossRef] [PubMed]
89. Harada, A.; Takashima, Y.; Nakahata, M. Supramolecular Polymeric Materials via Cyclodextrin-Guest Interactions. *Acc. Chem. Res.* **2014**, *47*, 2128–2140. [CrossRef] [PubMed]
90. Schneider, H.-J.; Yatsimirsky, A.K. Selectivity in supramolecular host-guest complexes. *Chem. Soc. Rev.* **2008**, *37*, 263–277. [CrossRef] [PubMed]
91. Schmidtchen, F.P. Reflections on the construction of anion receptors: Is there a sign to resign from design? *Coord. Chem. Rev.* **2006**, *250*, 2918–2928. [CrossRef]
92. Gokel, G.W.; Leevy, W.M.; Weber, M.E. Crown Ethers: Sensors for Ions and Molecular Scaffolds for Materials and Biological Models. *Chem. Rev.* **2004**, *104*, 2723–2750. [CrossRef] [PubMed]
93. Späth, A.; König, B. Molecular recognition of organic ammonium ions in solution using synthetic receptors. *Beilstein J. Org. Chem.* **2010**, *32*, 1–111. [CrossRef] [PubMed]
94. Zeng, F.; Han, Y.; Yan, Z.-C.; Liu, C.-Y.; Chen, C.-F. Supramolecular polymer gel with multi stimuli responsive, self-healing and erasable properties generated by host-guest interactions. *Polymer* **2013**, *54*, 6929–6935. [CrossRef]
95. Zhang, M.; Xu, D.; Yan, X.; Chen, J.; Dong, S.; Zheng, B.; Huang, F. Self-Healing Supramolecular Gels Formed by Crown Ether Based-Guest Interactions. *Angew. Chem.* **2012**, *124*, 7117–7121. [CrossRef]
96. Folch-Cano, C.; Yazdani-Pedram, M.; Claudio Olea-Azar, C. Inclusion and Functionalization of Polymers with Cyclodextrins: Current Applications and Future Prospects. *Molecules* **2014**, *19*, 14066–14079. [CrossRef] [PubMed]
97. Tan, S.; Ladewig, K.; Fu, Q.; Blencowe, A.; Qiao, G.G. Cyclodextrin-Based Supramolecular Assemblies and Hydrogels: Recent Advances and Future Perspectives. *Macromol. Rapid Commun.* **2014**, *35*, 1166–1184. [CrossRef] [PubMed]
98. Harada, A.; Hashidzume, A.; Yamaguchi, H.; Takashima, Y. Polymeric Rotaxanes. *Chem. Rev.* **2009**, *109*, 5974–6023. [CrossRef] [PubMed]
99. Nakahata, M.; Takashima, Y.; Yamaguchi, H.; Harada, A. Redox-responsive self-healing materials formed from host-guest polymers. *Nat. Commun.* **2011**, *2*, 511–516. [CrossRef] [PubMed]
100. Yan, Q.; Feng, A.; Zhang, H.; Yin, Y.; Yuan, J. Redox-switchable supramolecular polymers for responsive self-healing nanofibers in water. *Polym. Chem.* **2013**, *4*, 1216–1220. [CrossRef]
101. Wang, Y.-F.; Zhang, D.-L.; Zhou, T.; Zhang, H.-S.; Zhang, W.-Z.; Luo, L.; Zhang, A.-M.; Li, B.-J.; Zhang, S. A reversible functional supramolecular material formed by host-guest inclusion. *Polym. Chem.* **2014**, *5*, 2922–2927. [CrossRef]
102. Chuo, T.-W.; Wei, T.-C.; Liu, Y.-L. Electrically driven self-healing polymers based on reversible guest-host complexation of β-cyclodextrin and ferrocene. *J. Polym. Sci. A Polym. Chem.* **2013**, *51*, 3395–3403. [CrossRef]
103. Yu, C.; Wang, C.-F.; Chen, S. Robust Self-Healing Host-Guest Gels from Magnetocaloric Radical Polymerization. *Adv. Funct. Mater.* **2014**, *24*, 1235–1242. [CrossRef]
104. Chen, H.; Ma, X.; Wu, S.; Tian, H. A Rapidly Self-Healing Supramolecular Polymer Hydrogel with Photostimulated Room-Temperature Phosphorescence Responsiveness. *Angew. Chem. Int. Ed.* **2014**, *53*, 14149–14152. [CrossRef] [PubMed]

105. Zhang, Y.H.; Zhu, X.X. Polymers made from cholic acid derivatives: Selected properties. *Macromol. Chem. Phys.* **1996**, *197*, 3473–3482. [CrossRef]
106. Le Dévédec, F.; Fuentealba, D.; Strandman, S.; Bohne, C.; Zhu, X.X. Aggregation Behavior of Pegylated Bile Acid Derivatives. *Langmuir* **2012**, *28*, 13431–13440. [CrossRef] [PubMed]
107. Shao, Y.; Jia, Y.-G.; Shi, C.; Luo, J.; Zhu, X.X. Block and Random Copolymers Bearing Cholic Acid and Oligo(ethylene glycol) Pendant Groups: Aggregation, Thermosensitivity, and Drug Loading. *Biomacromolecules* **2014**, *15*, 1837–1844. [CrossRef] [PubMed]
108. Strandman, S.; Zhu, X.X. Biodegradable Shape Memory Polymers for Biomedical Applications. In *Shape Memory Polymers for Biomedical Applications*; Yahia, L.H., Ed.; Woodhead Publishing: Cambridge, UK, 2015; Volume Chapter 11, pp. 219–245.
109. Jia, Y.-G.; Zhu, X.X. Thermoresponsiveness of Copolymers Bearing Cholic Acid Pendants Induced by Complexation with β-Cyclodextrin. *Langmuir* **2014**, *30*, 11770–11775. [CrossRef] [PubMed]
110. Li, G.; Wu, J.; Wang, B.; Yan, S.; Zhang, K.; Ding, J.; Yin, J. Self-Healing Supramolecular Self-Assembled Hydrogels Based on Poly(L-glutamic acid). *Biomacromolecules* **2015**, *16*, 3508–3518. [CrossRef] [PubMed]
111. Miao, T.; Fenn, S.L.; Charron, P.N.; Oldinski, R.A. Self-Healing and Thermoresponsive Dual-Cross-Linked Alginate Hydrogels Based on Supramolecular Inclusion Complexes. *Biomacromolecules* **2015**, *16*, 3740–3750. [CrossRef] [PubMed]
112. Nakahata, M.; Takashima, Y.; Harada, A. Highly Flexible, Tough, and Self-Healing Supramolecular Polymeric Materials Using Host-Guest Interaction. *Macromol. Rapid Commun.* **2016**, *37*, 86–92. [CrossRef] [PubMed]
113. Miyamae, K.; Nakahata, M.; Takashima, Y.; Harada, A. Self-Healing, Expansion-Contraction, and Shape-Memory Properties of a Preorganized Supramolecular Hydrogel through Host-Guest Interactions. *Angew. Chem.* **2015**, *127*, 9112–9115. [CrossRef]
114. Kakuta, T.; Takashima, Y.; Nakahata, M.; Otsubo, M.; Yamaguchi, H.; Harada, A. Preorganized Hydrogel: Self-Healing Properties of Supramolecular Hydrogels Formed by Polymerization of Host-Guest-Monomers that Contain Cyclodextrins and Hydrophobic Guest Groups. *Adv. Mater.* **2013**, *25*, 2849–2853. [CrossRef] [PubMed]
115. Barrow, S.J.; Kasera, S.; Rowland, M.J.; del Barrio, J.; Scherman, O.A. Cucurbituril-Based Molecular Recognition. *Chem. Rev.* **2015**, *115*, 12320–12406. [CrossRef] [PubMed]
116. Rauwald, U.; Scherman, O.A. Supramolecular Block Copolymers with Cucurbit[8]uril in Water. *Angew. Chem. Int. Ed.* **2008**, *47*, 3950–3953. [CrossRef] [PubMed]
117. Rauwald, U.; Biedermann, F.; Deroo, S.; Robinson, C.V.; Scherman, O.A. Correlating Solution Binding and ESI-MS Stabilities by Incorporating Solvation Effects in a Confined Cucurbit[8]uril System. *J. Phys. Chem. B* **2010**, *114*, 8606–8615. [CrossRef] [PubMed]
118. Kim, K.; Selvapalam, N.; Ko, Y.H.; Park, K.M.; Kim, D.; Kim, J. Functionalized cucurbiturils and their applications. *Chem. Soc. Rev.* **2007**, *36*, 267–279. [CrossRef] [PubMed]
119. Assaf, K.I.; Nau, W.M. Cucurbiturils: From synthesis to high-affinity binding and catalysis. *Chem. Soc. Rev.* **2015**, *44*, 394–418. [CrossRef] [PubMed]
120. Appel, E.A.; Biedermann, F.; Rauwald, U.; Jones, S.T.; Zayed, J.M.; Scherman, O.A. Supramolecular Cross-Linked Networks via Host-Guest Complexation with Cucurbit[8]uril. *J. Am. Chem. Soc.* **2010**, *132*, 14251–14260. [CrossRef] [PubMed]
121. Appel, E.A.; Loh, X.J.; Jones, S.T.; Biedermann, F.; Dreiss, C.A.; Scherman, O.A. Ultrahigh-Water-Content Supramolecular Hydrogels Exhibiting Multistimuli Responsiveness. *J. Am. Chem. Soc.* **2012**, *134*, 11767–11773. [CrossRef] [PubMed]
122. Janeček, E.-R.; McKee, J.R.; Tan, C.S.Y.; Nykänen, A.; Kettunen, M.; Laine, J.; Ikkala, O.; Scherman, O.A. Hybrid Supramolecular and Colloidal Hydrogels that Bridge Multiple Length Scales. *Angew. Chem. Int. Ed.* **2015**, *54*, 5383–5388. [CrossRef] [PubMed]
123. Xiao, X.; Sun, N.; Qi, D.; Jiang, J. Unprecedented cucurbituril-based ternary host-guest supramolecular polymers mediated through included alkyl chains. *Polym. Chem.* **2014**, *5*, 5211–5217. [CrossRef]
124. Kulkarni, S.G.; Prucková, Z.; Rouchal, M.; Dastychová, L.; Vícha, R. Adamantylated trisimidazolium-based tritopic guests and their binding properties towards cucurbit[7]uril and β-cyclodextrin. *J. Incl. Phenom. Macrocycl. Chem.* **2015**, 1–10. [CrossRef]

125. Walsh, Z.; Janeček, E.-R.; Hodgkinson, J.T.; Sedlmair, J.; Koutsioubas, A.; Spring, D.R.; Welch, M.; Hirschmugl, C.J.; Toprakcioglu, C.; Nitschke, J.R.; *et al.* Multifunctional supramolecular polymer networks as next-generation consolidants for archaeological wood conservation. *Proc. Natl. Acad. Sci. USA* **2014**, *111*, 17743–17748. [CrossRef] [PubMed]
126. Brunsveld, L.; Folmer, B.J.B.; Meijer, E.W.; Sijbesma, R.P. Supramolecular Polymers. *Chem. Rev.* **2001**, *101*, 4071–4098. [CrossRef] [PubMed]
127. Chirila, T.V.; Lee, H.H.; Oddon, M.; Nieuwenhuizen, M.M.L.; Blakey, I.; Nicholson, T.M. Hydrogen-bonded supramolecular polymers as self-healing hydrogels: Effect of a bulky adamantyl substituent in the ureido-pyrimidinone monomer. *J. Appl. Polym. Sci.* **2014**, *131*, 39932. [CrossRef]
128. Van Gemert, G.M.L.; Peeters, J.W.; Söntjens, S.H.M.; Janssen, H.M.; Bosman, A.W. Self-Healing Supramolecular Polymers in Action. Macromol. *Chem. Phys.* **2012**, *213*, 234–242.
129. Dankers, P.Y.W.; Hermans, T.M.; Baughman, T.W.; Kamikawa, Y.; Kieltyka, R.E.; Bastings, M.M.C.; Janssen, H.M.; Sommerdijk, N.A.J.M.; Larsen, A.; van Luyn, M.J.A.; *et al.* Hierarchical Formation of Supramolecular Transient Networks in Water: A Modular Injectable Delivery System. *Adv. Mater.* **2012**, *24*, 2703–2709. [CrossRef] [PubMed]
130. Kieltyka, R.E.; Pape, A.C.H.; Albertazzi, L.; Nakano, Y.; Bastings, M.M.C.; Voets, I.K.; Dankers, P.Y.W.; Meijer, E.W. Mesoscale Modulation of Supramolecular Ureidopyrimidinone-Based Poly(ethylene glycol) Transient Networks in Water. *J. Am. Chem. Soc.* **2013**, *135*, 11159–11164. [CrossRef] [PubMed]
131. Weiss, R.G., Terech, P., Eds.; *Molecular Gels. Materials with Self-Assembled Fibrillar Networks*; Springer: Dordrecht, The Netherlands, 2006.
132. Weiss, R.G. The Past, Present, and Future of Molecular Gels. What Is the Status of the Field, and Where Is It Going? *J. Am. Chem. Soc.* **2014**, *136*, 7519–7530. [CrossRef] [PubMed]
133. Brizard, A.M.; Stuart, M.C.A.; van Esch, J.H. Self-assembled interpenetrating networks by orthogonal self assembly of surfactants and hydrogelators. *Faraday Discuss.* **2009**, *143*, 345–357. [CrossRef] [PubMed]
134. Strandman, S.; Le Dévédec, F.; Zhu, X.X. Self-Assembly of Bile Acid–PEG Conjugates in Aqueous Solutions. *J. Phys. Chem. B* **2013**, *117*, 252–258. [CrossRef] [PubMed]
135. Cui, J.; Campo, A.D. Multivalent H-bonds for self-healing hydrogels. *Chem. Commun.* **2012**, *48*, 9302–9304. [CrossRef] [PubMed]
136. Lin, Y.; Li, G. An intermolecular quadruple hydrogen-bonding strategy to fabricate self-healing and highly deformable polyurethane hydrogels. *J. Mat. Chem. B* **2014**, *2*, 6878–6885. [CrossRef]
137. Phadke, A.; Zhang, C.; Arman, B.; Hsu, C.-C.; Mashelkar, R.A.; Lele, A.K.; Tauber, M.J.; Arya, G.; Varghese, S. Rapid self-healing hydrogels. *Proc. Natl. Acad. Sci. USA* **2012**, *109*, 4383–4388. [CrossRef] [PubMed]
138. Varghese, S.; Lele, A.; Mashelkar, R. Metal-ion-mediated healing of gels. *J. Polym. Sci. A Polym. Chem.* **2006**, *44*, 666–670. [CrossRef]
139. Dai, X.; Zhang, Y.; Gao, L.; Bai, T.; Wang, W.; Cui, Y.; Liu, W. A Mechanically Strong, Highly Stable, Thermoplastic, and Self-Healable Supramolecular Polymer Hydrogel. *Adv. Mater.* **2015**, *27*, 3566–3571. [CrossRef] [PubMed]
140. Zhang, L.; Brostowitz, N.R.; Cavicchi, K.A.; Weiss, R.A. Perspective: Ionomer Research and Applications. *Macromol. React. Eng.* **2014**, *8*, 81–99. [CrossRef]
141. Lin, X.; Grinstaff, M.W. Ionic Supramolecular Assemblies. *Israel J. Chem.* **2013**, *53*, 498–510. [CrossRef]
142. Lin, X.; Navailles, L.; Nallet, F.; Grinstaff, M.W. Influence of Phosphonium Alkyl Substituents on the Rheological and Thermal Properties of Phosphonium-PAA-Based Supramolecular Polymeric Assemblies. *Macromolecules* **2012**, *45*, 9500–9506. [CrossRef]
143. Sun, J.-Y.; Zhao, X.; Illeperuma, W.R.K.; Chaudhuri, O.; Oh, K.H.; Mooney, D.J.; Vlassak, J.J.; Suo, Z. Highly stretchable and tough hydrogels. *Nature* **2012**, *489*, 133–136. [CrossRef] [PubMed]
144. Bai, T.; Liu, S.; Sun, F.; Sinclair, A.; Zhang, L.; Shao, Q.; Jiang, S. Zwitterionic fusion in hydrogels and spontaneous and time-independent self-healing under physiological conditions. *Biomaterials* **2014**, *35*, 3926–3933. [CrossRef] [PubMed]
145. Liu, H.; Xiong, C.; Tao, Z.; Fan, Y.; Tang, X.; Yang, H. Zwitterionic copolymer-based and hydrogen bonding-strengthened self-healing hydrogel. *RSC Adv.* **2015**, *5*, 33083–33088. [CrossRef]
146. Kostina, N.Y.; Sharifi, S.; de los Santos Pereira, A.; Michalek, J.; Grijpma, D.W.; Rodriguez-Emmenegger, C. Novel antifouling self-healing poly(carboxybetaine methacrylamide-co-HEMA) nanocomposite hydrogels with superior mechanical properties. *J. Mater. Chem. B* **2013**, *1*, 5644–5650. [CrossRef]

147. Haraguchi, K.; Ning, J.; Li, G. Changes in the Properties and Self-Healing Behaviors of Zwitterionic Nanocomposite Gels across Their UCST Transition. *Macromol. Symp.* **2015**, *358*, 182–193. [CrossRef]
148. Wang, Q.; Mynar, J.L.; Yoshida, M.; Lee, E.; Lee, M.; Okuro, K.; Kinbara, K.; Aida, T. High-water-content mouldable hydrogels by mixing clay and a dendritic molecular binder. *Nature* **2010**, *463*, 339–343. [CrossRef] [PubMed]
149. Wei, H.; Du, S.; Liu, Y.; Zhao, H.; Chen, C.; Li, Z.; Lin, J.; Zhang, Y.; Zhang, J.; Wan, X. Tunable, luminescent, and self-healing hybrid hydrogels of polyoxometalates and triblock copolymers based on electrostatic assembly. *Chem. Commun.* **2014**, *50*, 1447–1450. [CrossRef] [PubMed]
150. Zhang, L.; Qian, J.; Fan, Y.; Feng, W.; Tao, Z.; Yang, H. A facile CO_2 switchable nanocomposite with reversible transition from sol to self-healable hydrogel. *RSC Adv.* **2015**, *5*, 62229–62234. [CrossRef]
151. Haraguchi, K.; Uyama, K.; Tanimoto, H. Self-healing in Nanocomposite Hydrogels. *Macromol. Rapid Commun.* **2011**, *32*, 1253–1258. [CrossRef] [PubMed]
152. Han, D.; Yan, L. Supramolecular Hydrogel of Chitosan in the Presence of Graphene Oxide Nanosheets as 2D Cross-Linkers. *ACS Sustain. Chem. Eng.* **2014**, *2*, 296–300. [CrossRef]
153. Cong, H.-P.; Wang, P.; Yu, S.-H. Stretchable and Self-Healing Graphene Oxide-Polymer Composite Hydrogels: A Dual-Network Design. *Chem. Mater.* **2013**, *25*, 3357–3362. [CrossRef]
154. Guvendiren, M.; Lu, H.D.; Burdick, J.A. Shear-thinning hydrogels for biomedical applications. *Soft Matter* **2012**, *8*, 260–272. [CrossRef]
155. Desai, M.S.; Lee, S.-W. Protein-based functional nanomaterial design for bioengineering applications. *WIREs Nanomed. Nanobiotechnol.* **2015**, *7*, 69–97. [CrossRef] [PubMed]
156. Song, P.A.; Xu, Z.; Guo, Q. Bioinspired Strategy to Reinforce PVA with Improved Toughness and Thermal Properties via Hydrogen-Bond Self-Assembly. *ACS Macro Lett.* **2013**, *2*, 1100–1104. [CrossRef]
157. Li, G.; Yan, Q.; Xia, H.; Zhao, Y. Therapeutic-Ultrasound-Triggered Shape Memory of a Melamine-Enhanced Poly(vinyl alcohol) Physical Hydrogel. *ACS Appl. Mater. Interfaces* **2015**, *7*, 12067–12073. [CrossRef] [PubMed]
158. Li, G.; Zhang, H.; Fortin, D.; Xia, H.; Zhao, Y. Poly(vinyl alcohol)-Poly(ethylene glycol) Double-Network Hydrogel: A General Approach to Shape Memory and Self-Healing Functionalities. *Langmuir* **2015**, *31*, 11709–11716. [CrossRef] [PubMed]
159. Golinska, M.D.; Włodarczyk-Biegun, M.K.; Werten, M.W.T.; Stuart, M.A.C.; de Wolf, F.A.; de Vries, R. Dilute Self-Healing Hydrogels of Silk-Collagen-Like Block Copolypeptides at Neutral pH. *Biomacromolecules* **2014**, *15*, 699–706. [CrossRef] [PubMed]
160. Nowak, A.P.; Breedveld, V.; Pakstis, L.; Ozbas, B.; Pine, D.J.; Pochan, D.; Deming, T.J. Rapidly recovering hydrogel scaffolds from self-assembling diblock copolypeptide amphiphiles. *Nature* **2002**, *417*, 424–428. [CrossRef] [PubMed]
161. Ghoorchian, A.; Simon, J.R.; Bharti, B.; Han, W.; Zhao, X.; Chilkoti, A.; López, G.P. Bioinspired Reversibly Cross-linked Hydrogels Comprising Polypeptide Micelles Exhibit Enhanced Mechanical Properties. *Adv. Funct. Mater.* **2015**, *25*, 3122–3130. [CrossRef]
162. Sano, K.-I.; Kawamura, R.; Tominaga, T.; Oda, N.; Ijiro, K.; Osada, Y. Self-Repairing Filamentous Actin Hydrogel with Hierarchical Structure. *Biomacromolecules* **2011**, *12*, 4173–4177. [CrossRef] [PubMed]
163. Sathaye, S.; Mbi, A.; Sonmez, C.; Chen, Y.; Blair, D.L.; Schneider, J.P.; Pochan, D.J. Rheology of peptide- and protein-based physical hydrogels: Are everyday measurements just scratching the surface? *WIREs Nanomed. Nanobiotechnol.* **2015**, *7*, 34–68. [CrossRef] [PubMed]
164. Jacob, R.S.; Ghosh, D.; Singh, P.K.; Basu, S.K.; Jha, N.N.; Das, S.; Sukul, P.K.; Patil, S.; Sathaye, S.; Kumar, A.; *et al.* Self healing hydrogels composed of amyloid nano fibrils for cell culture and stem cell differentiation. *Biomaterials* **2015**, *54*, 97–105. [CrossRef] [PubMed]
165. Cai, L.; Dewi, R.E.; Heilshorn, S.C. Injectable Hydrogels with In Situ Double Network Formation Enhance Retention of Transplanted Stem Cells. *Adv. Funct. Mater.* **2015**, *25*, 1344–1351. [CrossRef] [PubMed]
166. Murphy, S.V.; Atala, A. 3D bioprinting of tissues and organs. *Nat. Biotechnol.* **2014**, *32*, 773–785. [CrossRef] [PubMed]
167. Kirchmajer, D.M.; Gorkin, R., III; in het Panhuis, M. An overview of the suitability of hydrogel-forming polymers for extrusion-based 3D-printing. *J. Mater. Chem. B* **2015**, *3*, 4105–4117. [CrossRef]
168. Wu, G.-H.; Hsu, S.-H. Review: Polymeric-Based 3D Printing for Tissue Engineering. *J. Med. Biol. Eng.* **2015**, *35*, 285–292. [CrossRef] [PubMed]

169. Jungst, T.; Smolan, W.; Schacht, K.; Scheibel, T.; Groll, J. Strategies and Molecular Design Criteria for 3D Printable Hydrogels. *Chem. Rev.* **2015**. [CrossRef] [PubMed]
170. Highley, C.B.; Rodell, C.B.; Burdick, J.A. Direct 3D Printing of Shear-Thinning Hydrogels into Self-Healing Hydrogels. *Adv. Mater.* **2015**, *27*, 5075–5079. [CrossRef] [PubMed]
171. Rodell, C.B.; Mealy, J.E.; Burdick, J.A. Supramolecular Guest-Host Interactions for the Preparation of Biomedical Materials. *Bioconjug. Chem.* **2015**, *26*, 2279–2289. [CrossRef] [PubMed]
172. Ozbolat, I.T. Bioprinting scale-up tissue and organ constructs for transplantation. *Trends Biotechnol.* **2015**, *33*, 395–400. [CrossRef] [PubMed]
173. Yu, L.; Ding, J. Injectable hydrogels as unique biomedical materials. *Chem. Soc. Rev.* **2008**, *37*, 1473–1481. [CrossRef] [PubMed]
174. Bae, K.H.; Wang, L.-S.; Kurisawa, M. Injectable biodegradable hydrogels: Progress and challenges. *J. Mater. Chem. B* **2013**, *1*, 5371–5388. [CrossRef]
175. Pape, A.C.H.; Bakker, M.H.; Tseng, C.C.S.; Bastings, M.M.C.; Koudstaal, S.; Agostoni, P.; Chamuleau, A.A.J.; Dankers, P.Y.W. An Injectable and Drug-loaded Supramolecular Hydrogel for Local Catheter Injection into the Pig Heart. *J. Vis. Exp.* **2015**, *100*, 52450. [CrossRef] [PubMed]
176. Gaffey, A.C.; Chen, M.H.; Venkataraman, C.M.; Trubelja, A.; Rodell, C.B.; Dinh, P.V.; Hung, G.H.; MacArthur, J.W.; Soopan, R.V.; Burdick, J.A.; *et al.* Injectable shear-thinning hydrogels used to deliver endothelial progenitor cells, enhance cell engraftment, and improve ischemic myocardium. *J. Thorac. Cardiovasc. Surg.* **2015**, *150*, 1268–1277. [CrossRef] [PubMed]
177. Yu, B.; Wang, C.; Ju, Y.M.; West, L.; Harmon, J.; Moussy, Y.; Moussy, F. Use of hydrogel coating to improve the performance of implanted glucose sensors. *Biosens. Bioelectron.* **2008**, *23*, 1278–1284. [CrossRef] [PubMed]
178. Goodman, S.B.; Yao, Z.; Keeney, M.; Yang, F. The Future of Biologic Coatings for Orthopaedic Implants. *Biomaterials* **2013**, *34*, 3174–3183. [CrossRef] [PubMed]
179. Agarwal, R.; García, A.J. Biomaterial strategies for engineering implants for enhanced osseointegration and bone repair. *Adv. Drug Deliv. Rev.* **2015**, *94*, 53–62. [CrossRef]
180. Zhang, S.; Bellinger, A.M.; Glettig, D.L.; Barman, R.; Lee, Y.-A.L.; Zhu, J.; Cleveland, C.; Montgomery, V.A.; Gu, L.; Nash, L.D.; *et al.* A pH-responsive supramolecular polymer gel as an enteric elastomer for use in gastric devices. *Nat. Mater.* **2015**, *14*, 1065–1071. [CrossRef] [PubMed]
181. Luo, F.; Sun, T.L.; Nakajima, T.; Kurokawa, T.; Ihsan, A.B.; Li, X.; Guo, H.; Gong, J.P. Free Reprocessability of Tough and Self-Healing Hydrogels Based on Polyion Complex. *ACS Macro Lett.* **2015**, *4*, 961–964. [CrossRef]
182. Cameron, A.R.; Frith, J.E.; Gomez, G.A.; Yap, A.S.; Cooper-White, J.J. The effect of time-dependent deformation of viscoelastic hydrogels on myogenic induction and Rac1 activity in mesenchymal stem cells. *Biomaterials* **2014**, *35*, 1857–1868. [CrossRef] [PubMed]

6

Hydrogel Microparticles as Sensors for Specific Adhesion: Case Studies on Antibody Detection and Soil Release Polymers

Alexander Klaus Strzelczyk [1], Hanqing Wang [1], Andreas Lindhorst [2], Johannes Waschke [2], Tilo Pompe [2], Christian Kropf [3], Benoit Luneau [3] and Stephan Schmidt [1,*]

1 Institute for Organic and Macromolecular Chemistry, Heinrich-Heine-Universität, Universitätsstrasse 1, 40225 Düsseldorf, Germany; Alexander.Strzelczyk@uni-duesseldorf.de (A.K.S.); hanqing.wang@hhu.de (H.W.)

2 Institute for Biochemistry, Universität Leipzig, Johannisallee 21-23, 04103 Leipzig, Germany; Andreas_Lindhorst@web.de (A.L.); jowaschke@cbs.mpg.de (J.W.); tilo.pompe@uni-leipzig.de (T.P.)

3 Henkel AG & Co KGaA, Henkelstr 67, D-40589 Düsseldorf, Germany; christian.kropf@henkel.com (C.K.); benoit.luneau@henkel.com (B.L.)

* Correspondence: stephan.schmidt@hhu.de

Abstract: Adhesive processes in aqueous media play a crucial role in nature and are important for many technological processes. However, direct quantification of adhesion still requires expensive instrumentation while their sample throughput is rather small. Here we present a fast, and easily applicable method on quantifying adhesion energy in water based on interferometric measurement of polymer microgel contact areas with functionalized glass slides and evaluation via the Johnson–Kendall–Roberts (JKR) model. The advantage of the method is that the microgel matrix can be easily adapted to reconstruct various biological or technological adhesion processes. Here we study the suitability of the new adhesion method with two relevant examples: (1) antibody detection and (2) soil release polymers. The measurement of adhesion energy provides direct insights on the presence of antibodies showing that the method can be generally used for biomolecule detection. As a relevant example of adhesion in technology, the antiadhesive properties of soil release polymers used in today's laundry products are investigated. Here the measurement of adhesion energy provides direct insights into the relation between polymer composition and soil release activity. Overall, the work shows that polymer hydrogel particles can be used as versatile adhesion sensors to investigate a broad range of adhesion processes in aqueous media.

Keywords: biomimetic hydrogel; biointerface; elastic solids; contact mechanics; poly(ethylene glycol) (PEG); soft colloidal probe; reflection interference contrast microscopy (RICM)

---

1. Introduction

Adhesive processes in aqueous environment are of great importance, e.g., in biology where crucial cellular functions are controlled by biological interfaces decorated with carbohydrates, proteins or lipids. Controlling adhesive interaction under water is also very important in technology, e.g., mussel adhesion on surfaces of ships as well as for cleaning processes or controlling the haptic sensation of surfaces [1,2]. On the molecular level, adhesion phenomena are typically indirectly studied using methods like quartz crystal microbalance, surface plasmon resonance or fluorescence microscopy [3,4]. These methods simply relate adhesion to the adsorbed amount of the molecules of interest. However, without direct measurement of adhesive interactions these methods provide only limited insight into the underlying mechanisms. Alternatively, direct force measurements by atomic force microscopy

(AFM), surface force apparatus are suitable to directly study adhesion phenomena between material surfaces [5–7]. In addition, such adhesion assays offer information on the nature of the interaction; for example, discrimination between specific and non-specific binding. However, these methods are often expensive, not easy to operate and slow when compared to other analytic techniques. Therefore, we have developed a fast, easy-to-use method that can be adapted to investigate and quantify a broad range of adhesion phenomena in aqueous environment. The method uses soft hydrogel microparticles (soft colloidal probes, SCPs) as adhesion sensors on glass surfaces functionalized with a suitable binding partner. The working principle of adhesion measurements based on hydrogel SCPs is schematically explained in Figure 1. When adhering to a functionalized glass coverslip surface, the SCPs undergo mechanical deformation due to their soft, gel-like structure [8,9]. The mechanical deformation can be related to the adhesion energy of the SCP with the surface by means of the Johnson–Kendall–Roberts (JKR) model of adhesion [10]:

$$W_{adh} = \frac{E_{eff}\, a^3}{6\pi\, R^2} \tag{1}$$

where a is the radius of contact, R radius of the SCP and $E_{eff} = [4E/3(1 - \nu^2)]$ its effective elastic modulus, with ν the Poisson ratio and E the elastic modulus of the SCP. Technically, the read-out of adhesion energies is straightforward and fast. First, E can be determined batch-wise for a large amount of SCPs by force-indentation measurements using an atomic force microscope. Assuming volume conservation ν can be taken as 0.5 in order to determine W_{adh}. SCPs of well-known mechanical properties are then subject to adhesion on various sample surfaces where a and R can be directly measured by an interferometric technique (reflection interference contrast microscopy, RICM) [11]. An additional advantage of this method is that PEG hydrogel particles as adhesion sensors show reduced non-specific binding to material surfaces and impurities due to the strong hydration of the PEG matrix. In addition, by functionalizing the PEG matrix with various molecules, including proteins [12], peptides [13] or carbohydrates [14], the assay can be adapted in a straightforward fashion to study adhesion phenomena in different contexts and application areas. Due to their biomimetic properties, SCPs have recently been successfully used as AFM probes to investigate interactions of cells [15] and biomaterial surfaces [16,17].

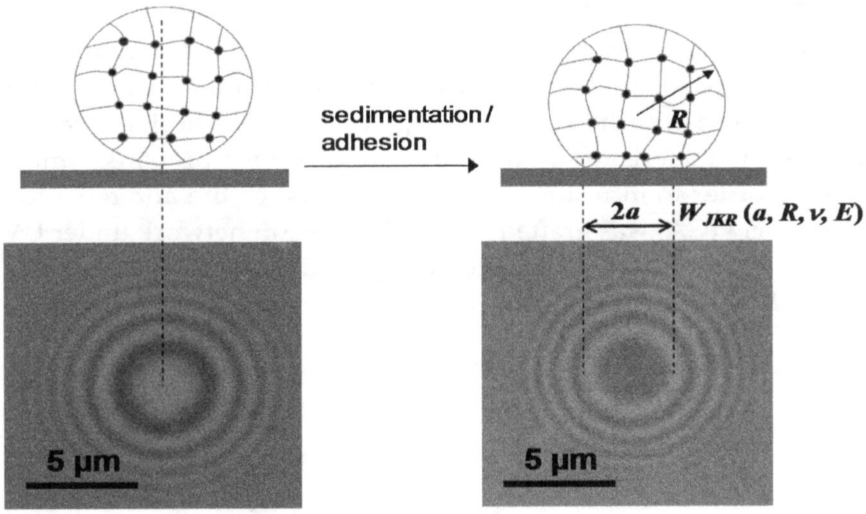

Figure 1. Principle of the Johnson-Kendall-Roberts (JKR) adhesion measurements with colloidal probes and typical reflection interference contrast microscopy (RICM) images (**bottom**) right before and after SCP adhesion. The dark area in the middle signifies the soft colloidal probe (SCP) contact area with the solid support.

Here we now focus on testing the applicability of the method in two applied areas, (1) detection of antibodies and (2) investigation of soil release processes of polymeric detergent additives. Given the commercial importance of these two areas, new ways of studying the underlying molecular interaction measurements is important. For instance, antibody detection is of great importance in medical diagnostics. Increased levels of antibodies in blood generally indicate exposure to certain antigens. Medical diagnostics routinely quantifies antibody concentrations in blood, which can be related to the presence of certain pathogenic antigens such as human immunodeficiency virus (HIV), measles or hepatitis [18]. More recent developments in antibody detection are directed towards diagnosis of cancer or autoimmune diseases [19,20]. The enzyme linked immunosorbent assay (ELISA) is the most commonly used antibody detection method due to its comparatively high sensitivity and selectivity. However, extensive cleaning procedures are required in order to achieve sufficiently high selectivity, which often hampers routine application of ELISA in medical diagnostics due to economic reasons. Therefore, here we aim at antibody detection by measuring the adhesion between antigen-coated hydrogel SCPs that capture antibodies and surfaces functionalized with protein A, an antibody binding protein. The potential advantage of this approach is that the extensive cleaning procedures usually required in ELISA would be significantly reduced. Sensitivity and selectivity are shown by determining the detection limit of the method as well as performing the assay in presence of interfering protein impurities. In the second application of the SCP adhesion assay, we investigate the 'soil release' activity of various polymer samples. Soil release polymers are often present in modern consumer detergent for fabric cleaning. They are supposed to fulfill two basic functions: (1) hindering redeposition of the soil (e.g., fats) from the washing solution to the fabric during washing, (2) formation of a protective coating on the fabric right after washing to enhance release of acquired hydrophobic soil in the next washing step. For both processes, it could be stated that soil release polymers act as antiadhesives. There is a range of polymers that have been empirically identified as soil release active. However, the soil release process and the underlying working principle of the active polymers so far have not been studied systematically. Therefore, we set out to mimic the adhesion of cotton fabrics on oily materials using functionalized SCPs and then quantitatively study the antiadhesive properties of various soil release polymers.

2. Results and Discussion

2.1. Synthesis and Functionalization of SCPs with Adhesion Molecules

As outlined in Figure 2, SCPs with a hydrogel matrix were prepared by crosslinking of poly(ethylene glycol) (8 kDa) diacrylamide (PEG_{8kDa}-dAAm) macromonomer droplets in 1 M Na_2SO_4 in water following previously established protocols [8]. For the two adhesion assays, antibody detection and soil release polymer characterization, two types of SCPs were prepared: antigen and cellobiose functionalized SCPs. In order to introduce coupling groups for the antigen and cellobiose for the adhesion assay, crotonic acid (CA) was grafted in the PEG-dAAm network under UV irradiation using benzophenone as active photophore [21,22]. Accordingly, PEG-CA SCPs with a CA functionalization degree on the order of 120 μmol CA per 1 g PEG-CA were obtained as measured via Toluidine blue O (TBO) titration.

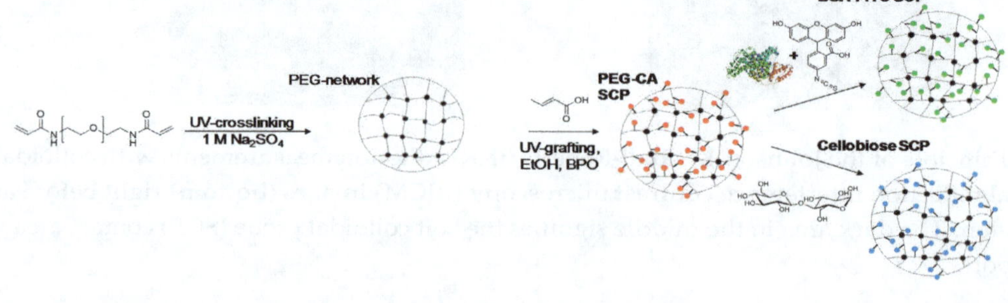

Figure 2. Sythetic route toward bovine serum albumin-fluoresceine isothiocyanate (BSA-FTIC) SCPs and cellubiose SCPs based on PEG-dAAm microgels.

For the preparation of antigen functionalized systems, fluoresceinisothiocyanate (FITC) was chosen as a model antigen. In order to functionalize SCPs with FITC, we used bovine serum albumin (BSA) as a carrier. First, BSA was coupled to the SCPs by activating the CA groups with ethyl dimethylaminopropyl carbodiimide (EDC) and *N*-hydroxysuccinimide (NHS) before binding the protein to the SCPs. Then the SCPs were cleaned and directly reacted with FITC that binds to the nucleophilic amino acid side chains of BSA, resulting in BSA-FITC SCPs. Successful functionalization of the SCPs could be readily confirmed by fluorescence microscopy detecting the presence of FITC on the SCPs (Figure 3). For the preparation of cellobiose-functionalized systems mimicking cotton fabrics, PEG-CA SCPs were reacted with the carbodiimide, activating the carboxylic acid groups to then form an ester with cellobiose. The degree of cellobiose functionalization was measured by an additional titration step with toluidine blue that essentially yielded the reduction of carboxylic acids due to esterification upon cellobiose coupling. The conversion was on the order of 90%, i.e., functionalization degrees on the order of 100 μmol g^{-1} were achieved. Taking into account the molecular weight of PEG-dAAm chains of 8000 kDa, this means that circa 0.8 cellobiose units per macromonomer were bound to the SCPs. Successful functionalization could be additionally confirmed by optical microscopy, where the TBO-labeled PEG-CA SCPs show a strong blue color that becomes significantly less intense in the case of cellobiose SCPs (Figure 4), indicating cellobiose functionalization. The elastic modulus of the final SCPs was determined by AFM force-indentation measurements (Figure S1) and evaluation with a recently introduced model [23]. This model considers that the SCPs deform at the indenter site and the contact site with the solid substrate. The model further assumes linear elasticity, which is justified by the fact that the deformation during force indentation measurements does not exceed 0.2% with respect to the particle's diameter. Both the BSA-FITC SCPs and cellobiose SCPs showed an elastic modulus of around 60 kPa.

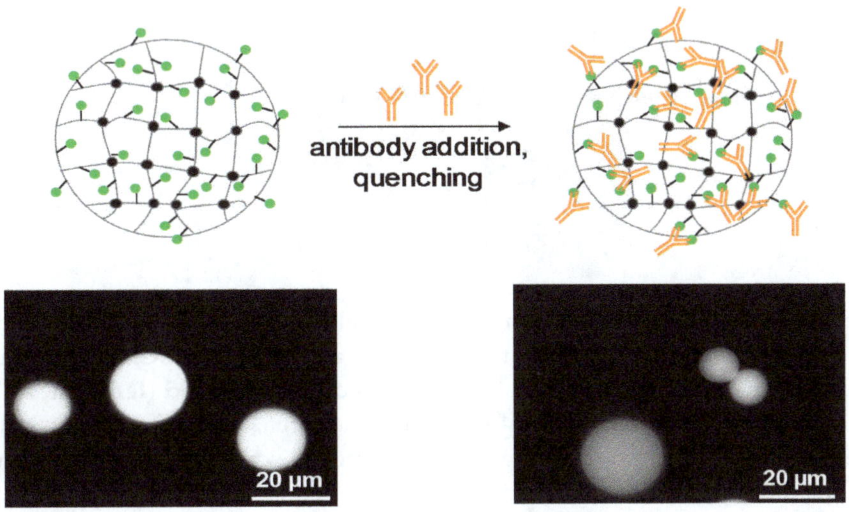

Figure 3. Fluorescence microscopy of BSA-FITC SCPs without addition of antibodies (**left**) and after addition of antibodies (**right**). Reduction on fluorescence intensity is due to quenching upon antibody binding and signifies specific interaction of the antibody with the SCPs.

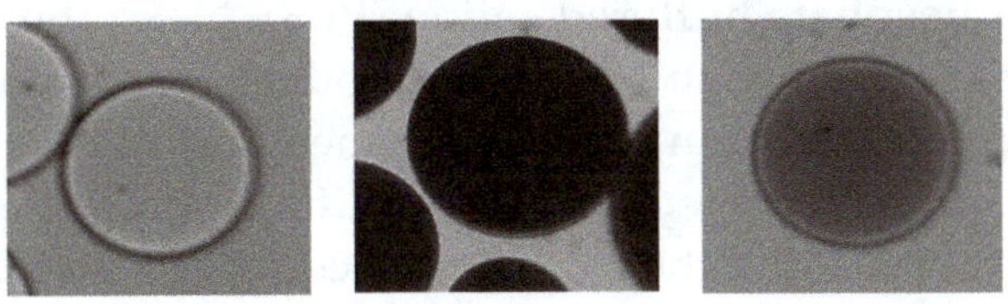

Figure 4. Toluidine blue (TBO) stained PEG SCPs before functionalization with crotonic acid (CA) (**left**). After CA functionalization PEG-CA SCPs bind TBO and acquire a dark color (**middle**). Reduced take-up of TBO after functionalization of CA groups with cellobiose (**right**).

2.2. Detection of Antibodies with a Combined SCP Pull-Down and Adhesion Assay

The FITC represent the model antigen in the adhesion assay with BSA-FITC SCPs. The overall concept is to capture the antibody from solution with the SCPs (pull-down assay) and then to detect the presence of the antibody by the actual SCP adhesion measurement (Figure 5). The FITC antibody used in this work belongs to one of the most important antibody subclasses-IgG (immunglobulin G)-as most clinic applications with antibodies rely on this subtype. IgG antibodies are composed of two identical heavy chains and two identical light chains which form the characteristic Y-shaped quaternary structure by intramolecular disulfide bonding. The two antigen binding sites are each formed by a light and heavy chain. The third specific binding site is formed only by the heavy chain, the so-called Fc-region at the base of the "Y". The specific affinity to the respective antigens is due to the large variability of amino acid sequences of the light chain, whereas the heavy chains are always identical for each antibody subtype [18]. Therefore, all antibodies belonging to the IgG subtype can be detected by specific binding at the Fc site. Here we use protein A as specifically interacting species binding to the Fc region in order to detect the antibody by means of adhesion. Additionally, for a preliminary test, the specific binding of the antibody to BSA-FITC SCPs can be visualized by fluorescence microscopy. This is due to fluorescence quenching of FITC when binding to an antibody [24]. The fluorescence microcopy measurements confirm binding of the antibody to the FITC-BSA SCPs, as can be seen by the ~30% reduction in fluorescence intensity (Figure 4). The reduction in intensity at the SCPs is lower as compared to experiments with FITC in solution [24], possibly because of a dense functionalization of BSA with FTIC.

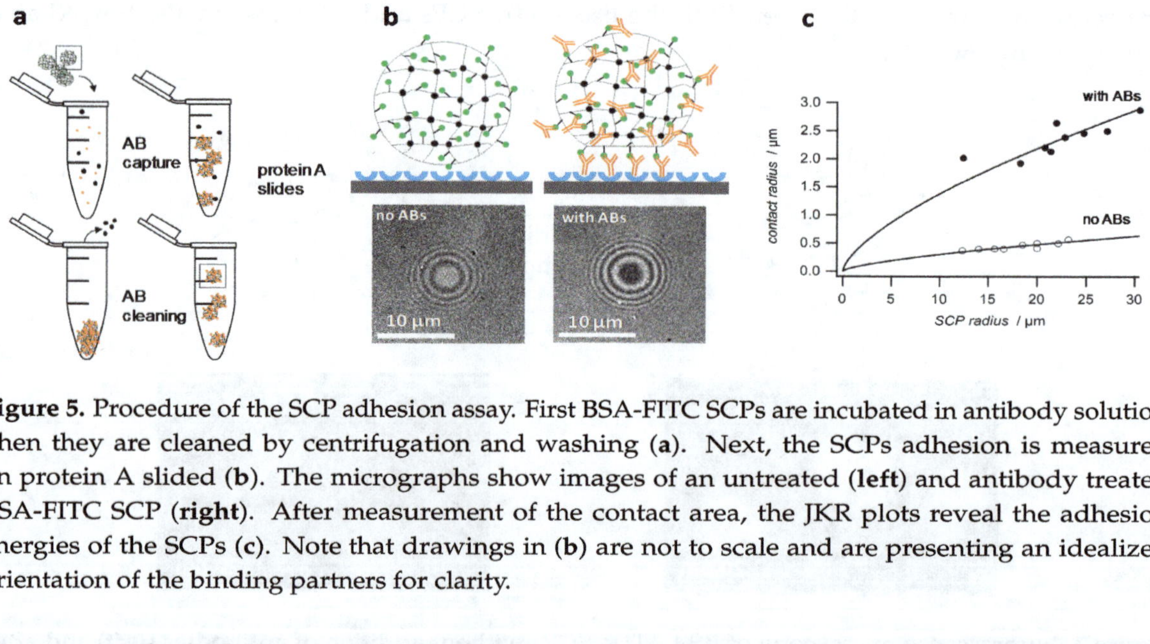

Figure 5. Procedure of the SCP adhesion assay. First BSA-FITC SCPs are incubated in antibody solution. Then they are cleaned by centrifugation and washing (**a**). Next, the SCPs adhesion is measured on protein A slided (**b**). The micrographs show images of an untreated (**left**) and antibody treated BSA-FITC SCP (**right**). After measurement of the contact area, the JKR plots reveal the adhesion energies of the SCPs (**c**). Note that drawings in (**b**) are not to scale and are presenting an idealized orientation of the binding partners for clarity.

Having confirmed the specific interactions of the IgG antibody with the BSA-FTIC SCPs, the adhesion energies of SCPs were measured on a protein A slide as a means to detect the antibodies (Figure 5b). In a first study, the BSA-FITC SCPs were incubated for two hours in a 0.1 mg mL^{-1} solution of the antibody in phosphate buffered saline (PBS). Next, the SCPs were centrifuged and washed with PBS three times to remove unbound antibodies. The protein A surface was prepared simply by allowing physisorption of protein A on the cleaned hydrophilic glass coverslips followed by physisorption of BSA on the coverslip to reduce non-specific binding sites. Next, the actual adhesion measurement was conducted by adding the SCPs to the coverslips in PBS. As expected, the contact areas of the SCPs with the protein A surface were significantly larger for SCPs treated with the antibody as compared to the negative control without antibodies due to the binding of the Fc-region with the protein A slides (Figure 5b). The contact areas versus SCP radius data could be fitted by equation 1 in order to obtain the adhesion energies of the SCPs (Figure 5c).

The adhesion energy of antibody-treated SCPs was roughly 20 times larger as compared to SCPs without antibody (Figure 6a) Next, the antibody detection limit was studied by charging the SCPs in solutions with antibody at concentrations ranging from 1 ng to 1 mg (Figure 6b). For antibody concentrations below 1 μg mL^{-1}, the adhesion energies did not increase compared to the negative control. Therefore, for the SCPs system established here, the detection limit is approximately 1 μg mL^{-1}. For comparison, the majority of commercial ELISA kits for protein analysis claim detection limits in the sub ng mL^{-1} range. Advanced ELISA methods may even reach detection limits in the lower pg mL^{-1} regime [25–27]. In unfavorable cases of low affinity between antibody and antigen in the micromolar range, the ELISA detection limit is increased up to 100 ng mL^{-1} [28]. Nevertheless, with a detection limit of 1 μg mL^{-1}, the SCP assay proved to be inferior to typical ELISA assays. The main reason could be the requirement of having both the protein A and the antibody at the glass surface and the SCP orientated in the right direction, so that they face each other in order to bind the antibodies Fc site. In case of the protein A surface, optimizing the orientation should be less important. The protein possesses five homologous binding sites for the FC-region of which a sufficient number is always likely to be accessible even if protein A is bound to a surface. However, the antibody is firmly bound to the crosslinked SCP network presenting the antigens, which means that the orientation of the Fc site is fixed, such that only a fraction of antibodies is able to bind to the protein A surface. Such hindrance is not present in classic ELISA assays where the analyte antibody is bound to a planar antigen surface from a freely dissolved state and then detected by reporter antibodies from solution. Therefore, in order to improve the detection limit of the SCP-antibody assays the elastic modulus of the SCPs should be decreased, e.g., by reducing crosslinking. This would allow larger (thermal) fluctuations of the network, and the SCP-bound antibodies would have more degrees of freedom and spatial reach to "find" protein A binding sites. Both protein A and antibodies have hydrodynamic diameters on the order of 10 nm. Thermal fluctuations of the PEG network should be of similar magnitude in order to allow sufficient probability that the orientation of the binding sites match. According to theory [29], this could be realized by polymer networks with an elastic modulus below 10 kPa, whereas the PEG-SCPs network used in this work had an elastic modulus of 60 kPa providing only fluctuations on the order of 3 nm. Another advantage of using softer PEG networks would be that the SCPs contact areas with the glass slide increase upon adhesion, thereby improving the sensitivity and detection limit of the method.

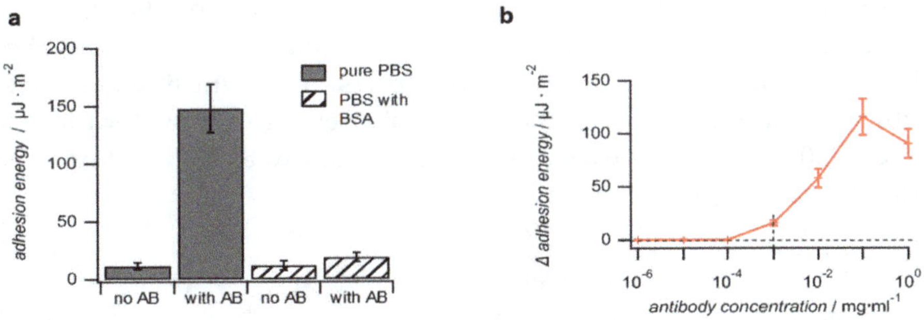

Figure 6. Results for SCPs adhesion assays for antibody detection. (**a**) Adhesion energies of BSA-FITC SCPs after incubation with antibodies (AB) on protein A slides. Measurements without antibody treatment were conducted as negative control. Measurements in presence and absence of 50 mg mL^{-1} BSA were conducted to investigate the selectivity of the method. (**b**) Measurement of BSA-FITC SCPs treated in different concentrations of antibody solution show that the detection limit is on the order of 1 μg mL^{-1} antibody.

In another assay, the SCPs were charged with antibodies in presence of 50 mg mL^{-1} BSA to test the feasibility of SCP-based antibody detection from impure solutions (selectivity). The applied amount of BSA reflects typical concentrations of serum albumin in blood. Evaluation of the contact areas showed that the overall adhesion of antibody-containing SCPs was drastically reduced (factor of ten) as compared to SCPs charged with pure antibody solution. Nevertheless, the adhesion energies of SCPs prepared from antibody-containing solutions were still larger by a factor of two as compared to the negative control, SCPs treated with 50 mg mL^{-1} BSA solution containing no antibodies (Figure 6a). The significant reduction in adhesion when incubating the SCPs in presence of BSA is most likely due to non-specific binding of BSA to the SCPs, which then interferes with the specific antibody binding. Unfortunately, the mere use of PEG as carrier material for antigens is not sufficient in order to significantly reduce non-specific protein interaction, although PEG hydrogels are widely believed to be protein-repellent materials. Our finding is in line with recent studies showing that the protein-repellent properties attributed to PEG are in fact due to a protein corona that forms around PEG in blood [3]. This means that PEG does interact non-specifically with serum albumin. Therefore, in order to reduce non-specific interactions, treating the antigen/antibody surface with detergents like Tween® is still required as is also the case for classic affinity assays including ELISA and the like.

Overall, we have successfully shown that the adhesion-based SCP assay allowed the detection of antibodies. For a proof-of-principle experiment, sensitivity and selectivity were not expected to match the performance of well-established methods that evolved and improved over decades like ELISA. Nevertheless, by further reducing non-specific binding and using significantly softer SCPs with increased sensitivity, further improvements in terms of sensitivity and specificity seem possible.

2.3. Characterization of Soil Release Polymers by SCP Adhesion Assays

As discussed above, adhesion phenomena are ubiquitous in nature and technology. Improved understanding of these processes is best obtained by developing assays that allow mimicking and quantifying the underlying adhesion phenomena. Therefore, here we adapted the SCP adhesion assay to study soil release polymers that are used as antiadhesives and antiredeposition agents in laundry processes. The adhesion between cellobiose functionalized SCPs and a hydrophobic surface is investigated to mimic a typical laundry situation, i.e., cotton fabrics soiled with oily substances. Cellobiose is composed of two glucose units, the same building block that constitutes cellulose and used here as soluble cellulose substitute. As a mimic for an oily substance, glass slides were functionalized with a hydrophobic trichloro(octadecyl)silane by chemical vapor deposition. The contact angle of these surfaces was larger than 90° indicating the high hydrophobicity of the surface and successful silanization. Next, we studied the adhesion energy of the cellobiose SCPs on the hydrophobic surfaces by RICM contact area measurements and evaluation with the JKR model (Figure S3). The adhesion energy of the cellobiose SCP with the hydrophobic surface (Figure 7a) was used here as a reference to test the antiadhesive properties of soil release polymers. The adhesive energy was on the order of 1700 μJ m^{-2}, which is in the expected range of adhesion energies in water for hydrophilic gels on hydrophobic surfaces [30]. The soil release effect was studied with four different polymers that are potential candidates as washing additive (Table 1).

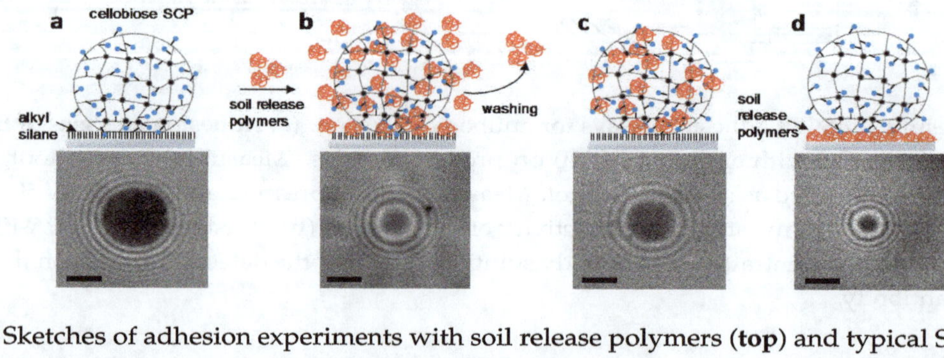

Figure 7. Sketches of adhesion experiments with soil release polymers (**top**) and typical SCP contact areas (**bottom**). (**a**) adhesion of bare cellobiose SCPs on hydrophobic glass as reference; (**b**) in presence of polymer samples (antiredepostion experiment); (**c**) after removal of the polymers by centrifugation and washing (antiadhesive coating experiment); (**d**) direct binding of cellobiose SCPs on polymer surfaces (direct binding experiment). Scale bars: 2 μm.

Table 1. Soil release polymer samples and adhesion energy from the three different assays. [a] Direct binding experiment between cellobiose SCP and hydrophobic surface. [b] Polymer was physisorbed on the glass/glymo slide.

Polymer	Adhesion Energies [μJ m^{-2}]		
	Antiredeposition	**Antiadhesive Coating**	**Direct Binding**
none/reference	1700 [a]		
Poly(propylene terephthalate)-*co*-Poly(ethylene glycol) (nonionic) PPT-*co*-PEG	41	1038	38 [b]
Copolymer A: cationic/neutral hydrophilic ratio 22:78	163	480	91
Copolymer B: cationic/neutral hydrophilic ratio 70:30	306	1020	126
Poly(acrylamide)	1142	1429	243

Soil release polymers usually consist of two building blocks. The first one is the driver for the adsorption of the polymer on the textile, the second one provides hydrophilicity and prevents hydrophobic stains to deeply penetrate the fibers. For example, for polyester fabrics, common soil release polymers are polyesters of terephthalic acid, propylene glycol and poly(ethylene glycol). The polymer structure mimics the poly(ethylene terephthalate) (PET) garment chemistry while the PEG component hydrophilizes the surface [31]. On the other hand, for cotton textiles a copolymer consisting of a quaternary ammonium-bearing monomer and a neutral hydrophilic monomer would be better suited, where the cationic monomer acts as an anchor group for the slight anionically charged cotton fiber while the neutral comonomer provides hydrophilicity to the surface. To test the proposed mechanisms, three different experiments were conducted: (1) adhesion of cellobiose SCPs on hydrophobic surface in presence of soil release polymers in the solution (antiredeposition experiment, Figure 7b); (2) adhesion to the hydrophobic surface with cellobiose SCPs that were pre-treated and washed with soil release polymers. This experiment reflects the antiadhesive properties of soil release polymer coatings post washing (antiadhesive coating experiment, Figure 7c); (3) direct binding of cellobiose SCPs on soil release polymer surfaces (direct binding experiment, Figure 7d). A selection of the SCP adhesion data and JKR-evaluation is shown in Figure S2.

For the first experiment with the soil release polymers, a cellobiose SCP solution containing 1 mg mL^{-1} soil release polymer was allowed to adhere on the hydrophobic surfaces. In this experiment, the polymer is present in solution during adhesion and thus allows studying the activity of the soil release agents during the washing step (Figure 7b). As expected, in presence of the soil release polymers, we observed a reduction of the adhesive energy between cellobiose SCPs and the hydrophobic glass slide. The overall trend was that non-ionic hydrophilic copolymer PPT-*co*-PEG achieved the strongest reduction in adhesive energy, whereas copolymers combining cationic groups and neutral hydrophilic monomer resulted in slightly increased adhesion (Table 1). This confirms the extraordinary antiadhesive properties of the PEG part, which could be attributed to the comparatively firm hydration shell around PEG resulting in excluded volume effects [32]. Although generally considered as a hydrophilic polymer, polyacrylamide is known to show strong interactions with hydrophobic surfaces which was also confirmed in this study [33]. Overall, the significant differences in adhesion of the hydrophilic polymers investigated to hydrophobic materials are quite intriguing and still not well understood [34,35].

In the second experiment, the adhesion measurement was performed with polymer-treated SCPs but in absence of polymer in the solution during adhesion (antiadhesive coating experiment, Figure 7c). Therefore, the polymer-treated SCPs were centrifuged and washed with water to remove the polymers from the solution. As expected, the overall adhesion energies increased compared to experiments with polymers present in solution. Interestingly, most polymer samples showed comparatively large adhesion energies in the range of 1000 $\mu J\ m^{-2}$ close to the results for SCPs without polymers. The exception is the weakly cationic copolymer A that showed a strong reduction in adhesion energy. Interestingly, the comparative copolymer B with a higher content of cationic groups did not achieve strong reduction in adhesion energy. It could be argued that, due to its strong polyelectrolyte character this highly charged polymer shows lateral repulsion and therefore reduced binding to the SCPs. The fact that the PEG-containing polymer did not achieve a strong reduction in adhesion energy, as was the case for the first assay, could be explained by its low tendency to form a stable antiadhesive layer on the cellobiose SCPs. As a result, it was removed from the SCPs during the centrifugation and washing steps, which then resulted in adhesion energies that were comparable to the pure cellobiose on the hydrophobic glass.

Finally, the adhesion energy of cellobiose SCPs on polymer surfaces was directly measured (direct binding experiment, Figure 7d). This experiment reflects the interaction of cotton fabrics directly with the soil release polymers. Here the cationic samples showed significant binding that increased with the number of cationic groups. This could be explained by the attractive interactions of polycations with cellulose [36], which also makes them potent flocculation and retention aids for cellulose in the paper industry. The interaction was lowest in the case of the non-ionic copolymer PPT-*co*-PEG. This is expected, as PPT-*co*-PEG does not present groups that would specifically interact with cellulose. This explains the absence of an antiadhesive coating and the large adhesion energies observed in the second experiment (Figure 7c) for this polymer. Interestingly, polyacrylamide showed the largest direct binding to cellobiose SCPs, indicating that the polymer forms a stable layer on the cellobiose SCPs. However, this coating still shows strong interactions with hydrophobic materials as confirmed by the first two experiments (Figure 7b,c).

Overall, the adhesion assay with cellobiose SCPs treated with various soil release polymers indicate that nonionic poly(ethylene glycol)-containing polymer provides high antiredeposition activity when used in the aqueous phase during the washing step. Here cationic copolymers show reduced activity; however, these polymers are more likely to form a stable film on cotton fabrics in order to form an antiadhesive layer against hydrophobic materials post washing. This confirms the generally believed mechanism of polymeric soil release: for cellulose-specialized soil release polymers, cationic groups of copolymers can bind to the slightly negatively charged cellulose, whereas the neutral part provides hydrophilicity and steric repulsion to the surface in order to reduce adhesive interactions.

3. Conclusions

In summary, a novel adhesion assay based on soft PEG hydrogel particles was established to investigate specific interactions of antibodies and antigens as well as cellobiose on hydrophobic surfaces in presence of soil release polymers. We tested the applicability of these assays in two commercially relevant areas, medical diagnostics of antibodies and detergent additives. In both cases, the adhesion assay could provide statistically significant amount of data in a short timescale compared to classic adhesion assay by means of AFM or surface force apparatus. It should be taken into account, however, that the adhesion method still requires the synthesis of the soft hydrogel particles for sensing, whereas in the case of more established methods like AFM the basic sensors (e.g., silicon cantilevers) are already available. In addition, regarding antibody detection, the SCP adhesion-based assay is still inferior in terms of sensitivity compared to established methods like ELISA. The sensitivity of antibody detection via adhesion of hydrogels sensors could be principally hampered due to the requirement of having the biomolecules properly oriented on the adhering surfaces. One of the strengths of the adhesion assays with soft, gel-like sensor particles is that it can mimic adhesive processes in nature and technology. This was in particular useful for the analysis of the soil release polymers, where several types of adhesion experiments could be established to decipher the mode of action of soil release polymers. It was found that the nonionic PEG-containing polymer is very active in antiredeposition of fats in aqueous media, whereas polymers containing cationic groups are more suited as antiadhesive coating on cellulose. The work shows that the presented SCP-RICM technique is well suited to study a large range of adhesion phenomena in nature and technology.

4. Experimental Section

4.1. Materials

Soil release polymer samples were provided by Henkel AG & Co KGaA and used without further purification (purity > 95%) The fluorescein antibody (polyclonal, type IgG) was produced by Rockland Immunochemicals (Limerick, PA, USA) and obtained from Biomol GmbH (Hamburg, Germany). All other chemicals were obtained from Sigma-Aldrich (Darmstadt, Germany). All water used here was produced by purification system with a resistivity higher than 18.2 MΩ·cm at 25 °C and UV treatment to break down organic impurities.

4.2. Soft Colloidal Probe Preparation

PEG SCPs were synthesized by crosslinking a dispersion of PEG-dAAm macromonomer droplets in a similar manner as described previously [8,14]. PEG-dAAm (Mn = 8000 Da) (50 mg, 6.3 µmol) was dispersed in a 1 M sodium sulfate/PBS solution (10 mL). Crotonic acid (400 µmol) were added to adjust the elastic modulus of the SCPs to about 30 kPa. Then the UV photoinitiator Irgacure 2959 (1 mg, 4.5 µmol) was added to the dispersion and vigorously shaken and photopolymerized under UV light. The PEG SCPs were washed with water and stored in water. The resulting particles were 10–70 µm in diameter. Next, the PEG SCPs were grafted with crotonic acid (CA) [21]. Briefly, water was exchanged by ethanol and benzophenone (250 mg, 1.4 mmol) and CA (1.5 g, 17.4 mmol) were added. Then the mixture was flushed with nitrogen for 30 s and irradiated with UV light for 900 s. The PEG-CA SCPs were washed with ethanol. FITC functionalized SCPs were prepared by first reacting the PEG-CA SCPs in 0.1 M 2-(*N*-morpholino)ethanesulfonic acid (MES) buffer pH 5.5 containing 0.1 M *N*-Hydroxysuccinimide and 0.1 M *N*-Ethyl-*N*′-(3-dimethylaminopropyl)carbodiimide (EDC). Next, the activated PEG-CA SCPs were centrifuged, washed and added to a 0.5 mg L^{-1} BSA solution in 0.1 M phosphate buffer (pH 8.0). Finally, 30 µL of a 0.5 mg mL^{-1} solution of FITC in DMSO were added to 2 mL of BSA-SCPs to form FITC-BSA SCPs followed by centrifugation and washing. Cellobiose functionalized SCPs were prepared by incubating the PEG-CA SCPs in 0.1 M MES buffer (pH 4.5) containing 0.1 M EDC and 1 mg mL^{-1} cellobiose for 2 h followed by washing and washing.

4.3. SCP Characterization

AFM force spectroscopy with a NanoWizard 3 system (JPK instruments AG, Berlin, Germany) was performed to determine the elastic modulus of the microparticles. As AFM probe a glass bead with a diameter 4.75 µm (cellobiose SCPs) or 10 µm (FITC-BSA SCPs) was glued with an epoxy glue onto a tipless, non-coated cantilever (spring constant 0.32 N m^{-1}; CSC12, NanoAndMore GmbH, Wetzlar, Germany). Several force curves were recorded for different SCPs and analyzed with an appropriate contact model developed by Glaubitz et al. [23].

4.4. Surface Preparation

Round glass coverslips (35 mm #1, Menzel Gläser, Braunschweig, Germany) were cleaned in a mixture of ammonia hydrogen peroxide (30%) and water (1:1:5, RCA protocol) at 70 °C. Protein A coated surface, glass slides were immersed in 0.5 mg mL^{-1} protein A (PBS pH7.4), flushed with RICM measurement solution (PBS pH7.4). For soil release polymer coating, glass slides were immersed in a mixture of 182.4 mL ethanol, 9.6 mL water, 192 µL acetic acid, and 1920 µL GLYMO, shaken for 120 min, flushed with ethanol, followed by annealing for 120 min at 90 °C. Before RICM measurement, the GLYMO slides were immersed in measurement solution (PBS pH7.4) containing soil release polymers a concentration of 1 mg mL^{-1}, shaken for 60 min, and flushed with RICM measurement solution. Hydrophobic surfaces mimicking contact to oily soils were prepared by exposing cleaned coverlips to an atmosphere of trichloro(octadecyl)silane; 5 mL of the silane were cast in a glass petri dish and placed in a vacuum desiccator with the coverslips. After adjusting a pressure of 20 mbar by a vacuum pump, the valves were closed and the coverslips were left overnight in the desiccator. Finally, the surfaces were rinsed with isopropanol and water followed by curing for two hours at 150 °C in a drying oven.

4.5. Determination of Functionalization Degree via TBO Titration

A quantity of 0.5 mL carboxylic group functionalized SCPs dispersion was washed with ethanol and dried under vacuum at 50 °C for 5 h until constant weight was reached. 1 mL toluidine blue O (TBO) aqueous solution with a concentration of 312.5 μM at pH 10–11 was added to the dry SCPs and shaken in the dark overnight to stain the SCPs. The stained SCP dispersion was centrifuged for 30 min at 4400 rpm. 0.3 mL of the supernatant was diluted to 2 mL with water. The absorbance at 633 nm of this solution was measured by UV-VIS spectroscopy and compared to the absorbance of a TBO standard (312.5 μM TBO in aqueous solution at pH 10–11 and 1.7 mL). The carboxylic group functionalization degree of this group of SCPs was calculated with the following equation $D_{CGF} = N_R(1 - A_S/A_R)/W_{Dry}$, where D_{CGF} is the carboxylic group functionalization degree, A_S and A_R is the UV-VIS absorbance of sample and reference, W_{Dry} is the dry weight of 0.5 mL SCPs, N_R is the amount of TBO in the reference in units of μmol. For each group of SCPs, the TBO titration experiment was repeated three times and the average carboxylic group functionalization degree of the three experiments was used as the carboxylic group functionalization degree of this group of SCPs.

4.6. Reflection Interference Contrast (RICM) Measurements

RICM on an IX 73 inverted microscope (Olympus, Tokyo, Japan) was used to obtain the contact area between the SCPs and the glass coverslip surfaces. For illumination, an Hg-vapor arc lamp was used with a green monochromator (546 nm). An UPlanFL N 60×/0.90 dry objective (Olympus Corporation, Japan), and uEye CMOS camera (IDS Imaging Development Systems GmbH, Obersulm, Germany) were used to image the RICM patterns. To conduct the JKR measurements of the adhesion energies, both the contact radius and the particle radius were measured. Images with RICM patterns were read out using self-written image analysis software, contact areas and particle profiles were evaluated using scripted peak finding algorithms (IgorPro Wavemetrics, Lake Oswego, OR, USA).

Acknowledgments: The authors acknowledge funding by the German Research foundation (DFG) in the project SCHM 2748/3-1 and by Henkel AG & Co KGaA.

Author Contributions: Alexander Klaus Strzelczyk, Hanqing Wang, Andreas Lindhorst collected, analyzed and interpreted the data. Johannes Waschke provided software tools for image analysis. Tilo Pompe, Christian Kropf, Benoit Luneau, designed parts of the work and revised the article. Stephan Schmidt designed the work and drafted the article. Henkel AG & Co KGaA as funding sponsor took part in designing the adhesion studies with soil release polymers. Apart from that, the founding sponsors had no role in the design of the study; in the collection, analyses, or interpretation of data; in the writing of the manuscript, and in the decision to publish the results.

References

1. Silverman, H.G.; Roberto, F.F. Understanding marine mussel adhesion. *Mar. Biotechnol.* **2007**, *9*, 661–681. [CrossRef] [PubMed]
2. Max, E.; Häfner, W.; Wilco Bartels, F.; Sugiharto, A.; Wood, C.; Fery, A. A novel AFM based method for force measurements between individual hair strands. *Ultramicroscopy* **2010**, *110*, 320–324. [CrossRef] [PubMed]
3. Ferreira, G.N.M.; da-Silva, A.-C.; Tomé, B. Acoustic wave biosensors: Physical models and biological applications of quartz crystal microbalance. *Trends Biotechnol.* **2009**, *27*, 689–697. [CrossRef] [PubMed]
4. Daniels, J.S.; Pourmand, N. Label-free impedance biosensors: Opportunities and challenges. *Electroanalysis* **2007**, *19*, 1239–1257. [CrossRef] [PubMed]

5. Dufrene, Y.F.; Pelling, A.E. Force nanoscopy of cell mechanics and cell adhesion. *Nanoscale* **2013**, *5*, 4094–4104. [CrossRef] [PubMed]
6. Lee, C.K.; Wang, Y.M.; Huang, L.S.; Lin, S.M. Atomic force microscopy: Determination of unbinding force, off rate and energy barrier for protein-ligand interaction. *Micron* **2007**, *38*, 446–461. [CrossRef] [PubMed]
7. Koehler, J.A.; Ulbricht, M.; Belfort, G. Intermolecular forces between a protein and a hydrophilic modified polysulfone film with relevance to filtration. *Langmuir* **2000**, *16*, 10419–10427. [CrossRef]
8. Pussak, D.; Behra, M.; Schmidt, S.; Hartmann, L. Synthesis and functionalization of poly(ethylene glycol) microparticles as soft colloidal probes for adhesion energy measurements. *Soft Matter* **2012**, *8*, 1664–1672. [CrossRef]
9. Moy, V.T.; Jiao, Y.K.; Hillmann, T.; Lehmann, H.; Sano, T. Adhesion energy of receptor-mediated interaction measured by elastic deformation. *Biophys. J.* **1999**, *76*, 1632–1638. [CrossRef]
10. Johnson, K.L.; Kendall, K.; Roberts, A.D. surface energy and contact of elastic solids. *Proc. R. Soc. Lond. Ser. Math. Phys. Sci.* **1971**, *324*, 301. [CrossRef]
11. Limozin, L.; Sengupta, K. Quantitative Reflection Interference Contrast Microscopy (RICM) in Soft Matter and Cell Adhesion. *Chem. Phys. Chem.* **2009**, *10*, 2752–2768. [CrossRef] [PubMed]
12. Martin, S.; Wang, H.; Hartmann, L.; Pompe, T.; Schmidt, S. Quantification of protein-materials interaction by soft colloidal probe spectroscopy. *Phys. Chem. Chem. Phys. PCCP* **2015**, *17*, 3014–3018. [CrossRef] [PubMed]
13. Schmidt, S.; Reinecke, A.; Wojcik, F.; Pussak, D.; Hartmann, L.; Harrington, M.J. Metal-mediated molecular self-healing in histidine-rich mussel peptides. *Biomacromolecules* **2014**, *15*, 1644–1652. [CrossRef] [PubMed]
14. Pussak, D.; Ponader, D.; Mosca, S.; Ruiz, S.V.; Hartmann, L.; Schmidt, S. Mechanical Carbohydrate Sensors Based on Soft Hydrogel Particles. *Angew. Chem. Int. Ed.* **2013**, *52*, 6084–6087. [CrossRef] [PubMed]
15. Martin, S.; Wang, H.; Rathke, T.; Anderegg, U.; Möller, S.; Schnabelrauch, M.; Pompe, T.; Schmidt, S. Polymer hydrogel particles as biocompatible AFM probes to study CD44/hyaluronic acid interactions on cells. *Polymer* **2016**, *102*, 342–349. [CrossRef]
16. Pussak, D.; Ponader, D.; Mosca, S.; Pompe, T.; Hartmann, L.; Schmidt, S. Specific adhesion of carbohydrate hydrogel particles in competition with multivalent inhibitors evaluated by AFM. *Langmuir* **2014**, *30*, 6142–6150. [CrossRef] [PubMed]
17. Helfricht, N.; Doblhofer, E.; Bieber, V.; Lommes, P.; Sieber, V.; Scheibel, T.; Papastavrou, G. Probing the adhesion properties of alginate hydrogels: A new approach towards the preparation of soft colloidal probes for direct force measurements. *Soft Matter* **2017**, *13*, 578–589. [CrossRef] [PubMed]
18. Berg, J.M.; Tymoczko, J.L.; Stryer, L. *Biochemistry*; W.H. Freeman: New York, NJ, USA, 2010.
19. Lei, Q.Q.; Liu, J.W.; Zheng, H. Potential role of anti-p53 antibody in diagnosis of lung cancer: Evidence from a bivariate meta-analysis. *Eur. Rev. Med. Pharmacol. Sci.* **2013**, *17*, 3012–3018. [PubMed]
20. Simon, J.A.; Cabiedes, J.; Ortiz, E.; Alcocer-Varela, J.; Sanchez-Guerrero, J. Anti-nucleosome antibodies in patients with systemic lupus erythematosus of recent onset. Potential utility as a diagnostic tool and disease activity marker. *Rheumatology* **2004**, *43*, 220–224. [CrossRef] [PubMed]
21. Schmidt, S.; Wang, H.; Pussak, D.; Mosca, S.; Hartmann, L. Probing multivalency in ligand-receptor-mediated adhesion of soft, biomimetic interfaces. *Beilstein J. Org. Chem.* **2015**, *11*, 720–729. [CrossRef] [PubMed]
22. Dorman, G.; Nakannura, H.; Pulsipher, A.; Prestwich, G.D. The Life of Pi Star: Exploring the Exciting and Forbidden Worlds of the Benzophenone Photophore. *Chem. Rev.* **2016**, *116*, 15284–15398. [CrossRef] [PubMed]
23. Glaubitz, M.; Medvedev, N.; Pussak, D.; Hartmann, L.; Schmidt, S.; Helm, C.A.; Delcea, M. A novel contact model for AFM indentation experiments on soft spherical cell-like particles. *Soft Matter* **2014**, *10*, 6732–6741. [CrossRef] [PubMed]
24. Watt, R.M.; Voss, E.W. Mechanism of quenching of fluorescein by anti-fluorescein igg antibodies. *Immunochemistry* **1977**, *14*, 533–541. [CrossRef]
25. Liu, M.Y.; Jia, C.P.; Huang, Y.Y.; Lou, X.H.; Yao, S.H.; Jin, Q.H.; Zhao, J.L.; Xiang, J.Q. Highly sensitive protein detection using enzyme-labeled gold nanoparticle probes. *Analyst* **2010**, *135*, 327–331. [CrossRef] [PubMed]
26. Jia, C.P.; Zhong, X.Q.; Hua, B.; Liu, M.Y.; Jing, F.X.; Lou, X.H.; Yao, S.H.; Xiang, J.Q.; Jin, Q.H.; Zhao, J.L. Nano-ELISA for highly sensitive protein detection. *Biosens. Bioelectron.* **2009**, *24*, 2836–2841. [CrossRef] [PubMed]
27. Zhou, F.; Wang, M.M.; Yuan, L.; Cheng, Z.P.; Wu, Z.Q.; Chen, H. Sensitive sandwich ELISA based on a gold nanoparticle layer for cancer detection. *Analyst* **2012**, *137*, 1779–1784. [CrossRef] [PubMed]

28. Zhang, S.Y.; Garcia-D'Angeli, A.; Brennan, J.P.; Huo, Q. Predicting detection limits of enzyme-linked immunosorbent assay (ELISA) and bioanalytical techniques in general. *Analyst* **2014**, *139*, 439–445. [CrossRef] [PubMed]
29. De Gennes, P.G. *Scaling Concepts in Polymer Physics*; Cornell University Press: Ithaca, NJ, USA, 1979.
30. Erath, J.; Schmidt, S.; Fery, A. Characterization of adhesion phenomena and contact of surfaces by soft colloidal probe AFM. *Soft Matter* **2010**, *6*, 1432–1437. [CrossRef]
31. O'Lenick, A.J. Soil release polymers. *J. Surfactants Deterg.* **1999**, *2*, 553–557. [CrossRef]
32. Sheth, S.R.; Leckband, D. Measurements of attractive forces between proteins and end-grafted poly(ethylene glycol) chains. *Proc. Natl. Acad. Sci. USA* **1997**, *94*, 8399–8404. [CrossRef] [PubMed]
33. Sedeva, I.G.; Fornasiero, D.; Ralston, J.; Beattie, D.A. The Influence of Surface Hydrophobicity on Polyacrylamide Adsorption. *Langmuir* **2009**, *25*, 4514–4521. [CrossRef] [PubMed]
34. Schottler, S.; Becker, G.; Winzen, S.; Steinbach, T.; Mohr, K.; Landfester, K.; Mailander, V.; Wurm, F.R. Protein adsorption is required for stealth effect of poly(ethylene glycol)-and poly(phosphoester)-coated nanocarriers. *Nat. Nanotechnol.* **2016**, *11*, 372–377. [CrossRef] [PubMed]
35. Israelachvili, J. The different faces of poly(ethylene glycol). *Proc. Natl. Acad. Sci. USA* **1997**, *94*, 8378–8379. [CrossRef] [PubMed]
36. Podsiadlo, P.; Choi, S.Y.; Shim, B.; Lee, J.; Cuddihy, M.; Kotov, N.A. Molecularly engineered nanocomposites: Layer-by-layer assembly of cellulose nanocrystals. *Biomacromolecules* **2005**, *6*, 2914–2918. [CrossRef] [PubMed]

7

Gels Obtained by Colloidal Self-Assembly of Amphiphilic Molecules

Paula Malo de Molina [1,*] and Michael Gradzielski [2,*]

1 Centro de Física de Materiales (CSIC, UPV/EHU) and Materials Physics Center MPC, Paseo Manuel de Lardizabal 5, E-20018 San Sebastián, Spain
2 Institut für Physikalische & Theoretische Chemie—Stranski Laboratorium, Technische Universität Berlin, Straße des 17. Juni 124, 10623 Berlin, Germany
* Correspondence: paulamalodemolina@gmail.com (P.M.d.M.); michael.gradzielski@tu-berlin.de (M.G.)

Abstract: Gelation in water-based systems can be achieved in many different ways. This review focusses on ways that are based on self-assembly, i.e., a bottom-up approach. Self-assembly naturally requires amphiphilic molecules and accordingly the systems described here are based on surfactants and to some extent also on amphiphilic copolymers. In this review we are interested in cases of low and moderate concentrations of amphiphilic material employed to form hydrogels. Self-assembly allows for various approaches to achieve gelation. One of them is via increasing the effective volume fraction by encapsulating solvent, as in vesicles. Vesicles can be constructed in various morphologies and the different cases are discussed here. However, also the formation of very elongated worm-like micelles can lead to gelation, provided the structural relaxation times of these systems is long enough. Alternatively, one may employ amphiphilic copolymers of hydrophobically modified water soluble polymers that allow for network formation in solution by self-assembly due to having several hydrophobic modifications per polymer. Finally, one may combine such polymers with surfactant self-assemblies and thereby produce interconnected hybrid network systems with corresponding gel-like properties. As seen here there is a number of conceptually different approaches to achieve gelation by self-assembly and they may even become combined for further variation of the properties. These different approaches are described in this review to yield a comprehensive overview regarding the options for achieving gel formation by self-assembly.

Keywords: gels; self-assembly; surfactants; amphiphilic polymers; rheology; colloids; micelles; microemulsions; vesicles

---

1. Introduction

Self-assembly of amphiphilic molecules in solution can lead to a large variety of different colloidal structures [1], where these structures can have a profound effect on the macroscopic properties of these solutions. In this review we will, in particular, focus on systems which form gels by self-assembly of corresponding amphiphilic compounds. Of course, at very high concentrations all amphiphilic systems will form gel-type structures simply due to dense packing, where typically very stiff hexagonal, cubic or lamellar phases are formed [2,3]. However, we will explicitly not discuss such liquid crystalline phases of dense packing, but rather, focus in our review on more dilute systems in which basically the self-assembled structures lead to a gel-like behaviour of the systems. We will also exclude surfactant gels in which the surfactant is present in crystallized form. Especially for longer chain surfactants this can be achieved relatively easily and often leads to gelation, typically via formation of fibres [4] but also for the case of vesicular or lamellar structures, as for instance, known for the case of phospholipids ("gel phase") [5]. Accordingly, our review is concerned with gels formed by reversible dynamic assemblies, where the properties depend largely on the molecular architecture of the amphiphilic molecules, which in turn, control the structure and dynamics of these self-assembled systems.

Here it has to be noted that the definition of what is a gel is not necessarily undisputed and an authoritative source for a definition according to the International Union of Pure and Applied Chemistry (IUPAC) is that a gel is a "Non-fluid colloidal network or polymer network that is expanded throughout its whole volume by a fluid." In addition, it is stated that "a gel has a finite, usually rather small, yield stress" [6]. However, the experimental confirmation of a yield stress is nothing straightforward [7], as practically that amounts to the situation of having a structural relaxation time τ_{str} longer than the experimental observation time window.

For instance, the viscosity of self-assembled systems can become largely increased upon the formation of rodlike or wormlike micelles and once these are sufficiently long and/or concentrated viscoelastic surfactant solutions are formed [8,9]. Accordingly such systems can have zero-shear viscosities several orders of magnitude higher than the solvent itself, and this already at concentrations well below 1% [10]. However, such systems should be viscoelastic but not gels due to the expected finite structural relaxation time τ_{str}. Their elastic properties are based on entanglements and for wormlike surfactant micelles, different to polymer networks, there is a finite reptation time. In addition, wormlike micelles are dynamic chains, that break with a characteristic time scale [11]. Nonetheless such systems (as well as fibres) have recently been discussed to have a gel-like collective response that arises from these topological interactions (entanglements) [12]. The crucial parameter here is the effective structural relaxation time τ_{str}. Systems with an infinite (or at least substantially longer than the observation window) τ_{str} may be defined as self-assembled gels and they constitute interesting systems for formulations as they allow to exercise rheological control in a simple fashion. Accordingly systems with a finite but sufficiently long relaxation times (τ_{str} >> s) may for practical purposes be considered as gels, as in the relevant time and frequency range they respond similarly, which means mainly elastic and only to a much lesser extent viscous.

In general, the shear modulus G_0 and the zero-shear viscosity η_0 are directly related to each other via the structural relaxation time τ_{str} via Equation (1).

$$\eta_0 = G_0 \cdot \tau_{str} \tag{1}$$

As stated before, dense packing of micelles [3] or vesicles [13] also leads to systems with pronounced rheological properties which typically have a yield stress, i.e., do not flow at all if not subjected to a minimum external stress. There are many principal ways of achieving gel-like behaviour by self-assembly and a larger number of them have been well established for surfactant assemblies [14]. Gel formation can also be achieved by appropriate surfactant mixtures and/or employing polymeric amphiphiles [15], or combinations thereof. Here in particular block copolymers of the Pluronic type (PEO-PPO-PEO; PEO: poly(ethylene oxide), PPO: poly(propylene oxide)) are frequently employed as gelating systems, which have the capacity to form gels already at rather low concentrations due to the fact that their large PEO head groups can bind a substantially larger volume of water than their PEO chains would have themselves. This enhanced effective hard sphere volume explains their facile gelation [16], where it has been noted that such gels disappear again upon the admixture of low Mw surfactants that dissolve the copolymer micelles [17]. In principle, they are just densely packed micelles, often in a liquid crystalline cubic arrangement, but the main practical difference to most conventional surfactants is the large amount of bound water, thereby facilitating gelation already at rather low surfactant concentrations (of 15–25 wt %). Furthermore, Pluronics are attractive systems from the point of applicability, as they have permission to be employed in almost any field of pharmaceutical or cosmetical applications [18].

So far we just focussed on the situation of gelation due to self-assembly. However, particularly interesting in that respect are naturally systems which are responsive to external parameters, such as pH, ionic strength, temperature, magnetic and electric fields, shear fields etc., as they allow to control the rheological properties externally and to construct smart systems that adapt correspondingly to such external stimuli.

As seen from Equation (1) a main parameter is the shear modulus G_0 that is directly related to the structural arrangement of the colloidal systems, as determined by the mesoscopic structure. From a simple network theory G_0 is given by [19]:

$$G_0 = \nu \cdot {}^1N \cdot k \cdot T \tag{2}$$

where ${}^1N$ is the number of cross-linking network points. In this theory it is simply assumed that each such network point can store one kT as elastic energy, in analogy with the energy stored per degree of freedom in an harmonic oscillator. ν is a parameter of the order one associated with a given specific structural arrangement. For an ideal network $\nu = 1$ for an affine network and $\nu = (1 - 2/f)$ for phantom networks, where f is the cross-link functionality.

However, as stated above similarly important is the structural relaxation time τ_{str} that determines how long lived a given structural arrangement will be, which then is the key property that controls viscosity. Again, in a very simplified fashion this structural relaxation time can be approximated by:

$$\tau_{str} = A \cdot e^{E_a/k \cdot T} \tag{3}$$

where E_a is the activation energy required to break a cross-linking point, and A is the fastest possible break-up time (which is given by the inverse natural oscillation frequency of the network, for instance the movement of a hydrophobic sticker, and therefore typically is in the range of 10^{-10} s). E_a is the energetic effort for breaking up a given self-assembled connection, which in water is typically related to transferring a hydrophobic chain out of its environment in the hydrophobic assembly into the aqueous surrounding. This is known to be about 1.2 kT per CH_2 group [20] and similarly values are also known for other hydrophobic moieties.

In the following, we will discuss in the various chapters different typical approaches to achieve colloidal gels by self-assembly of amphiphilic molecules, which are based on wormlike micelles, densely packed vesicles, self-assembling polymers, or bridging of surfactant structures by amphiphilic copolymers.

2. Viscoelastic Networks of Wormlike Micelles

Self-assembly of surfactants in form of micellar aggregates can lead to the formation of surfactant gels, which are an interesting class of molecular gels, without having to be of crystalline nature [21]. The formation of viscoelastic surfactant solutions may occur directly upon dissolution of a surfactant in aqueous solution, but is also often observed upon addition of an additive, e.g., of salt, hydrophobic counterions or cosurfactant to ionic surfactant solutions [22]. This empiric observation has been around for a long time and may already occur for surfactant concentrations well below 1 wt % [8]. Initially the structural origin of this interesting rheological behaviour was unclear but became clarified by intense research more than 30 years ago. It could be attributed to the formation of overlapping long wormlike micelles and was then also directly imaged by transmission electron microscopy (TEM) already more than 30 years ago for the cases of dimethyloleylamine oxide [23] or cetyltrimethylammonium bromide (CTAB)/sodium salicylate (NaSal) [24] (Figure 1A). Such systems for both in salt-free water but also in the presence of larger concentrations of salt, like shown in Figure 1B for the case of 100 mM NaCl. An interesting question here has been the branching of such long wormlike micelles but more recently it has been shown by cryo-TEM [25] (Figure 1C) that branching does occur in wormlike micelles and also has a profound effect on the rheological properties of these networks as the appearance of branching points increases the shear modulus G_0 [26].

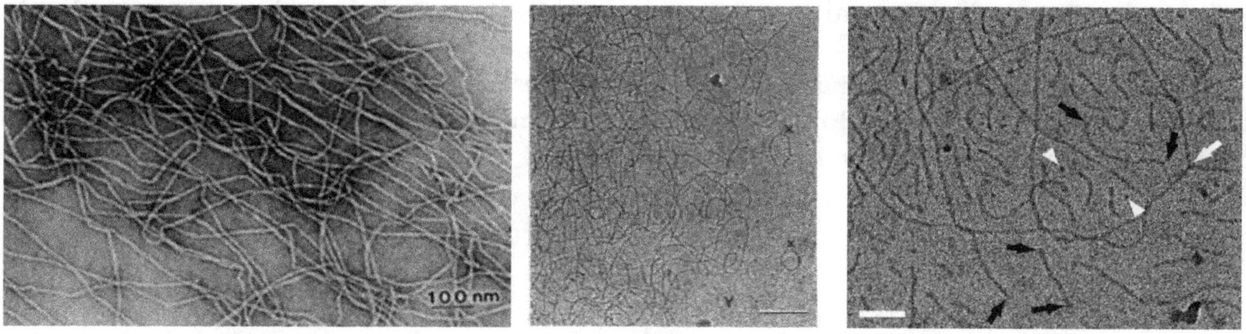

(A) (B) (C)

Figure 1. Electron micrograph of (**A**) a sample of 1 mM CTAB/1 mM NaSal [24] (With permission of Springer); (**B**) 50 mM CTAC/50 mM NaSal in 100 mM NaCl (scale bar: 100 nm) [27]; (**C**) NaOleate solution containing 15 wt % octyltrimethyl ammonium bromide (OTAB) (scale bar: 50 nm), white arrows indicate branching points and black arrows the end-caps [25].

Viscoelastic and gel-like systems have also been intensely studied for mixtures of cationic surfactants, such as alkyltrimethylammonium or alkylpyridinium, with hydrophobic counterions such as benzoates, salicylates, or naphthoates. Similarly, anionic surfactants of the alkylcarboxylate or alkylsulfate type form very viscous solutions with counterions such as tetraalkylammonium salts [8,10]. However, it might be noted that often also the addition of simple salts to ionic surfactants can lead to a substantial enhancement of the viscoelastic properties of a given surfactant solution. This is simply due to the reduced head group size as the electrostatic screening increases. This then increases the packing parameter of the surfactant and thereby one has a shift to more elongated micelles (the packing parameter p is given as: $p = v/(a_0 \cdot l_c)$; where v is the volume of the hydrophobic part of the surfactant, a_0 the head group area, and l_c a critical length which is roughly equal but less than the fully extended length of the hydrocarbon chain of the surfactant [28]). This effect is typically more pronounced for multi-valent counterions, as for instance demonstrated for sodium dodecyl trioxyethylene sulfate (SDTES) where the efficiency of the ions follows the rule: $Al^{3+} > Mg^{2+} > Ca^{2+} > Na^{+}$ [29]. However, also the addition of simple NaCl to palmitylamido-sulfobetaine (PDAS) has been shown to help gelation properties where here in particular a thermoreversible gelation is observed which takes place upon heating from 30 to 40 °C and is linked to a transition of globular to wormlike micelles but only at very high surfactant concentrations of 1 mol/L [30].

The rheological behaviour of such systems of entangled wormlike micelles in oscillatory experiments can to a first order often be described by Maxwellian behaviour, as given by Equation (4) for the frequency dependence of the storage modulus G′ and the loss modulus G″, but at higher frequencies typically marked deviations are observed. These can be attributed to the fact that the wormlike micelles have a finite life time and, depending on the detailed molecular composition, will have a characteristic breaking time τ_{break} [31], which determines at which frequency one will observe deviations from the picture expected for simple wormlike objects (as they are present in polymer solutions).

$$G' = G_0 \cdot \frac{\omega^2 \cdot \tau_{str}{}^2}{1 + \omega^2 \cdot \tau_{str}{}^2} \tag{4a}$$

$$G'' = G_0 \cdot \frac{\omega \cdot \tau_{str}}{1 + \omega^2 \cdot \tau_{str}{}^2} \tag{4b}$$

Typically, one observes that both, shear modulus G_0 and zero-shear viscosity η_0 follow a power law above a certain concentration c_0:

$$G_0 = A \cdot \left(\frac{c - c_0}{c_0}\right)^{\gamma} \tag{5a}$$

$$\eta_0 = B \cdot \left(\frac{c - c_0}{c_0}\right)^{\beta} \tag{5b}$$

where c_0 is an effective overlap concentration, A and B some system dependent pre-factors, and β and γ some system dependent power law exponents. γ depends mostly on the structural interconnection and typically is in the range of 1.5–3 while β may vary much more widely (between 1.5 to 8.5) as according to Equation (1) it also depends on the power law that applies to τ_{str}. Typically β increases substantially with increasing electrostatic interaction of the micelles, being smallest for neutral systems and highest for unscreened charged systems [10].

As already discussed before normal solutions of wormlike micelles have a finite structural relaxation time τ_{str}, which means they flow rather quickly under gravity. However, τ_{str} scales with the length of the hydrophobic chain of the surfactant, which normally is directly related to the kinetic exit time of the hydrophobic chain from the micelle and as the activation energy per CH_2 unit is about 1.2 kT having 2 additional CH_2 groups results in an increase by a factor ~10 (see Equation (3)). Accordingly, for long chain systems τ_{str} may move out of the experimental observation window. For instance one observes gel formation for erucyl dimethyl amidopropyl betaine (EDAB) for concentrations already above 10 mM and the shear modulus follows a power law of $G_0 \sim c(\text{EDAB})^{1.8}$ for the surfactant concentration. Such a scaling is also in good agreement with theoretical predictions [32]. Of course, this behaviour is temperature dependent and upon heating to 60 °C one observes normal viscoelastic behaviour again [33], since τ_{str} depends strongly on temperature (scaling according to Equation (3)). This concept of having very long hydrophobic chains to enhance the elastic properties of wormlike micelles has also been extended to pseudo gemini surfactants composed of *N*-erucamido-*N*,*N*-dimethylamine ($UC_{22}AMPM$) and maleic acid with a molar ratio of 2:1. This system was found to be quite temperature-sensitive and to be showing pronounced elastic properties already at concentration of 25 mM and, not surprisingly, these properties increase substantially upon increasing the surfactant concentration further. The most interesting aspect here is that it is quite sensitive to pH in the range of 6 and 7.5 [34], thereby indicating the importance of the charging conditions here.

In general, it is well known that gemini surfactants have a pronounced tendency for forming very elongated wormlike micelles [35]. As they have automatically two hydrophobic chains anchored within the wormlike micelle their structural relaxation times τ_{str} are much larger than those of the corresponding single chain surfactants. Accordingly, here one also produces easily viscoelastic surfactant solutions, where it has been found that the presence of a hydroxy group in the head group enhances viscosity and elastic properties, as seen in particular for the comparison of 2-hydroxyl-propanediyl-α,ω-*bis*-(dimethyldodecylammonium bromide) (12-3(OH)-12) and propanediyl-α,ω-bis(dimethyldodecylammonium bromide) (12-3-12) [36]. The same study also demonstrated the importance of the alkyl chain length as for 12-3(OH)-12 τ_{str} was in the range of s, while for 14-3(OH)-14 it moved into the range of hundreds or thousands of s, i.e., allowing for tube inversion and real gel formation (again this is not a surprising finding as one introduces 4 CH_2 groups into the hydrophobic surfactant moiety and according to our above statements would expect to see an increase of τ_{str} by a factor ~100).

In summary, it can then be stated that viscoelastic surfactant systems based on wormlike micelles allow for the formation of viscoelastic fluids, that in practice can mutate to gels. The main control parameter here is the structural relaxation time τ_{str}, which in turn depends mainly on the length of the hydrophobic moiety.

3. Densely Packed Vesicle Gels

Another way of obtaining highly viscous or even gel-like systems is by having densely packed vesicles. This applies to unilamellar or multi-lamellar vesicles (ULVs, MLVs) that are closed surfactant bilayers (with one or many shells) (Figure 2). The rheology of vesicle systems has been reviewed some while ago [37], and they show typically enhanced viscosity compared to micellar systems and a shear thinning behaviour. Their enclosing of solvent allows to have densely packed systems of spherical objects, with the amphiphilic volume fraction typically in the range of 2–15 wt % of amphiphilic substance. In general, one may expect low volume fractions for ULVs and higher volume fractions with correlated layers in the MLVs.

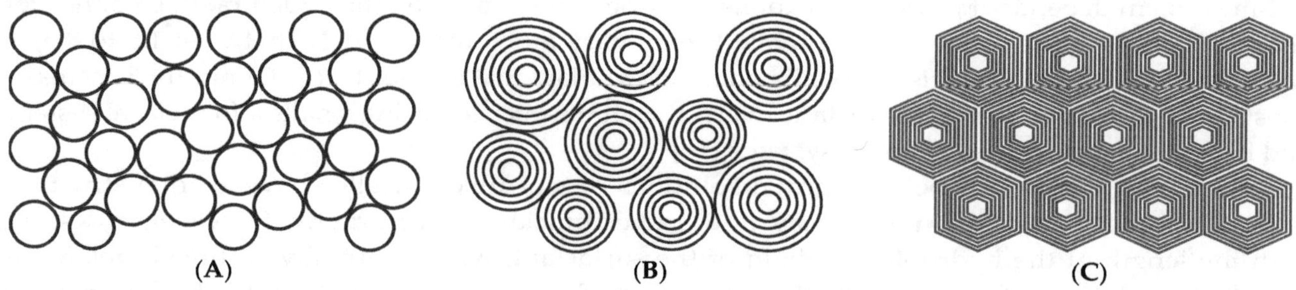

Figure 2. Sketch of different types of densely packed vesicle gels. (**A**) made up from unilamellar vesicles (ULVs); (**B**) made up from multilamellar vesicles (MLVs); (**C**) densely packed deformed vesicles at high concentration.

3.1. Vesicle Gels Based on Unilamellar Vesicles (ULVs)

We may first consider vesicle gels formed by ULVs. Such gels were already reported as early as 1968 by Fontell and Ekwall for densely packed vesicles observed in the system NaOleate/decanol/water [38]. However, it remained largely unnoticed, as the authors did not emphasize that interesting point in their study, which instead focussed on other aspects of surfactant self-assembly and phase behaviour. These investigations were taken up on basically similar systems in the 1990's by Hoffmann et al. who not only studied the mesoscopic structure of these systems, but also their rheological properties in some detail [39,40]. These vesicle gels simply form spontaneously by diffusion of the cosurfactant into the oleate solution. Thereby the initial micellar oleate solution becomes first transformed into increasingly longer wormlike micelles with a corresponding marked increase of viscosity. This process occurs within the first 1–2 min and then is followed somewhat later by a transition from wormlike micelles into well-defined ULVs, where this process is typically completed after about 15–20 min and is accompanied by a gelation of the system, i.e., it possesses a yield stress then of ~200 Pa [41]. The shear modulus G_0 is in the range of 1000–10,000 Pa, increasing with increasing surfactant concentration. It can be well explained via Equation (2), as now one has more vesicles since they become reduced in radius from 24 to 14 nm, in order to retain the packing volume fraction while having larger amounts of amphiphilic bilayer to disperse. The dense and highly ordered packing of vesicles can be seen well in the FF-TEM shown in Figure 3A.

The observed structural progression can simply be explained by the change of the packing parameter p of the amphiphilic system that comes about by incorporating the octanol into the oleate system. The initially present globular micelles elongate into increasingly long rodlike micelles, which then finally transform to well-defined and rather monodisperse ULVs as followed and confirmed by time resolved small-angle neutron scattering (SANS) experiments [42]. The formed gel phase is isotropic and quite transparent and is found for oleate concentrations of ~150–400 mM and concentrations of 1-octanol of ~450–700 mM, which means that the amphiphilic film is largely composed of the cosurfactant 1-octanol and therefore also this bilayer is with ~2.2–2.5 nm rather thin. In addition, such vesicle gels can also be formed by adding other cosurfactants like heptanol, hexanol, or geraniol to an aqueous Na oleate solution, while shorter or longer alcohols lead to systems with much reduced elastic properties [40]. It might also be noted that here one is not restricted to oleate as surfactant, but the structurally related isostearate possesses a quite similar phase behaviour. It is interesting to note that subsequent NMR work on these well-defined ULV gels proved the existence of µm-size kind of "super-structure" or "grain-like structure" in these systems, which is several hundred times bigger than the individual vesicles, and which for instance for aspects of release of active agents from them should be of relevance [43]. A somewhat related investigation showed that NaOleate can also become transformed into a gel phase by addition of *N,N-bis*(carboxylatomethyl) glutamate (GLDA). However, in that case not the formation of ULVs is responsible for gelation but instead long and stiff fibrils of lamellae are at its origin [44].

Further work also showed that the structural features of the vesicle gel can be retained during silication, where the initial vesicle gel contained in addition tetraethyl orthosilicate (TEOS) as a silica source. The TEOS then hydrolyses more slowly than required for the vesicle gel formation to take place, therefore not interfering with it [45]. Interestingly the incorporation of the silica network leads to a reduction of the elastic properties, which becomes very pronounced beyond a certain critical TEOS concentration [45].

Some time ago, the formation of vesicle gels of strings of vesicles was reported for the case of gemini surfactants, where it was speculated that this percolating system of vesicles comes about by the protrusion of small chains from the vesicle surface (Figure 3B) [46]. This then renders the vesicles attractive to each other and the bridging of two vesicles by the gemini surfactant thus leads to the formation of strings that for high enough concentration yields a space-filling percolated network. Such behaviour was observed for a number of gemini surfactants, all having in common a large asymmetry with respect to the length of their two hydrocarbon chains and gel strength and yield stress were found to depend markedly on the molecular structure of the gemini surfactant.

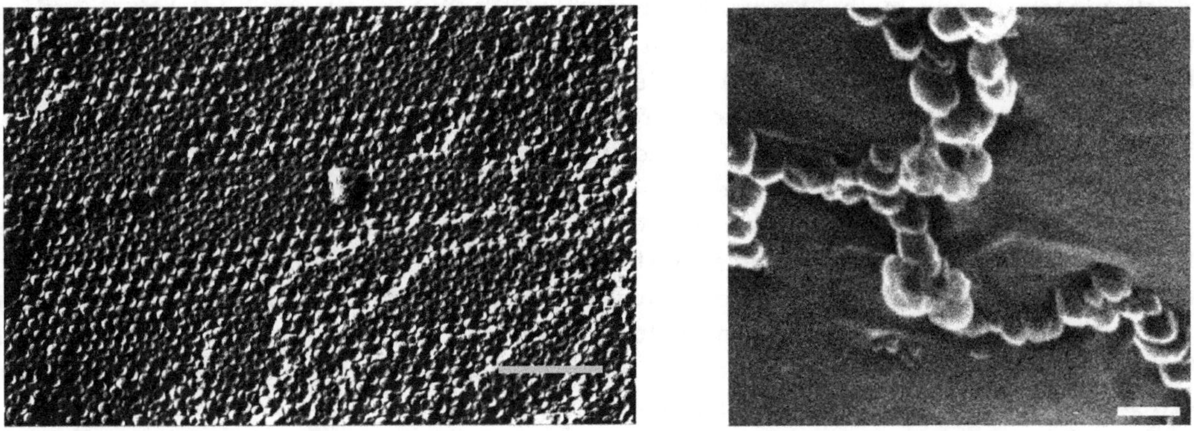

(A) (B)

Figure 3. (**A**) Freeze-fracture transmission electron microscopy (FF-TEM) image of densely packed ULV in the system 182 Na isostearate/567 mM 1-octanol (the aqueous solution contained 20 wt % glycrol to facilitate the FF preparation (size bar: 200 nm) [41]; (**B**) cryo scanning electron microscopy (cryo-SEM) image of a C_{18}–C_8 gemini vesicle gel (size bar: 66.7 nm) [46].

When getting more and more densely packed, one may expect that the charged vesicles escape from this crowded situation by deflation and formation of bi- or multilamellar vesicles. This mechanism actually has been described recently by theory and supported by experimental evidence [47].

3.2. *Vesicle Gels Based on Multilamellar Vesicles (MLVs)*

Of course, the concept of densely packed spherical colloids just described for ULVs can be extended to multilamellar vesicles (MLVs) and actually here the formation of such gellike and viscoelastic systems has been reported much more often. Such vesicle gels cannot only be formed by spherically shaped ULVs and MLVs (Figure 4A) but for higher concentrations of amphiphilic material the vesicles have to deviate from a spherical shape in order to allow for a more dense packing (see Figures 2C and 4). Such "deformed MLVs" then are of polyhedral shape, which allows for the correspondingly required more dense packing. Examples for such structures are depicted in Figure 4B. If made from phospholipid such liposome gels of MLVs are also of high interest for practical formulations in the context of delivery systems [48].

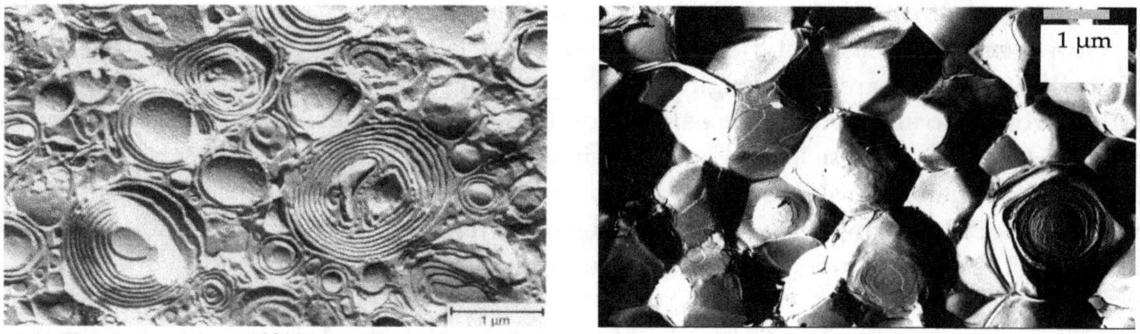

(A) (B)

Figure 4. FF-TEM micrographs of the systems: (**A**) 90 mM TDMAO/10 mM TTABr/220 mM 1-hexanol (Reproduced ("Adapted" or "in part") from [49] with permission of The Royal Society of Chemistry.); (**B**) 360 mM TDMAO/40 mM TTABr/780 mM 1-hexanol/700 mM NaCl.

An example for such densely packed MLV is given for the zwitterionic surfactant tetradecyl-dimethyl amine oxide (TDMAO) that by addition of a cosurfactant like hexanol or heptanol becomes transformed into a state of vesicles or lamellae. By protonation or substitution of the TDMAO by the cationic tetradecyltrimethyl ammonium bromide (TTAB) this system can be shifted further into the state of MLVs [50]. Already at a surfactant concentration of 100 mM (~2.5 wt %) the formation of MLVs results in pronounced elastic properties with a shear modulus G_0 in the range of 10–100 Pa and the formed systems even exhibit a yield stress. This is due to the fact that one has here μm sized onion-type MLVs that are densely packed, as seen in Figure 4A [49]. An interesting observation is that G_0 is very sensitive to the charging of the systems. While the uncharged system shows basically no gel-like behaviour, already the presence of 1 mol % charged surfactant leads to viscoelastic properties and raising this value to 4 mol % leads to a substantial increase of G_0 to ~40 Pa, while further charging then has no effect and G_0 remains constant thereafter. It is also interesting to note that the rheological behaviour is almost the same whether one charges the system by protonation or by substituting TDMAO by TTAB, indicating that it is a purely electrostatic effect. Of course, as the rheological properties here depend so strongly on electrostatics the addition of salt then leads to a substantial reduction of the viscoelastic properties again. Upon increasing the surfactant concentration by a factor 4 and having a rather high salinity of 700 mM NaCl one observes the formation of densely packed multifacetted vesicles as depicted in Figure 4B (the high salinity here reduces the electrostatic repulsion between the bilayers and thereby facilitates their dense packing).

It might be noted that this MLV TDMAO/TTAB/1-hexanol system can elegantly be changed in its structure by application of shear forces, where with increasing shear rate one reduces the number of lamellae in the MLVs until at very high shear rates of several 1000 s^{-1} finally unilamellar vesicles are present (Figure 5A). Of course, at the same time the number of vesicles increases substantially and this also leads to an increase of the shear modulus G_0 (Figure 5B) which is in agreement with Equation (2) and demonstrates that here the number of effective network points has increased [51]. It might be noted that this transformation of vesicle morphology is basically irreversible (i.e., no relaxation process to any of the other structures was observed), but therefore it also remains unclear which state here is really thermodynamically preferred. Of course, it should be noted that size control of MLVs by shear had already reported before [52] but not in the context of parallel rheological control.

A somewhat related system is also based on TDMAO but uses the fact that TDMAO is a weak base, which can become protonated by a strong acid. This was accomplished by mixing it with perfluorolauric acid (PFLA) and this catanionic surfactant system forms birefringent gels for surfactant concentrations higher than 50 mM and a molar content of PFLA of 80–90%. 100 mM systems have shear moduli of ~1000 Pa and the structural investigations show the presence of MLVs here, but ones that are in the crystalline state at room temperature and only melt around 50 °C, as also the pure PFLA melts at 55 °C [53].

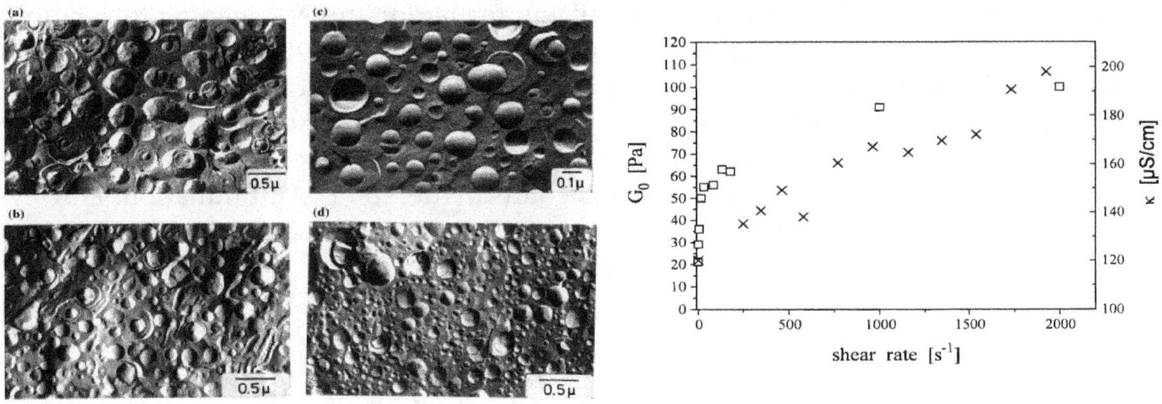

(A) (B)

Figure 5. (**A**): FF-TEM micrographs of the system 90 mM TDMAO/10 mM TTABr/220 mM 1-hexanol: (**a**) immediately after shearing the sample for 1.5 h at a shear rate of 200 s^{-1}; (**b**) 2000 s^{-1}; (**c**) 4000 s^{-1}; (**d**) after allowing the system depicted in (**c**) to relax under stirring for 12 days; (**B**): Shear modulus G_0 (□) and electric conductivity during shear in vorticity direction (×) versus shear rate of the pre-shear for the same system. The vesicle solution was sheared at the given shear rate until the apparent shear viscosity indicated a steady state. Then, shearing was stopped and the modulus was measured in an oscillation experiment (Original in [51]).

Another interesting observation on the TDMAO/cosurfactant system was that one can also induce the formation of a densely packed MLV gel by the addition of a hydrocarbon (here decane) to a concentrated solution of TDMAO and benzyl alcohol, where the structure and rheological properties are controlled by the amount of decane contained [54].

As indicated the formation of MLVs is often linked to the presence of a cosurfactant. Accordingly, similar systems have also been described for the case of the classical surfactant SDS and cetyl alcohol. The reason for the gellike behaviour here could be attributed to jammed packing of uni- or multi-lamellar vesicles as determined mainly a combination of ^{1}H and ^{13}C-NMR [55]. For phospholipids such gels are formed from lecithin upon the addition of sodium deoxycholate, sodium cholate, sodium taurodeoxycholate, or sodium taurocholate, where the rheological properties depend on the precise ratio of bile salt and lecithin. Robust gels are formed already around molar ratios for bile salt/lecithin of ~0.2 and lecithin concentration of 400 mM [56].

In another type of system based on a classical nonionic surfactant $C_{12}E_4$, that is well known to form vesicles [57] gelation was induced and controlled by the addition of cationic dodecyltrimethyl ammonium bromide DTAB. Yield stress and elastic modulus increase with increasing content of DTAB, where a maximum is already achieved around 4–5 mol % substitution. Of course, the elastic properties can further be controlled by the total concentration of surfactant as shown in Figure 6 [58]. A similar phase and rheological behaviour was observed for another nonionic surfactant of somewhat longer chain length and in addition containing ethylhexylglyceride as cosurfactant. Here the addition of SDS resulted in the formation of a vesicle gel with a yield stress, which could be explained by a simple electrostatic model for the bending constant of the bilayers [59]. Later work showed a similar behaviour upon admixing the anionic surfactant sodium *bis*(2-ethyl hexyl)sulfosuccinate (AOT) [60].

Furthermore, it could also be shown that charging of the $C_{12}E_4$ bilayers by means of adding anionic perfluorolauric acid (PFLA) leads to the formation of vesicle gels for concentrations around 10 wt %, where yield stress and elastic modulus increase substantially with increasing content of ionic surfactant PFLA [61]. In a related later study the charging of the $C_{12}E_4$ system was done by adding the amphiphilic anionic dye sodium 4-phenylazobenzoic acid (AzoNa). The obtained vesicle gels were responsive to temperature, pH and light, increasing in elastic properties with increasing temperature in a reversible way and being stable in the pH range of 7 to 11, while losing their gel properties outside this pH range [62]. Illumination by UV light initiates a transition from trans to cis conformation of the AzoNa which promotes bilayer formation thereby leading to gelation of the system. This process then can be switched back by illuminating with visible light [62], i.e., this is a light-responsive self-assembled gel. It might be added that a similar light responsive formation of a hydrogel could also be achieved for the case of the cationic surfactant, alkyltrimethylammonium bromide (C_nTAB, $n = 12, 14, 16$, and 18) via the addition of sodium azobzenzene 4,4′-dicarboxylic acid ($AzoNa_2$), which can be switched from cis to trans conformation by UV illumination. However, the strong gels formed there are not due to vesicle formation but the reason is the formation of very long (many μm) fibers in the presence of the cis-$AzoNa_2$ [63].

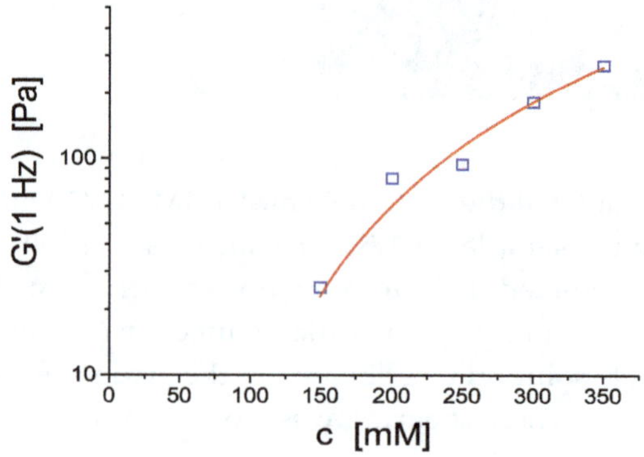

Figure 6. Storage modulus G′ as a function of total concentration for vesicles gels composed of Brij30 (technical grade $C_{12}E_4$) and 4 mol % DTAB, solid line: $G'(1\ Hz) = A \times (c - c_0)^x$; $c_0 = 76$ mM, x = 1.87.

A classical way of forming vesicles is the mixture of cationic and anionic surfactants, i.e., for catanionic surfactants [64]. By a variation on that theme it has been shown that the mixture of tetradecyl or dodecyl trimethyl ammonium hydroxide (TTAOH/DTAOH) with the 2-hydroxy-1-carboxy-naphthoate (HCN). This leads to salt free systems where HCN constitutes the hydrophobic counterion. For both surfactants (DTA and TTA) one observes already at 100 mM concentration around equimolar mixing or some HCN excess the formation of MLVs which form a weak gel with about 20 Pa storage modulus and exhibiting a yield stress [65]. A similar behaviour had been observed before when employing CTAOH as cationic surfactant and here it was interesting to note that for lower amounts of admixed HCN a viscoelastic phase of wormlike micelles is formed, where this viscoelastic solutions show a very similar storage modulus at high frequency as the MLV gels, but not having their yield stress [66]. In that context, it is also interesting to note that these pronounced elastic properties of a typical gel system are only observed for the salt-free C_nTA/HCN systems, while mixing their salts (which produces the equimolar amount of salt, e.g., NaBr) leads to a reduction of the elastic modulus by a factor 100 while at the same time one observes the formation of MLV but together with other locally lamellar structures [67]. Apparently the elastic properties of this system are largely controlled by the electrostatic interaction.

It might be noted that catanionic surfactants are most known for their spontaneous formation of well-defined ULVs but at higher concentration will often form MLVs, which then are in dense packing forming gels, as for instance described for the case of CTAT/SDBS [68].

4. Hydrophobically Modified Polymers

So far we have considered gels that are formed by surfactant self-assembly due to the aggregate structure, i.e., long worm-like micelles and densely packed micelles. However, in general, one may achieve similar self-assembled systems by employing amphiphilic copolymers. A particularly interesting class of polymers in that context are hydrophobically modified water-soluble polymers (see Figure 7). Here the hydrophobic modification is typically an alkyl chain of similar length as in surfactants and, therefore, has a tendency to associate with other such chains. Depending on the number of hydrophobic modifications along the polymer backbone, their concentration and the flexibility of the polymer, these hydrophobic chains may self-assemble like a micelle or yield connecting hydrophobic contact points. The structure and rheological properties differ significantly for the different polymer architectures. Hydrophobically modified water-soluble polymers can be classified into three main groups (Figure 7A):

1. Telechelic polymers, which are linear polymers end-capped with two stickers, alkyl chains or short hydrophobic blocks;
2. Low functionality multisticker polymers;
3. Multisticker grafted polymer chains with randomly distributed pendant hydrophobes along the hydrophilic chain (comb polymers).

Of course, beyond the architecture here it is also interesting to have systems where this hydrophilic/hydrophobic balance depends on external parameters, such as temperature, pressure, pH, ionic strength, etc., i.e., stimuli responsive systems.

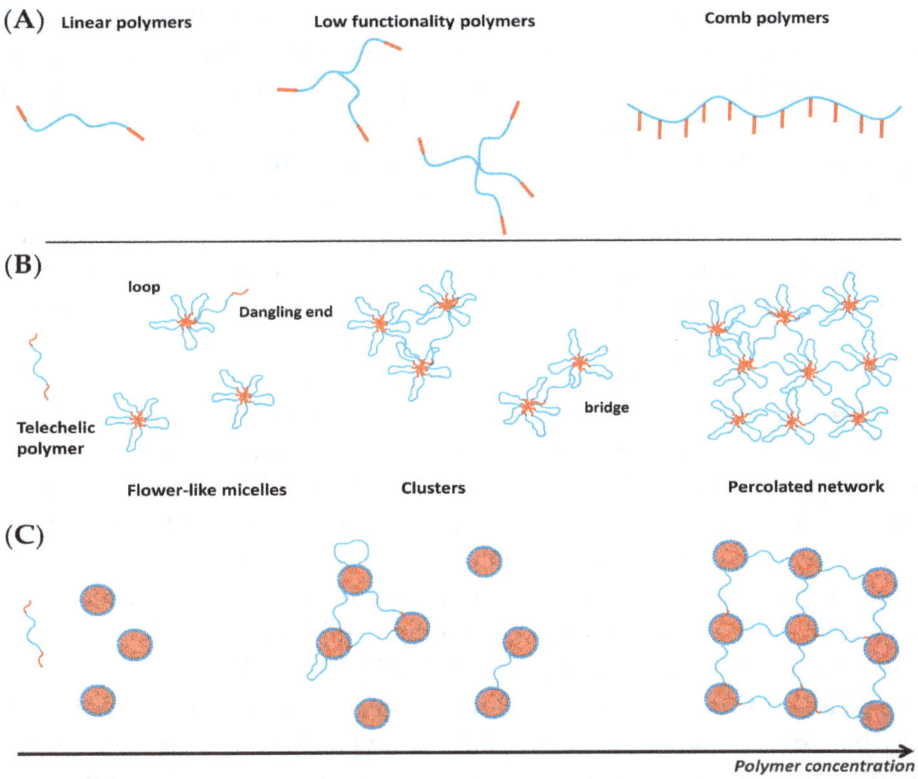

Figure 7. Schematic representation of (**A**) the polymer architecture of linear and low functionality telechelic polymers and comb-type hydrophobically modified polymers; association of telechelic polymers in (**B**) aqueous solutions and (**C**) with microemulsions as a function of the polymer concentration.

4.1. Telechelic Polymers

The self-assembly and mechanical properties of telechelic polymers, or ABA triblock copolymers (with A being hydrophobic and B being hydrophilic), is relatively well understood. The hydrophilic block most widely used is poly(ethylene oxide) [69–72] but there are examples with other chemistries, such as poly *N,N'*-dimethylacrylamide [73,74]. The hydrophobic end-group is typically a hydrocarbon alkyl chain [72,74,75]. However, also fluorocarbon alkyl chains [76] and hydrophobic polymer blocks (for example, polystyrene [77] and polybutadiene [78]) have been explored. Of course, one may also have a similar situation for ABC copolymers with A and C being two different hydrophobic units, which has also been discussed and studied in some detail [78]. The current understanding of the self-assembly and mechanical properties of telechelic polymers has been collected in a recent review by Chassenieux et al. [79]. Telechelic polymers aggregate at low concentrations into flower-like micelles [70], where different techniques yield different micellar aggregation numbers for a given system. For instance, fluorescence gives lower numbers than those obtained by static light or neutron scattering, and these in turn are slightly lower than the ones from dynamic light scattering (DLS) and viscosimetry [72,80,81]. However, they all agree that, compared to surfactant micelles, associative polymer micelles have lower aggregation numbers ($N_{agg} < 50$). Theoretically, the aggregation number of flower-like micelles results from balancing the interfacial energy, configurational entropy, and excluded volume interactions in the corona against the deformation energy of the hydrophobic chains in the core. The polydispersity of the micelles arises from thermal fluctuations [82]. For long hydrophilic chains, the loop formation does not affect significantly the free energy and micelles of telechelic chains theoretically are predicted to have the same aggregation number as a solution of double the number of chains with one hydrophobic end-cap group [69]. This result has been in fact found experimentally by Sérero et al. [70].

At higher polymer concentrations, as the number density of micelles increases, they come closer together and polymer chains are able to reversibly form bridges between micelles leading to the formation of clusters [83] (Figure 7B). The cluster size grows with increasing concentration until the percolation concentration, where one cluster spans the entire volume and a transient network is formed. The rheological properties of these networks have been studied in detail [71,84–86]. Linear oscillatory shear measurements exhibit viscoelastic behaviour with one relaxation time and a high frequency modulus that can be described by a Maxwell model (Equation (4)) with one single relaxation time τ_{str} and a plateau elastic modulus G_0 (Equation (2)). The structural relaxation time τ_{str} is related to the residence time of the hydrophobic sticker in the micelle. Thus, the experimental structural times are very similar to the relaxation times determined for micellar kinetics for surfactants with the same chain length and increases strongly with increasing chain length of hydrophobic sticker [85]. Variation of the end-group chemistry consequently affects the sticker residence time in the micelles, i.e., hydrophobically modified polymers with fluorocarbon chains have longer lifetimes of the bridges than polymers end-capped with alkyl chains of the same length [76]. The elastic modulus in case of flexible and unentangled chains is expected to depend on the number of bridges. In the simplest case of rubber theory each bridge contributes 1 kT to the elasticity (see Equation (2)). Thus, predicting the value of G_0 depends on the ability to estimate the fraction of bridges versus loops and dangling ends [87]. In terms of the non-linear rheology, they often exhibit shear thickening behaviour prior to a sharp decrease in viscosity of several orders of magnitude [71,88,89].

The structure of flower-like micelles fits well to a model for star polymers [90]. With this model one can well describe the small angle scattering data of isolated micelles at low concentrations. In the concentrated regime micelles are experimentally found to interact repulsively [70,74,91], due to the excluded volume of the bridging chains. However, theoretically an attractive component due to bridging is expected [92] (Figure 8). The strength of the attractive interaction has been predicted to be about 1 k_BT per chain, regardless of the lifetime of the bridges. At constant concentration, an increase in the strength of the attraction leads to phase separation [92].

4.2. Amphiphilic Polymers with Multiple Hydrophobic Stickers

Polymers with many hydrophobic groups attached to a hydrophilic backbone associate in more complicated structures since the hydrophobes may associate with other hydrophobes of the same polymer molecule or of different molecules. Depending on the polymer architecture and concentration, these polymers also form transient networks. They show a strong viscosity increase in the semi-dilute regime that is more pronounced if the grafting density is higher or the length of the stickers is increased. These solutions exhibit viscoelastic behavior. Their rheological response presents a broader distribution of relaxation times compared to the HM end-capped linear polymers [73]. A theoretical work proposed by Rubinstein and Semenov [93] predicted that the dynamics in the dilute regime is mainly controlled by intramolecular association. With increasing concentration, the polymer chain dynamics can be described by a sticky Rouse model for unentangled polymers and a sticky reptation model for entangled polymers. Experimental data of HM neutral polymers, such as polyacrylamide [94] or poly(*N*,*N′*-dimethylacrylamide) [73], agree with the theoretical predictions.

In the case of just a low number of stickers per polymer, i.e., more than 2 but less than 10, considerably less work can been found in the literature. Associative star block copolymers—such as poly(acrylic acid)-block-polystyrene $(PAA\text{-}b\text{-}PS)_4$ [77], poly(ethylene glycol)-*b*-poly(*N*-isoprylacryl-amide) (PEO-*b*-PNiPAAM)*x*, with $x = 2$–8 [95] or (PEG-*b*-PLLA) [96,97]—aggregate in aqueous solutions und undergo a sol-gel transition at a critical concentration in the same way bifunctional polymers do. In general, polymers with higher number of associative groups form networks more effectively due to lower intramolecular association. However, more systematic studies are needed in order to understand the effect of the polymer architecture on the network properties. Recent investigations on 3-arm and 4-arm end-capped polymers showed that the polymer functionality impacts substantially the rheological properties of the network in terms of the network elasticity. Higher functionality leads to higher connectivity and thus to higher plateau moduli. The viscoelastic behaviour is still almost of Maxwell type with one relaxation time given by the sticker length [74].

The interactions of associative groups that give rise to the transient junctions can be other than hydrophobic interactions. Sophisticated end-functionalization of polymers has led to a library of associating polymers that bond through noncovalent physical interactions such as metal−ligand coordination [98–100], hydrogen bonding, [101] and host−guest interactions [102]. Compared to the classical telechelic polymers, the multiplicity of the network links is given by the type of physical bonding and it is generally less than 5.

Also, although most of the work has been done on water based systems, polymers that associate in non-polar solvents havebeen studied. They form micelles and networks with the same governing physics as in the water-based systems. For instance, triblock copolymers with a middle block soluble in paraffin oil and insoluble polystyrene end-blocks are able to associate into interconnected micelles in paraffin oil [103,104]. Analogously, block copolymers with an oil soluble middle block and water soluble end-groups are able to bridge and network reverse swollen micelles in oil [105,106] in the same way that occurs in the water based systems described below.

4.3. Stimuli Responsive Copolymers

All the aspects discussed above apply similarly to systems where the hydrophilicity of one block becomes switched on or off by an external parameter, such as temperature, pressure, pH, or ionic strength—and, of course, such switchable systems are very interesting. Accordingly, here many concepts have been presented and in our review we want to focus purely on such where by the change of an external parameter the hydrophilicity/hydrophobicity can be switched and thereby the rheological properties of the systems. As an example for the case of temperature sensitive systems this means that one of the blocks has to possess a lower (LCST) or an upper (UCST) critical solution temperature [107]. As the field of stimuli responsive polymers and their effect on macroscopic properties is a very wide one we want in the following only discuss some exemplary cases relevant for gelation, for instance arising from interconnection of self-assembled entities.

Most frequently ABA triblock copolymers have been employed where the block A is the switchable block. A good example for such a system are copolymers with A = 2-(diisopropyl-amino)ethyl methacrylate), DPA or 2-(diethylamino)ethyl methacrylate), DEA; B = 2-metha-cryloyloxyethyl phosphorylcholine, MPC, where DEA and DPA possess a LCST, which means that they become hydrophobic above a certain temperature and then form hydrophobic domains (see Figure 7B). This means that above this temperature and for concentrations about 10 wt % gelation takes place. These gels then dissolve again at lower pH due to protonation of the A block [108], i.e., are in addition pH-sensitive. A slight variation of this system then was done with PNIPAm as hydrophobic A block which forms gel-like systems at temperatures above 35 °C [109]. An interesting extension of this work then lead to thermogelling systems that contain PNIPAm as LCST A block. In addition, these polymers contained a S–S bond in the center of the copolymer, which then allows for chemical disintegration of the gel by a reduction reaction, that for instance can be done under very mild conditions by the tripeptide glutathione [110]. A similar redox-responsive system has been constructed for the case of NIPAM-*b*-PDMA-*b*-PNIPAM or PDEGA-*b*-PDMA-*b*-PDEGA copolymers, which were obtained by the RAFT procedure and correspondingly contain a central trithiocarbonate unit. The initially formed gel can be broken by aminolysis and the formed thiol capped copolymer micelles can be cross-linked reversibly by oxidation [111]. Of course, the gelation concentration depends on the detailed molecular architecture of the ABA copolymer and for PLGA-PEO-PLGA was shown to be mainly dependent on the block lengths [112].

Of course, in addition to just having temperature responsiveness one may also switch hydrophobicity by pH. This was for instance demonstrated for the case of ($PDEAEM_{25}$-*b*-PEO_{100}-*b*-PPO_{65}-*b*-PEO_{100}-*b*-$PDEAEM_{25}$) pentablock copolymers. By SANS experiments it could be shown that above the LCST at 70 °C one has at pH 7.4 micellar aggregates that form a cross-linked gel upon raising the pH to 10.5 [113]. However, the pH-dependence can also arise from the center block, as demonstrated for the case of PMMA-PDMAEMA-PMMA, where above a certain polymer concentration gels are formed in aqueous solution due to the formation of bridged hydrophobic domains. However, this mechanism is only well working at intermediate pH of ~4, where the PDMAEMA chain is almost fully charged and thereby fully stretched. At higher pH the PDMAEMA chain becomes neutralized and therefore is no longer able to bridge and at lower pH one has automatically a substantial increase of the ionic strength which then screens the electrostatic repulsion within the PDMAEMA chain and therefore much reduced chain elongation. Accordingly, one can switch by pH from a gel state at pH ~4 to a sol state at higher or lower pH [114].

A coupled pH/temperature responsiveness of a sol-gel transition has been observed for linear triblock copolymer, poly(methoxydi(ethyleneglycol) methacrylate-*co*-methacrylic acid)-*b*-PEO-*b*-poly (methoxydi(ethylene glycol) methacrylate-*co*-methacrylic acid) (P(DEGMMA-*co*-MAA)-*b*-PEO-*b*-P(DEGMMA-*co*-MAA)), and by appropriately changing pH and temperature one can obtain successive sol-gel and gel-sol transitions in a narrow pH range [115]. However, here exist also other approaches for introducing thermoresponsiveness and for instance it can also be obtained by having hydrophobic dipeptides (dityrosine end groups) that end-cap a PEG chain. The dipeptides then form β-sheet fibrils that lead to a gel-sol transition near body-temperature [116].

Of course, there are many more ways to employ amphiphilic polymers for the formation of polymeric hydrogels and we have depicted here only some, which are more directly related to our general theme of surfactant based hydrogels. However, the interested reader may here be referred to recent reviews that focus on this topic of polymeric hydrogels [117,118].

5. Micellar Systems, Microemulsions or Vesicles Cross-linked by Amphiphilic Polymers

As seen in the chapter before amphiphilic copolymers can self-assemble into network gels by themselves, but such self-assembly can substantially be altered and strengthened by the presence of surfactant. This means that one may cross-link micelles or microemulsion droplets (which for simplicity one may consider as micelles swollen with a hydrophobic compound) by the amphiphilic copolymer and similarly vesicles may become cross-linked by such polymers. It might be mentioned that our review here is not complete with respect to the options existing for using mixtures of surfactant and polymer to achieve gel type systems, as that can also be achieved by mixtures of polyelectrolyes and surfactants as they have been reviewed recently [119,120]. A well-established case for such a system is hydroxyethyl cellulose (HEC), which can be cationically and hydrophobicially modified and by combining with oppositely charged anionic surfactants one can form highly viscous systems, which may exhibit gel-like behaviour already at very low concentrations (whereas the pure HEC and surfactant solutions are water viscous). At equimolar mixing of charges one may observe precipitation but upon approaching this two-phase region a very marked increase of viscosity by several orders of magnitude will take place [121], typically in a rather narrow concentration range. The formation of gels strongly depends on the constitution of the polymer and its concentration. Having oppositely charged groups present and also the presence of hydrophobic modification on the polymer reduces the concentration at which gelation is observed [122]. Without the electrostatic interaction no more pronounced interaction is observed and accordingly no gelation takes place.

5.1. *Surfactant Micelles Interacting with Amphiphilic Polymers*

A rather classical case of viscosity enhancement or gelation is observed in mixtures of surfactants with hydrophobically modified polymers, typically a water soluble polymer where alkyl chains are present as hydrophobic side-arms.

Non-ionic and ionic surfactants have a strong affinity for the hydrophobic domains formed by associative polymers. The zero shear viscosity as a function of the surfactant concentration has a maximum at a concentration close to the cmc of the pure surfactant solution. Assuming that the surfactant has no interaction with the hydrophilic chain, the addition of surfactant to a solution of HM-polymer of constant concentration results in the creation of mixed micelles. Their rheological properties arise mainly from two effects: (1) the lifetime of the stickers in the mixed micelles and (2) the number density of mixed micelles (cross-linking points) and hence the distance between them.

For aqueous solutions of telechelic polymers that are, as explained above, quasi-Maxwellian fluids, the changes on the network relaxation and the elastic modulus are easy to observe [123]. Higher surfactant concentrations increase the number density of the micelles in the system, which brings the micelles closer together, thereby enabling the polymer to form bridges. At the same time, there is a lower probability of having many stickers in the same micelle. The net effect of both opposing factors is a peak in the modulus as a function of surfactant concentration. At high polymer concentrations, where loops are less probable, the addition of surfactant has a rather limited influence on the formation of bridges, resulting in an almost immediate drop of the modulus without appreciable prior increase [123].

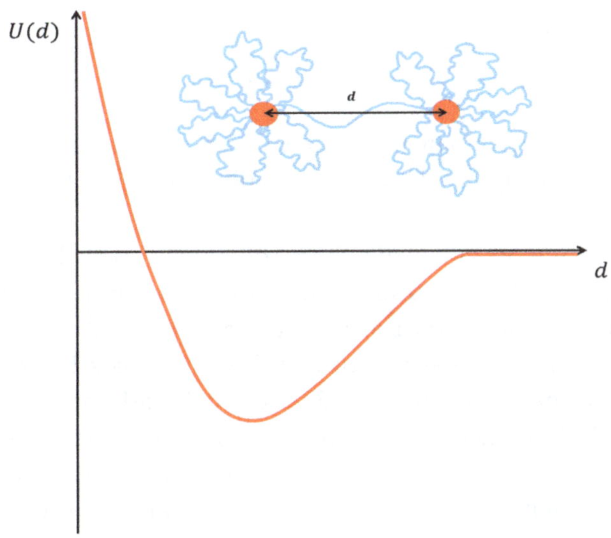

Figure 8. Schematic representation of the interaction potential between micelles (microemulsions) that are decorated and bridged by a telechelic polymer. The interaction has an effective attractive interaction between the micelles, due to the bridging, and a repulsive interaction, due to the steric repulsion between the micelles induced by the presence of the water soluble polymer that decorates the micelles.

Now let's consider the other case, where telechelic polymer is added to a surfactant solution of constant concentration with already formed micelles. In this case, the addition of polymer leads to the decoration followed by the subsequent interconnection of micelles with the corresponding increase in viscosity and elastic modulus. In terms of the structure, subsequent addition of telechelic polymer to a micellar solution doesn't significantly change its structure but induces repulsive interactions between them [124], as well as attractive ones due to the capacity for bridging micelles. A typical curve for the potential energy between two micelles (or similarly microemulsion droplets) in the presence of telechelic polymer is given in Figure 8.

In the case of HM-grafted polymers, the viscosity and the relaxation time undergo a maximum as a function of the added surfactant concentration. Compared to the case of regular telechelic polymers, the increase of the number density of micelles results in fewer cross-links and the corresponding decrease in the elasticity. Therefore, the non-monotonic variation of the viscosity arises from variations in the residence time in mixed micelles and will vary depending on the nature of the surfactant—it's length and head group. The concentration at the viscosity maximum correlates with the CMC of the surfactant [125,126] and increases substantially with increasing concentration of the hydrophobically modified polymer, as shown in Figure 9 for the case of hydrophobically modified polyacrylamide (HMPAM) with sodium dodecyl sulfate (SDS).

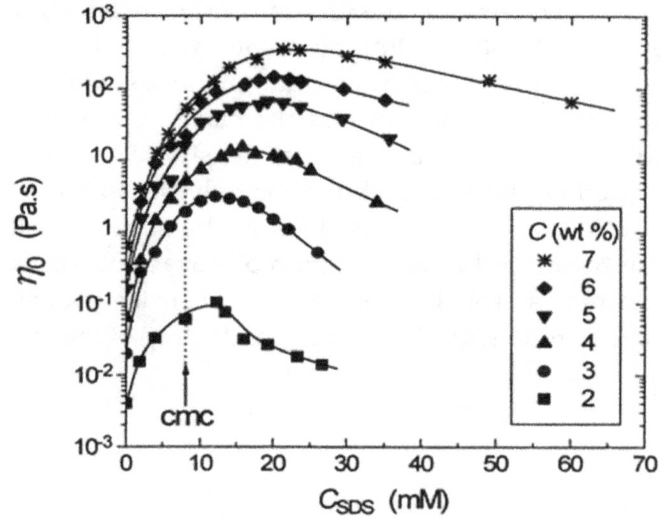

Figure 9. Effect of sodium dodecyl sulfate (SDS) concentration on the zero-shear viscosity of aqueous hydrophobically modified polyacrylamide (HMPAM) solutions of different concentration C [126].

However, micelles may have other shapes than spherical. The addition of telechelic polymers to worm-like micelles also leads to additional interconnection between the micelles. In this case, worm like micelles already have viscoelastic behaviour arising from the micelle´s interconnection. The addition of telechelic polymer results in a viscoelastic fluid with two coupled networks characterized by two relaxation times, one related to the worm-like micelle network and one related to the telechelic polymer bridging different micelles [127]. In such systems one observes two characteristic rheological relaxation times with two corresponding elastic moduli—to be described by two Maxwell fluids. The fast mode can be associated to the network of telechelic active chains that bridge two micelles, while the slow mode arises from the network of entangled wormlike micelles [128]. It has been demonstrated by SANS that, in a similar way as for spherical micelles, the telechelic polymer induces both an effective attractive interaction between the micelles, due to the bridging of the micelles, and a repulsive interaction, due to the steric repulsion between the micelles induced by the presence of the water soluble polymer that decorates the micelles [129].

5.2. Microemulsions Interacting with Polymers

Microemulsions (ME) are homogeneous, thermodynamically stable and finely dispersed mixtures of oil and water stabilized by a surfactant film [130]. MEs may occur in the form of oil-in-water

(O/W) and water-in-oil (W/O) droplets, or as bicontinuous structures [131]. O/W microemulsions are attractive formulations for encapsulation of hydrophobic active agents, substrates, or enzymes in aqueous environments. Dilute microemulsions have the viscosity of its continuous component (or the average of the both for the case of bicontinuous systems) [132] irrespective of their structure. An effective way to enhance the viscosity of droplet microemulsions is by the addition of telechelic polymers [133–136], which is interesting as for many applications of microemulsions an enhanced viscosity can be desirable. An example for the remarkable increase of viscosity that can be achieved by addition of telechelic, bridging polymers is shown in Figure 10. Here one sees an increase by about 4 orders of magnitude that takes place in a rather narrow range of polymer addition (where apparently the polymer concentration is sufficient to connect all the microemulsion droplets, which here is the case for having about 7-8 hydrophobic stearyl stickers per microemulsion droplet).

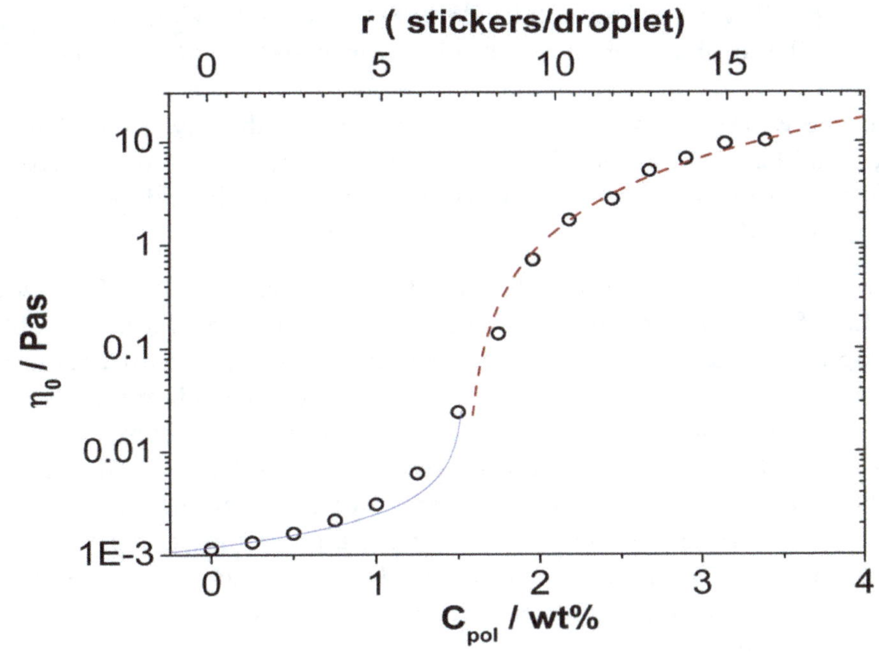

Figure 10. Zero-shear viscosity η_0 at 25 °C of the mixtures of a microemulsion (100 mM TDMAO/35 mM decane in water) as a function of the concentration of C_{18}-EO_{150}-C_{18} measured with a capillary viscometer until a concentration of 2 wt % and with the instrument AR-G2 above this concentration. Solid line: $\eta_0 = 0.0016((1.54 - c)/\text{wt}\,\%)^{-0.7}$. Dashed line: $\eta_0 = 3.6((c - 1.54)/\text{wt}\,\%)^{1.7}$ ([136]—Published by The Royal Society of Chemistry).

Another good reason for studying microemulsion networks is that they are good examples of model transient networks. The cross-linking points are the microemulsion droplets with typically very low polydispersity, that are located at a distance given by their number density and the telechelic polymer that connects them has a controlled length given by its molecular weight, persistence length and chain conformation. The stickers of the polymer solubilise into the microemulsion droplets uniformly [137]. If the drops are further apart than the chain length, the polymer forms loops with two ends localized in the same droplet. When the drops are closer than the end-to-end distance of the polymer, the polymer forms bridges between two droplets. The formation of bridges leads to the formation of clusters of droplets and, above the polymer percolation concentration, an infinite network of droplets spans the entire volume leading to a significant increase of the viscosity [136] (Figure 7C).

Experimental results with small-angle neutron and X-ray scattering (SANS/SAXS) show that the structure of the microemulsion droplets in terms of shape and size is not affected by the addition of telechelic polymer. However, the polymer changes the interaction between the droplets [133–136]. The interaction has three contributions (see also Figure 8):

(1) The interaction between the droplets without polymer (excluded volume [135] or Yukawa repulsion for charged surfactants [138]);
(2) an entropic attraction induced by the bridging polymer [139];
(3) a soft repulsion caused by the self-excluding polymer chains between the droplets [92,140].

Depending on the relative importance of these contributions, the net interaction is attractive or repulsive [141].

The system exhibits a phase separation between a fluid sol phase and a polymer rich network phase when the net interaction is attractive enough [75]. The attractive interaction that leads to phase separation has a purely entropic origin since the increase in polymer configurations overcomes the entropy loss due to the phase separation and the formation of a dense phase [139]. Thus, it only depends on the relative length of the polymer compared to the separation between the droplets and not on the sticker length. It was experimentally demonstrated that the end-group does not influence the phase behaviour [75], but phase separation can be suppressed by introducing additional repulsive interactions [138].

Microemulsion networks exhibit viscoelastic behaviour with only one characteristic relaxation process described by a Maxwell model [134,142]. The relaxation time of the network is given by the residence time of the end group in the microemulsion droplet and, thus, depends on the length of the hydrophobic end-group [143].

The effect of the polymer architecture on the structure and dynamics of ME networks has also been studied [141,143]. Low functionality telechelic star polymers with 3 and 4 arms are able to interconnect microemulsion droplets. Neutron scattering experiments show that the attraction induced between the drops is larger with higher functionality polymers, which leads to a larger phase separation area in the polymer concentration-droplet concentration space. The repulsive component of the interaction potential, however only depends on the volume fraction of the hydrophilic chains. This leads to the observation that the local structure of the microemulsion networks is the same for the same polymer concentration, regardless of its functionality. Linear rheology experiments show that below the percolation threshold the viscosity is more influenced by the volume fraction of the created clusters. Above the percolation concentration higher polymer functionalities lead to a higher connectivity and, thus a higher elastic modulus.

In summary, it can be stated that the addition of hydrophobically modified polymers (especially telechelic ones) is a very attractive way of controlling the viscosity and gelation properties of otherwise low viscous microemulsions. The obtained rheological properties are controlled via the length of the hydrophobic sticker, the ratio of stickers per microemulsion droplets, the length of the hydrophilic chain, and the architecture of the bridging polymer.

5.3. Vesicles Interacting with Polymers

There is less work done on the fundamental properties on vesicles compared to microemulsions with added HM polymers. Such hydrophobically modified polymers can interact with vesicles and interconnect them within vesicle solutions [144–148] (Figure 11A). However, the vesicle/polymer systems are more complicated because, unlike in the case of microemulsions and micelles, the structure of the vesicles do not necessarily remain intact upon the addition of hydrophobically modified polymers. Anchored polymers in vesicle membranes (see Figure 11B) can have two effects: (1) change the curvature; (2) change the membrane gel to fluid transition temperature. In any case, the addition of hydrophobically modified linear polymers to vesicle solutions leads to the formation of networks. These networks have higher elasticity with higher polymer concentration [144–146]. Here again, it is necessary that the hydrophilic part of the polymer is longer than the distance between vesicle membranes to form bridges.

Grafted polymers such as HM-chitosan also form vesicle networks [147,148]. This system is particularly advantageous because of its biocompatibility.

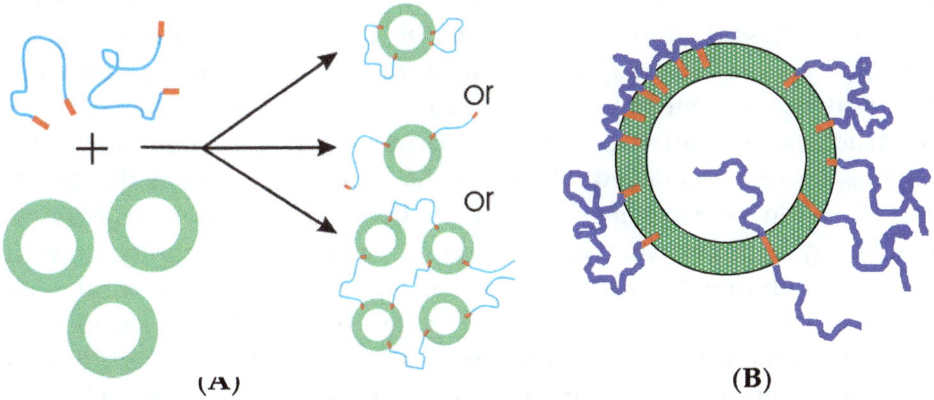

Figure 11. Schematic representation of the (**A**) association of telechelic polymers and vesicles leading to the formation of decorated vesicles and vesicle networks; and (**B**) anchoring of hydrophobically modified polymers of different architectures to vesicle membranes.

A very interesting case has been studied with catanionic vesicles composed of sodium dodecyl sulfate (SDS)/didodecyldimethylammonium bromide (DDAB) or sodium dodecylbenzenesulfonate (SDBS)/cetyltrimethylammonium tosylate (CTAT) in mixtures with hydrophobically modified sodium polyacrylate (hm-NaPA). When the vesicles are positively charged (as controlled by the SDS/DDAB or SDBS/CTAT ratio) they will form precipitates with the oppositely charged hm-NaPA but for the anionically charged vesicles one finds the formation of a gel already at as low concentrations of 1.25 wt % surfactant and 0.4 wt % hm-NaPA [149]. A similar behaviour has been reported for SDBS/CTAT vesicles when combined with hydrophobically modified (with *n*-dodecyl chains) chitosan, hm-chitosan, but interestingly no gelation was observed when adding the hm-chitosan to wormlike CTAT micelles. Here only an increase of viscosity and shift to a longer structural relaxation time takes place [147]. Gelation at higher polymer and surfactant concentration (but still in the 1 wt % regime) also has been reported for catanionic vesicles in the SDS/DDAB system upon admixing cationically modified cellulose (JR400 or LM200) [150]. Further studies then demonstrated that the hydrophobic modification present (a dodecyl chain employed for quaternization of the amine at the cellulose) in LM200 results in a more pronounced change of the rheological properties and the formation of more and more long-lived cross-links [144–146].

This principle of gelation was then also extended to phospholipids such as dipalmitoylphosphatidyl-choline (DPPC) and here similar gels with hm-chitosan can be formed which could be formulated as an injectable system for drug delivery as demonstrated with doxorubicin as model drug [151]. This type of vesicle/hm-chitosan gel then could also be transformed into a photoresponsive system by substituting the cationic surfactant by p-octyloxydiphenyl-iodonium hexafluoroantimonate (ODPI). ODPI can be regarded as a cationic surfactant but upon illumination with UV-light becomes decomposed into uncharged hydrophobic products, which leads to a transformation of the initially present vesicles into micelles, which macroscopically is seen as gel to sol transition [152].

6. Conclusions

This review discusses the structure and mechanical properties of gels formed by self-assembly of amphiphilic molecules at low and moderate concentration, i.e., well below dense packing of the molecules. There are several ways of achieving this. One way is by increasing the effective volume fraction via the enclosure of large volumes of solvent (the case of vesicles) or through the formation of supramolecular structures (like worm-like micelles) and network formation. The viscoelastic properties of these materials are given by the structure and the lifetime of the network bonds. In case of densely packed vesicles that time would be given by the cage opening of neighbouring vesicles. This time is typically much longer than the experimental window and thus effectively the system behaves as an elastic solid with a yield stress. The structural properties of worm-like micelle solutions are described by the micelle´s length, persistence length and concentration. Worm-like micelles have two relaxing mechanisms: the breaking and reforming of the micelles and the entanglement of cross-links, both determining the structural relaxation time τ_{str}. In principle, worm-like micelles yield viscoelastic solutions, but once τ_{str} becomes sufficiently large one has effectively a gel as the flow becomes so slow that it is not observed within the experimental window. In the case of networks of associative polymers the number density of hydrophobic domains (polymeric or surfactant or microemulsions), the length and functionality of the polymer determine the structure, whereas the structural relaxation is mainly due to the residence time of the stickers in the hydrophobic domains (in the absence of entanglements). Especially the case of microemulsions is interesting as they are able to solubilise large amounts of active agents but for applications often a much enhanced viscosity is required as it can be obtained by the combination with amphiphilic polymers.

Great advances in the synthesis of polymers and surfactants with controlled architecture and chemistry have been making the field moving fast towards systems of higher complexity and stimuli-responsiveness, where typically on has a response to changes of pH, temperature, or ionic strength. Mostly such systems work by interconnecting hydrophobic domains in aqueous solution, where the interaction is mostly of hydrophobic or electrostatic nature but also the chain entropy of the polymers may play a role. Stimuli-responsive systems are particularly interesting due to the possibility of switching from a liquid state to a gel state upon an external trigger, the same way nature does to create biological function. Especially here further future research advancements are to be expected.

Therefore, in summary, it can be stated that it is possible to control the rheological properties of soft matter systems largely via the principles of self-assembly and this pertains also to the situation of forming practical gels, as they are often required for many practical formulations.

References

1. Evans, F.D.; Wennerstöm, H. *The Colloidal Domain: Where Physics, Chemistry, Biology, and Technology Meet (Advances in Interfacial Engineering)*; Wiley-VCH: Weinheim, Germany, 1996.
2. Tiddy, G. Surfactant-water liquid crystal phases. *Phys. Rep.* **1980**, *57*, 1–46. [CrossRef]
3. Laughlin, R.G. *The Aqueous Phase Behavior of Surfactants*; Academic Press: London, UK, 1994; Volume 6.
4. Estroff, L.A.; Hamilton, A.D. Water Gelation by Small Organic Molecules. *Chem. Rev.* **2004**, *104*, 1201–1218. [CrossRef] [PubMed]
5. Torchilin, V.; Weissig, V. *Liposomes: A Practical Approach*; Oxford University Press: Oxford, UK, 2003.
6. International Union of Pure and Applied Chemistry. *Compendium of Chemical Terminology Gold Book*, 2nd ed.; International Union of Pure and Applied Chemistry: Research Triangle Park, NC, USA, 2014.
7. Møller, P.C.F.; Mewis, J.; Bonn, D. Yield stress and thixotropy: On the difficulty of measuring yield stresses in practice. *Soft Matter* **2006**, *2*, 274–283. [CrossRef]
8. Rehage, H.; Hoffmann, H. Viscoelastic surfactant solutions: Model systems for rheological research. *Mol. Phys.* **1991**, *74*, 933–973. [CrossRef]
9. Zana, R.; Kaler, E.W. *Giant Micelles: Properties and Applications*; CRC Press: Boca Raton, FL, USA, 2007; Volume 140.
10. Hoffmann, H. Viscoelastic Surfactant Solutions. In *ACS Symposium Series, Vol. 578*; ACS Publications: Washington, DC, USA, 1994; pp. 2–31.
11. Cates, M.E. Dynamics of living polymers and flexible surfactant micelles: Scaling laws for dilution. *J. Phys.* **1988**, *49*, 1593–1600. [CrossRef]
12. Raghavan, S.R.; Douglas, J.F. The conundrum of gel formation by molecular nanofibers, wormlike micelles, and filamentous proteins: Gelation without cross-links? *Soft Matter* **2012**, *8*, 8539–8546. [CrossRef]
13. Gradzielski, M. Vesicles and vesicle gels—Structure and dynamics of formation. *J. Phys. Condens. Matter* **2003**, *15*, R655–R697. [CrossRef]

14. Hoffmann, H.; Ulbricht, W. Surfactant gels. *Curr. Opin. Colloid Interface Sci.* **1996**, *1*, 726–739. [CrossRef]
15. Schmolka, I.R. A comparison of block copolymer surfactant gels. *J. Am. Oil Chem. Soc.* **1991**, *68*, 206–209. [CrossRef]
16. Mortensen, K.; Brown, W.; Nordén, B. Inverse melting transition and evidence of three-dimensional cubatic structure in a block-copolymer micellar system. *Phys. Rev. Lett.* **1992**, *68*, 2340–2343. [CrossRef] [PubMed]
17. Hecht, E.; Hoffmann, H. Interaction of ABA block copolymers with ionic surfactants in aqueous solution. *Langmuir* **1994**, *10*, 86–91. [CrossRef]
18. Escobar-Chávez, J.J.; López-Cervantes, M.; Naik, A.; Kalia, Y.; Quintanar-Guerrero, D.; Ganem-Quintanar, A. Applications of thermo-reversible pluronic F-127 gels in pharmaceutical formulations. *J. Pharm. Pharm. Sci.* **2006**, *9*, 339–358. [PubMed]
19. Flory, P.J. Statistical Mechanics of Swelling of Network Structures. *J. Chem. Phys.* **1950**, *18*, 108–111. [CrossRef]
20. Tanford, C. *The Hydrophobic Effect: Formation of Micelles and Biological Membranes*, 2nd ed.; Wiley: Hoboken, NJ, USA, 1980.
21. Raghavan, S.R. Distinct character of surfactant gels: A smooth progression from micelles to fibrillar networks. *Langmuir* **2009**, *25*, 8382–8385. [CrossRef] [PubMed]
22. Berret, J.-F. Rheology of wormlike micelles: Equilibrium properties and shear banding transitions. In *Molecular Gels*; Springer: Berlin/Heidelberg, Germany, 2006; pp. 667–720.
23. Imae, T.; Kamiya, R.; Ikeda, S. Electron microscopic observation of rod-like micelles of dimethyloleylamine oxide regenerated from its aqueous solutions. *J. Colloid Interface Sci.* **1984**, *99*, 300–301. [CrossRef]
24. Sakaiguchi, Y.; Shikata, T.; Urakami, H.; Tamura, A.; Hirata, H. Electron microscope study of viscoelastic cationic surfactant systems. *Colloid Polym. Sci.* **1987**, *265*, 750–753. [CrossRef]
25. Ziserman, L.; Abezgauz, L.; Ramon, O.; Raghavan, S.R.; Danino, D. Origins of the viscosity peak in wormlike micellar solutions. 1. Mixed catanionic surfactants. A cryo-transmission electron microscopy study. *Langmuir* **2009**, *25*, 10483–10489. [CrossRef] [PubMed]
26. Oelschlaeger, C.; Schopferer, M.; Scheffold, F.; Willenbacher, N. Linear-to-branched micelles transition: A rheometry and diffusing wave spectroscopy (DWS) study. *Langmuir* **2008**, *25*, 716–723. [CrossRef] [PubMed]
27. Clausen, T.M.; Vinson, P.K.; Minter, J.R.; Davis, H.T.; Talmon, Y.; Miller, W.G. Viscoelastic micellar solutions: Microscopy and rheology. *J. Phys. Chem.* **1992**, *96*, 474–484. [CrossRef]
28. Israelachvili, J.N.; Mitchell, D.J.; Ninham, B.W. Theory of self-assembly of hydrocarbon amphiphiles into micelles and bilayers. *J. Chem. Soc. Faraday Trans. 2 Mol. Chem. Phys.* **1976**, *72*, 1525–1568. [CrossRef]
29. Mu, J.-H.; Li, G.-Z.; Jia, X.-L.; Wang, H.-X.; Zhang, G.-Y. Rheological Properties and Microstructures of Anionic Micellar Solutions in the Presence of Different Inorganic Salts. *J. Phys. Chem.* **2002**, *106*, 11685–11693. [CrossRef]
30. Chu, Z.; Feng, Y. Thermo-switchable surfactant gel. *Chem. Commun.* **2011**, *47*, 7191–7193. [CrossRef] [PubMed]
31. Turner, M.S.; Cates, M.E. Linear viscoelasticity of wormlike micelles: A comparison of micellar reaction kinetics. *J. Phys. II* **1992**, *2*, 503–519. [CrossRef]
32. Cates, M.E.; Candau, S.J. Statics and dynamics of worm-like surfactant micelles. *J. Phys. Condens. Matter* **1990**, *2*, 6869–6892. [CrossRef]
33. Kumars, R.; Kalur, G.C.; Ziserman, L.; Danino, D.; Raghavan, S.R. Wormlike micelles of a C22-tailed zwitterionic betaine surfactant: From viscoelastic solutions to elastic gels. *Langmuir* **2007**, *23*, 12849–12856. [CrossRef] [PubMed]
34. Feng, Y.; Chu, Z. pH-Tunable wormlike micelles based on an ultra-long-chain "pseudo" gemini surfactant. *Soft Matter* **2015**, *11*, 4614–4620. [CrossRef] [PubMed]
35. Bernheim-Groswasser, A.; Zana, R.; Talmon, Y. Sphere-to-cylinder transition in aqueous micellar solution of a dimeric (gemini) surfactant. *J. Phys. Chem.* **2000**, *104*, 4005–4009. [CrossRef]
36. Pei, X.; Zhao, J.; Ye, Y.; You, Y.; Wei, X. Wormlike micelles and gels reinforced by hydrogen bonding in aqueous cationic gemini surfactant systems. *Soft Matter* **2011**, *7*, 2953–2960. [CrossRef]
37. Gradzielski, M. The rheology of vesicle and disk systems—Relations between macroscopic behaviour and microstructure. *Curr. Opin. Colloid Interface Sci.* **2011**, *16*, 13–17. [CrossRef]
38. Fontell, K.; Mandell, L.; Ekwall, P. Some isotropic mesophases in systems containing amphiphilic compounds. *Acta Chem. Scand* **1968**, *22*, 3209–3223. [CrossRef]

39. Gradzielski, M.; Bergmeier, M.; Müller, M.; Hoffmann, H. Novel Gel Phase: A Cubic Phase of Densely Packed Monodisperse, Unilamellar Vesicles. *J. Phys. Chem.* **1997**, *101*, 1719–1722. [CrossRef]
40. Gradzielski, M.; Müller, M.; Bergmeier, M.; Hoffmann, H.; Hoinkis, E. Structural and Macroscopic Characterization of a Gel Phase of Densely Packed Monodisperse, Unilamellar Vesicles. *J. Phys. Chem.* **1999**, *103*, 1416–1424. [CrossRef]
41. Gradzielski, M.; Bergmeier, M.; Hoffmann, H.; Müller, M.; Grillo, I. Vesicle Gel Formed by a Self-Organization Process. *J. Phys. Chem.* **2000**, *104*, 11594–11597. [CrossRef]
42. Gradzielski, M.; Grillo, I.; Narayanan, T. Morphological Transitions in Amphiphilic Systems Probed by Small-Angle Scattering Techniques. In *Self-Assembly*; Robinson, B.H., Ed.; IOS Press: Amsterdam, The Netherland, 2003; pp. 410–421.
43. Lasič, S.; Åslund, I.; Oppel, C.; Topgaard, D.; Söderman, O.; Gradzielski, M. Investigations of vesicle gels by pulsed and modulated gradient NMR diffusion techniques. *Soft Matter* **2011**, *7*, 3947–3955. [CrossRef]
44. Jeong, Y.; Uezu, K.; Kobayashi, M.; Sakurai, S.; Masunaga, H.; Inoue, K.; Sasaki, S.; Shimada, N.; Takeda, Y.; Kaneko, K.; et al. Complex made from tetrasodium *N, N-bis* (carboxylatomethyl) glutamate and sodium oleate that forms a highly ordered lamella in gel phase. *Bull. Chem. Soc. Jpn.* **2007**, *80*, 410–417. [CrossRef]
45. Oppel, C.; Prévost, S.; Noirez, L.; Gradzielski, M. The use of highly ordered vesicle gels as template for the formation of silica gels. *Langmuir* **2011**, *27*, 8885–8897. [CrossRef] [PubMed]
46. Menger, F.M.; Peresypkin, A. V Strings of vesicles: Flow behavior in an unusual type of aqueous gel. *J. Am. Chem. Soc.* **2003**, *125*, 5340–5345. [CrossRef] [PubMed]
47. Seth, M.; Ramachandran, A.; Murch, B.P.; Leal, L.G. Origins of microstructural transformations in charged vesicle suspensions: The crowding hypothesis. *Langmuir* **2014**, *30*, 10176–10187. [CrossRef] [PubMed]
48. Diec, K.H.; Sokolowski, T.; Wittern, K.P.; Schreiber, J.; Meier, W. New liposome gels by self organization of vesicles and intelligent polymers. *Cosmet. Toilet.* **2002**, *117*, 55–62.
49. Hoffmann, H.; Thunig, C.; Schmiedel, P.; Munkert, U. Gels from surfactant solutions with densely packed multilamellar vesicles. *Faraday Discuss.* **1995**, *101*, 319–333. [CrossRef]
50. Hoffmann, H.; Thunig, C.; Schmiedel, P.; Munkert, U. Complex fluids with a yield value; their microstructures and rheological properties. *Nuovo Cim. D* **1994**, *16*, 1373–1390. [CrossRef]
51. Bergmeier, M.; Gradzielski, M.; Hoffmann, H.; Mortensen, K. Behavior of Ionically Charged Lamellar Systems under the Influence of a Shear Field. *J. Phys. Chem.* **1999**, *103*, 1605–1617. [CrossRef]
52. Diat, O.; Roux, D. Preparation of monodisperse multilayer vesicles of controlled size and high encapsulation ratio. *J. Phys.* **1993**, *3*, 9–14. [CrossRef]
53. Long, P.; Hao, J. A gel state from densely packed multilamellar vesicles in the crystalline state. *Soft Matter* **2010**, *6*, 4350–4356. [CrossRef]
54. Hufnagl, A.; Kinzel, S.; Gradzielski, M. Vesicles and Vesicle Gels—Structure and Solubilisation Properties. *Tenside Surfactants Deterg.* **2007**, *44*, 110–115. [CrossRef]
55. Grewe, F.; Ortmeyer, J.; Haase, R.; Schmidt, C. Colloidal Gels Formed by Dilute Aqueous Dispersions of Surfactant and Fatty Alcohol. In *Colloid Process Engineering*; Springer: Berlin/Heidelberg, Germany, 2015; pp. 21–43.
56. Cheng, C.-Y.; Wang, T.-Y.; Tung, S.-H. Biological Hydrogels Formed by Swollen Multilamellar Liposomes. *Langmuir* **2015**, *31*, 13312–13320. [CrossRef] [PubMed]
57. Lauger, J.; Linemann, R.; Richtering, W. Shear orientation of a lamellar lyotropic liquid crystal. *Rheol. Acta* **1995**, *34*, 132–136. [CrossRef]
58. Kinzel, S.; Gradzielski, M. Control of phase behavior and properties of vesicle gels by admixing ionic surfactants to the nonionic surfactant brij 30. *Langmuir* **2008**, *24*, 10123–10132. [CrossRef] [PubMed]
59. Zou, A.; Hoffmann, H.; Freiberger, N.; Glatter, O. Influence of ionic charges on the bilayers of lamellar phases. *Langmuir* **2007**, *23*, 2977–2984. [CrossRef] [PubMed]
60. Dong, R.; Zhong, Z.; Hao, J. Self-assembly of onion-like vesicles induced by charge and rheological properties in anionic–nonionic surfactant solutions. *Soft Matter* **2012**, *8*, 7812–7821. [CrossRef]
61. Dong, R.; Wu, J.; Dong, S.; Song, S.; Tian, F.; Hao, J. Interconvertible Self-Assembly and Rheological Properties of Planar Bilayers and Vesicle Gels in Anionic/Nonionic (CF/CH) Surfactant Solutions. *Chem. Asian J.* **2013**, *8*, 1863–1872. [CrossRef] [PubMed]

62. Wang, D.; Wei, G.; Dong, R.; Hao, J. Multiresponsive viscoelastic vesicle gels of nonionic $C_{12}EO_4$ and anionic AzoNa. *Chem. A Eur. J.* **2013**, *19*, 8253–8260. [CrossRef] [PubMed]
63. Wang, D.; Hao, J. Multiple-stimulus-responsive hydrogels of cationic surfactants and azoic salt mixtures. *Colloid Polym. Sci.* **2013**, *291*, 2935–2946. [CrossRef]
64. Kaler, E.W.; Murthy, A.K.; Rodriguez, B.E.; Zasadzinski, J.A.N. Spontaneous vesicle formation in aqueous mixtures of single-tailed surfactants. *Science* **1989**, *245*, 1371–1375. [CrossRef] [PubMed]
65. Abdel-Rahem, R.; Hoffmann, H. Novel viscoelastic systems from cationic surfactants and hydrophobic counter-ions: Influence of surfactant chain length. *J. Colloid Interface Sci.* **2007**, *312*, 146–155. [CrossRef] [PubMed]
66. Abdel-Rahem, R.; Gradzielski, M.; Hoffmann, H. A novel viscoelastic system from a cationic surfactant and a hydrophobic counterion. *J. Colloid Interface Sci.* **2005**, *288*, 570–582. [CrossRef] [PubMed]
67. Horbaschek, K.; Hoffmann, H.; Thunig, C. Formation and properties of lamellar phases in systems of cationic surfactants and hydroxy-naphthoate. *J. Colloid Interface Sci.* **1998**, *206*, 439–456. [CrossRef] [PubMed]
68. Coldren, B.A.; Warriner, H.; van Zanten, R.; Zasadzinski, J.A.; Sirota, E.B. Lamellar gels and spontaneous vesicles in catanionic surfactant mixtures. *Langmuir* **2006**, *22*, 2465–2473. [CrossRef] [PubMed]
69. Laflèche, F.; Durand, D.; Nicolai, T. Association of Adhesive Spheres Formed by Hydrophobically End-Capped PEO. 1. Influence of the Presence of Single End-Capped PEO. *Macromolecules* **2003**, *36*, 1331–1340.
70. Séréro, Y.; Aznar, R.; Porte, G.; Berret, J.-F.; Calvet, D.; Collet, A.; Viguier, M. Associating Polymers: From "Flowers" to Transient Networks. *Phys. Rev. Lett.* **1998**, *81*, 5584–5587. [CrossRef]
71. Berret, J.-F.; Séréro, Y.; Winkelman, B.; Calvet, D.; Collet, A.; Viguier, M. Nonlinear rheology of telechelic polymer networks. *J. Rheol.* **2001**, *45*, 477–492. [CrossRef]
72. Alami, E.; Almgren, M.; Brown, W.; François, J. Aggregation of Hydrophobically End-Capped Poly(ethylene oxide) in Aqueous Solutions. Fluorescence and Light-Scattering Studies. *Macromolecules* **1996**, *29*, 2229–2243. [CrossRef]
73. Cram, S.L.; Brown, H.R.; Spinks, G.M.; Hourdet, D.; Creton, C. Hydrophobically Modified Dimethylacrylamide Synthesis and Rheol. l Behavior. *Macromolecules* **2005**, *38*, 2981–2989. [CrossRef]
74. Herfurth, C.; Malo de Molina, P.; Wieland, C.; Rogers, S.; Gradzielski, M.; Laschewsky, A. One-step RAFT synthesis of well-defined amphiphilic star polymers and their self-assembly in aqueous solution. *Polym. Chem.* **2012**, *3*, 1606–1617. [CrossRef]
75. Filali, M.; Aznar, R.; Svenson, M.; Porte, G.; Appell, J. Swollen Micelles Plus Hydrophobically Modified Hydrosoluble Polymers in Aqueous Solutions: Decoration versus Bridging. A Small Angle Neutron Scattering Study. *J. Phys. Chem.* **1999**, *103*, 7293–7301.
76. Rufier, C.; Collet, A.; Viguier, M.; Oberdisse, J.; Mora, S. Asymmetric End-Capped Poly(ethylene oxide). Synthesis and Rheological Behavior in Aqueous Solution. *Macromolecules* **2008**, *41*, 5854–5862.
77. Hietala, S.; Mononen, P.; Strandman, S.; Järvi, P.; Torkkeli, M.; Jankova, K.; Hvilsted, S.; Tenhu, H. Synthesis and rheological properties of an associative star polymer in aqueous solutions. *Polymer* **2007**, *48*, 4087–4096. [CrossRef]
78. Taribagil, R.R.; Hillmyer, M.A.; Lodge, T.P. Hydrogels from ABA and ABC triblock polymers. *Macromolecules* **2010**, *43*, 5396–5404. [CrossRef]
79. Chassenieux, C.; Nicolai, T.; Benyahia, L. Rheology of associative polymer solutions. *Curr. Opin. Colloid Interface Sci.* **2011**, *16*, 18–26. [CrossRef]
80. Pham, Q.T.; Russel, W.B. Micellar Solutions of Associative Triblock Copolymers: The Relationship between Structure and Rheology. *Langmuir* **1999**, 5139–5146. [CrossRef]
81. Yekta, A.; Duhamel, J.; Brochard, P. A fluorescent probe study of micelle-like cluster formation in aqueous solutions of hydrophobically modified poly (ethylene oxide). *Macromolecules* **1993**, *26*, 1829–1836. [CrossRef]
82. Meng, X.X.; Russel, W.B. Structure and size of spherical micelles of telechelic polymers. *Macromolecules* **2005**, *38*, 593–600. [CrossRef]
83. Chassenieux, C.; Nicolai, T.; Durand, D. Association of Hydrophobically End-Capped Poly(ethylene oxide). *Macromolecules* **1997**, *30*, 4952–4958. [CrossRef]
84. Lundberg, D.J.; Brown, R.G.; Glass, J.E.; Eley, R.R. Synthesis, Characterization, and Solution Rheology of Model Hydrophobically-Modified, Water-Soluble Ethoxylated Urethanes. *Langmuir* **1994**, *10*, 3027–3034. [CrossRef]

85. Annable, T. The rheology of solutions of associating polymers: Comparison of experimental behavior with transient network theory. *J. Rheol.* **1993**, *37*, 695–726. [CrossRef]
86. Winnik, M.A.; Yekta, A. Associative polymers in aqueous solution. *Curr. Opin. Colloid Interface Sci.* **1997**, *2*, 424–436. [CrossRef]
87. Zhong, M.; Wang, R.; Kawamoto, K.; Olsen, B.D.; Johnson, J.A. Quantifying the impact of molecular defects on polymer network elasticity. *Science* **2016**, *353*, 1264–1268. [CrossRef] [PubMed]
88. Ma, S.X.; Cooper, S.L. Shear Thickening in Aqueous Solutions of Hydrocarbon End-Capped Poly (ethylene oxide). *Macromolecules* **2000**, *34*, 3294–3301. [CrossRef]
89. Pellens, L.; Gamez Corrales, R.; Mewis, J. General nonlinear rheological behavior of associative polymers. *J. Rheol.* **2004**, *48*, 379–393. [CrossRef]
90. Dozier, W.D.; Huang, J.S.; Fetters, L.J. Colloidal nature of star polymer dilute and semidilute solutions. *Macromolecules* **1991**, *24*, 2810–2814. [CrossRef]
91. Rufier, C.; Collet, A.; Viguier, M.; Oberdisse, J.; Mora, S. Influence of Surfactants on Hydrophobically End-Capped Poly(ethylene oxide) Self-Assembled Aggregates Studied by SANS. *Macromolecules* **2011**, *44*, 7451–7459. [CrossRef]
92. Semenov, A.N.; Joanny, J.F.; Khokhlov, A.R. Associating polymers: Equilibrium and linear viscoelasticity. *Macromolecules* **1995**, *28*, 1066–1075. [CrossRef]
93. Rubinstein, M.; Semenov, A.N. Dynamics of entangled solutions of associating polymers. *Macromolecules* **2001**, *34*, 1058–1068. [CrossRef]
94. Regalado, E.J.; Selb, J.; Candau, F. Viscoelastic behavior of semidilute solutions of multisticker polymer chains. *Macromolecules* **1999**, *32*, 8580–8588. [CrossRef]
95. Lin, H.-H.; Cheng, Y.-L. In-Situ Thermoreversible Gelation of Block and Star Copolymers of Poly(ethylene glycol) and Poly(*N*-isopropylacrylamide) of Varying Architectures. *Macromolecules* **2001**, *34*, 3710–3715. [CrossRef]
96. Park, S.Y.; Han, D.K.; Kim, S.C. Synthesis and Characterization of Star-Shaped PLLA−PEO Block Copolymers with Temperature-Sensitive Sol−Gel Transition Behavior. *Macromolecules* **2001**, *34*, 8821–8824. [CrossRef]
97. Nagahama, K.; Ouchi, T.; Ohya, Y. Temperature-Induced Hydrogels Through Self-Assembly of Cholesterol-Substituted Star PEG-*b*-PLLA Copolymers: An Injectable Scaffold for Tissue Engineering. *Adv. Funct. Mater.* **2008**, *18*, 1220–1231. [CrossRef]
98. Burnworth, M.; Tang, L.; Kumpfer, J.R.; Duncan, A.J.; Beyer, F.L.; Fiore, G.L.; Rowan, S.J.; Weder, C. Optically healable supramolecular polymers. *Nature* **2011**, *472*, 334–337. [CrossRef] [PubMed]
99. Holten-Andersen, N.; Harrington, M.J.; Birkedal, H.; Lee, B.P.; Messersmith, P.B.; Lee, K.Y.C.; Waite, J.H. pH-induced metal-ligand cross-links inspired by mussel yield self-healing polymer networks with near-covalent elastic moduli. *Proc. Natl. Acad. Sci. USA* **2011**, *108*, 2651–2655. [CrossRef] [PubMed]
100. Tang, S.; Habicht, A.; Li, S.; Seiffert, S.; Olsen, B.D. Self-Diffusion of Associating Star-Shaped Polymers. *Macromolecules* **2016**, *49*, 5599–5608. [CrossRef]
101. Guo, M.; Pitet, L.M.; Wyss, H.M.; Vos, M.; Dankers, P.Y.W.; Meijer, E.W. Tough stimuli-responsive supramolecular hydrogels with hydrogen-bonding network junctions. *J. Am. Chem. Soc.* **2014**, *136*, 6969–6977. [CrossRef] [PubMed]
102. Appel, E.A.; Forster, R.A.; Koutsioubas, A.; Toprakcioglu, C.; Scherman, O.A. Activation energies control the macroscopic properties of physically cross-linked materials. *Angew. Chem. Int. Ed.* **2014**, *53*, 10038–10043. [CrossRef] [PubMed]
103. Dürrschmidt, T.; Hoffmann, H. Organogels from ABA triblock copolymers. *Colloid Polym. Sci.* **2001**, *279*, 1005–1012. [CrossRef]
104. Monge, S.; Joly-Duhamel, C.; Boyer, C.; Robin, J.-J. Synthesis and Characterisation of Organogels from ABA Triblock Copolymers. *Macromol. Chem. Phys.* **2007**, *208*, 262–270. [CrossRef]
105. Meier, W.; Falk, A.; Odenwald, M.; Stieber, F. Microemulsion elastomers. *Colloid Polym. Sci.* **1996**, *274*, 218–226. [CrossRef]
106. Blochowicz, T.; Gögelein, C.; Spehr, T.; Müller, M.; Stühn, B. Polymer-induced transient networks in water-in-oil microemulsions studied by small-angle X-ray and dynamic light scattering. *Phys. Rev. E Stat. Nonlinear Soft Matter Phys.* **2007**, *76*, 1–9. [CrossRef] [PubMed]

107. Schmaljohann, D. Thermo-and pH-responsive polymers in drug delivery. *Adv. Drug Deliv. Rev.* **2006**, *58*, 1655–1670. [CrossRef] [PubMed]
108. Ma, Y.; Tang, Y.; Billingham, N.C.; Armes, S.P.; Lewis, A.L. Synthesis of biocompatible, stimuli-responsive, physical gels based on ABA triblock copolymers. *Biomacromolecules* **2003**, *4*, 864–868. [CrossRef] [PubMed]
109. Li, C.; Tang, Y.; Armes, S.P.; Morris, C.J.; Rose, S.F.; Lloyd, A.W.; Lewis, A.L. Synthesis and characterization of biocompatible thermo-responsive gelators based on ABA triblock copolymers. *Biomacromolecules* **2005**, *6*, 994–999. [CrossRef] [PubMed]
110. Li, C.; Madsen, J.; Armes, S.P.; Lewis, A.L. A New Class of Biochemically Degradable, Stimulus-Responsive Triblock Copolymer Gelators. *Angew. Chem. Int. Ed.* **2006**, *45*, 3510–3513. [CrossRef] [PubMed]
111. Vogt, A.P.; Sumerlin, B.S. Temperature and redox responsive hydrogels from ABA triblock copolymers prepared by RAFT polymerization. *Soft Matter* **2009**, *5*, 2347–2351. [CrossRef]
112. Yu, L.; Chang, G.T.; Zhang, H.; Ding, J.D. Injectable block copolymer hydrogels for sustained release of a PEGylated drug. *Int. J. Pharm.* **2008**, *348*, 95–106. [CrossRef] [PubMed]
113. Determan, M.D.; Guo, L.; Thiyagarajan, P.; Mallapragada, S.K. Supramolecular self-assembly of multiblock copolymers in aqueous solution. *Langmuir* **2006**, *22*, 1469–1473. [CrossRef] [PubMed]
114. Bossard, F.; Aubry, T.; Gotzamanis, G.; Tsitsilianis, C. pH-Tunable rheological properties of a telechelic cationic polyelectrolyte reversible hydrogel. *Soft Matter* **2006**, *2*, 510–516. [CrossRef]
115. O'Lenick, T.G.; Jiang, X.; Zhao, B. Thermosensitive Aqueous Gels with Tunable Sol–Gel Transition Temperatures from Thermo-and pH-Responsive Hydrophilic ABA Triblock Copolymer. *Langmuir* **2010**, *26*, 8787–8796. [CrossRef] [PubMed]
116. Hamley, I.W.; Cheng, G.; Castelletto, V. A Thermoresponsive Hydrogel Based on Telechelic PEG End-Capped with Hydrophobic Dipeptides. *Macromol. Biosci.* **2011**, *11*, 1068–1078. [CrossRef] [PubMed]
117. Appel, E.A.; del Barrio, J.; Loh, X.J.; Scherman, O.A. Supramolecular polymeric hydrogels. *Chem. Soc. Rev.* **2012**, *41*, 6195–6214. [CrossRef] [PubMed]
118. Voorhaar, L.; Hoogenboom, R. Supramolecular polymer networks: Hydrogels and bulk materials. *Chem. Soc. Rev.* **2016**, *45*, 4013–4031. [CrossRef] [PubMed]
119. Chiappisi, L.; Hoffmann, I.; Gradzielski, M. Complexes of oppositely charged polyelectrolytes and surfactants—Recent developments in the field of biologically derived polyelectrolytes. *Soft Matter* **2013**, *9*, 3896–3909. [CrossRef]
120. Lindman, B.; Antunes, F.; Aidarova, S.; Miguel, M.; Nylander, T. Polyelectrolyte-surfactant association from fundamentals to applications. *Colloid J.* **2014**, *76*, 585–594. [CrossRef]
121. Kästner, U.; Hoffmann, H.; Dönges, R.; Ehrler, R. Interactions between modified hydroxyethyl cellulose (HEC) and surfactants. *Colloids Surf. A Physicochem. Eng. Asp.* **1996**, *112*, 209–225. [CrossRef]
122. Hoffmann, H.; Kästner, U.; Dönges, R.; Ehrler, R. Gels from modified hydroxyethyl cellulose and ionic surfactants. *Polym. Gels Netw.* **1996**, *4*, 509–526. [CrossRef]
123. Annable, T.; Buscall, R.; Ettelaie, R. Influence of surfactants on the rheology of associating polymers in solution. *Langmuir* **1994**, *10*, 1060–1070. [CrossRef]
124. Appell, J.; Rawiso, M. Interactions between Nonionic Surfactant Micelles Introduced by a Telechelic Polymer A Small Angle Neutron Scattering Study. *Langmuir* **1998**, *14*, 4409–4414. [CrossRef]
125. Piculell, L.; Egermayer, M.; Sjöström, J. Rheology of mixed solutions of an associating polymer with a surfactant. Why are different surfactants different? *Langmuir* **2003**, *19*, 3643–3649.
126. Jiménez-Regalado, E.; Selb, J.; Candau, F. Effect of Surfactant on the Viscoelastic Behavior of Semidilute Solutions of Multisticker Associating Polyacrylamides. *Langmuir* **2000**, *16*, 8611–8621. [CrossRef]
127. Tabuteau, H.; Ramos, L.; Nakaya-Yaegashi, K.; Imai, M.; Ligoure, C. Nonlinear rheology of surfactant wormlike micelles bridged by telechelic polymers. *Langmuir* **2009**, *25*, 2467–2472. [CrossRef] [PubMed]
128. Nakaya-Yaegashi, K.; Ramos, L.; Tabuteau, H.; Ligoure, C. Linear viscoelasticity of entangled wormlike micelles bridged by telechelic polymers: An experimental model for a double transient network. *J. Rheol.* **2008**, *52*, 359–377. [CrossRef]
129. Ramos, L.; Ligoure, C. Structure of a new type of transient network: Entangled wormlike micelles bridged by telechelic polymers. *Macromolecules* **2007**, *40*, 1248–1251. [CrossRef]
130. Gradzielski, M. Recent developments in the characterisation of microemulsions. *Curr. Opin. Colloid Interface Sci.* **2008**, *13*, 263–269. [CrossRef]

131. Langevin, D. Microemulsions. *Acc. Chem. Res.* **1988**, *21*, 255–260. [CrossRef]
132. Gradzielski, M.; Langevin, D.; Sottmann, T.; Strey, R. Small angle neutron scattering near the wetting transition: Discrimination of microemulsions from weakly structured mixtures. *J. Chem. Phys.* **1996**, *104*, 3782–3787. [CrossRef]
133. Gradzielski, M.; Raucher, A.; Hoffmann, H. Hydrophobically cross-linked micellar solutions: Microstructure and properties of the solutions. *J. Phys. IV* **1993**, *3*, C1-65–C1-79. [CrossRef]
134. Bagger-Jörgensen, H.; Coppola, L.; Thuresson, K.; Olsson, U.; Mortensen, K. Phase Behavior, Microstructure, and Dynamics in a Nonionic Microemulsion on Addition of Hydrophobically End-Capped Poly(ethylene oxide). *Langmuir* **1997**, *13*, 4204–4218. [CrossRef]
135. Maccarrone, S.; Frielinghaus, H.; Allgaier, J.; Richtery, D.; Lindner, P. SANS study of polymer-linked droplets. *Langmuir* **2007**, *23*, 9559–9562. [CrossRef] [PubMed]
136. Malo de Molina, P.; Appavou, M.-S.; Gradzielski, M. Oil-in-water microemulsion droplets of TDMAO/decane interconnected by the telechelic C18-EO150-C18: Clustering and network formation. *Soft Matter* **2014**, *10*, 5072–5084. [CrossRef] [PubMed]
137. Hed, G.; Safran, S.A. The immunity of polymer-microemulsion networks. *Eur. Phys. J.* **2006**, *19*, 69–76. [CrossRef] [PubMed]
138. Porte, G.; Ligoure, C.; Appell, J.; Aznar, R. Bridging interactions due to telechelic linkers balanced by screened Coulombic repulsions. *J. Stat. Mech. Theory Exp.* **2006**, *2006*, P05005. [CrossRef]
139. Zilman, A.; Kieffer, J.; Molino, F.; Porte, G.; Safran, S.A. Entropic phase separation in polymer-microemulsion networks. *Phys. Rev. Lett.* **2003**, *91*, 15901. [CrossRef] [PubMed]
140. Bhatia, S.R.; Russel, W.B. End-Capped Associative Polymer Chains between Nanospheres: Attractions in Ideal Solutions. *Macromolecules* **2000**, *33*, 5713–5720. [CrossRef]
141. Malo de Molina, P.; Ihlefeldt, F.S.; Prévost, S.; Herfurth, C.; Appavou, M.-S.; Laschewsky, A.; Gradzielski, M. Phase Behavior of Nonionic Microemulsions with Multi-end-capped Polymers and Its Relation to the Mesoscopic Structure. *Langmuir* **2015**, *31*, 5198–5209. [CrossRef] [PubMed]
142. Michel, E.; Filali, M.; Aznar, R.; Porte, G.; Appell, J. Percolation in a Model Transient Network: Rheology and Dynamic Light Scattering. *Langmuir* **2000**, *16*, 8702–8711. [CrossRef]
143. Malo de Molina, P.; Herfurth, C.; Laschewsky, A.; Gradzielski, M. Structure and dynamics of networks in mixtures of hydrophobically modified telechelic multiarm polymers and oil in water microemulsions. *Langmuir* **2012**, *28*, 15994–16006. [CrossRef] [PubMed]
144. Meier, W.; Hotz, J.; GuntherAusborn, S. Vesicle and cell networks: Interconnecting cells by synthetic polymers. *Langmuir* **1996**, *12*, 5028–5032. [CrossRef]
145. Antunes, F.E.; Marques, E.F.; Gomes, R.; Thuresson, K.; Lindman, B.; Miguel, M.G. Network formation of catanionic vesicles and oppositely charged polyelectrolytes. Effect of polymer charge density and hydrophobic modification. *Langmuir* **2004**, *20*, 4647–4656. [PubMed]
146. dos Santos, T.; Medronho, B.; Antunes, F.E.; Lindman, B.; Miguel, M. How does a non-ionic hydrophobically modified telechelic polymer interact with a non-ionic vesicle? Rheological aspects. *Colloids Surf. A Physicochem. Eng. Asp.* **2008**, *319*, 173–179. [CrossRef]
147. Lee, J.H.; Gustin, J.P.; Chen, T.; Payne, G.F.; Raghavan, S.R. Vesicle-biopolymer gels: Networks of surfactant vesicles connected by associating biopolymers. *Langmuir* **2005**, *21*, 26–33. [CrossRef] [PubMed]
148. Ruocco, N.; Frielinghaus, H.; Vitiello, G.; D'Errico, G.; Leal, L.G.; Richter, D.; Ortona, O.; Paduano, L. How hydrophobically modified chitosans are stabilized by biocompatible lipid aggregates. *J. Colloid Interface Sci.* **2015**, *452*, 160–168. [CrossRef] [PubMed]
149. Ashbaugh, H.S.; Boon, K.; Prud'homme, R.K. Gelation of "catanionic" vesicles by hydrophobically modified polyelectrolytes. *Colloid Polym. Sci.* **2002**, *280*, 783–788. [CrossRef]
150. Marques, E.F.; Regev, O.; Khan, A.; Miguel, M.G.; Lindman, B. Interactions between catanionic vesicles and oppositely charged polyelectrolytes phase behavior and phase structure. *Macromolecules* **1999**, *32*, 6626–6637. [CrossRef]
151. Lee, J.-H.; Oh, H.; Baxa, U.; Raghavan, S.R.; Blumenthal, R. Biopolymer-Connected Liposome Networks as Injectable Biomaterials Capable of Sustained Local Drug Delivery. *Biomacromolecules* **2012**, *13*, 3388–3394. [CrossRef] [PubMed]
152. Oh, H.; Javvaji, V.; Yaraghi, N.A.; Abezgauz, L.; Danino, D.; Raghavan, S.R. Light-induced transformation of vesicles to micelles and vesicle-gels to sols. *Soft Matter* **2013**, *9*, 11576–11584. [CrossRef]

8

Controlling the Organization of Colloidal Sphero-Cylinders using Confinement in a Minority Phase

Niek Hijnen † and Paul S. Clegg *

School of Physics & Astronomy, University of Edinburgh, Peter Guthrie Tait Road, Edinburgh EH9 3FD, UK; Niek.Hijnen@akzonobel.com

* Correspondence: paul.clegg@ed.ac.uk

† Current address: AkzoNobel, Stoneygate Lane, Felling, Gateshead NE10 0JY, UK.

Abstract: We demonstrate experimentally that a phase-separating host solvent can be used to organize colloidal rods into different cluster and network states. The rods are silica sphero-cylinders which are preferentially wet by the water-rich phase of an oil–water binary liquid system. By beginning with the rods dispersed in the single-fluid phase and then varying the temperature to enter the demixed regime, a precisely chosen volume of water-rich phase can be created. We then show how this can be used to create independent clusters of rods, a percolating network, a network of clusters or a system that undergoes hindered phase separation. These different modes are selected by choosing the relative volumes of the rods and the water-rich phase and by the timing of the temperature change.

Keywords: colloid; wetting; capillarity

---

1. Introduction

Non-spherical colloids move, assemble, and percolate differently compared to standard spherical particles [1–3]. Because many natural and synthetic particles, e.g., mineral particles, bacteria, viruses, graphene, carbon nanotubes and fibres are non-spherical, understanding and controlling the new behaviour is valuable [4]. Much has already been achieved by tuning mutual interactions via surface charge or the addition of depletants [5]. Less explored is the approach of organizing non-spherical particles using a phase transition in the host solvent. Low concentrations of rods or platelets can be corralled into a percolating arrangement driven by the phase-separation kinetics. Here, we are considering the host solvent to be a binary mixture of polymers or low molecular weight liquids. The corral is created by the preferred fluid domain or by the interface between the two fluids. Interfacial trapping requires considerable control over the wettability of the colloids [6,7]; by contrast, confining rods or platelets to one of the fluid domains should be more straightforward.

We focus here on colloidal sphero-cylinders as model rod-shaped particles; we are interested in their behaviour as they are forced into a confining volume of solvent. Onsager showed that a population of rods of aspect ratio, A, will exhibit a nematic phase for volume fractions, Φ_{ons}, above $\Phi_{ons}A = 3.29$ due to the combined effect of the decrease in entropy when the rods align and the excluded volume associated with the relative orientation of the rods. This threshold is valid even down to modest aspect ratios [8,9]. As the volume fraction of rods is further increased, we expect to find [10,11] the maximum for amorphous packing, Φ_a, to be described by $\Phi_a A = 5.1$ and the maximum for ordered packing, Φ_{max}, at

$$\Phi_{max} = \left(\frac{\pi}{\sqrt{27}} + \frac{\pi}{\sqrt{12}}A\right) / \left(\sqrt{\frac{2}{3}} + A\right) \approx \frac{0.6 + 0.9A}{0.8 + A}. \tag{1}$$

At quite low volume fractions, a fine network of 'sticky' sphero-cylinders can form and percolate across the sample. A homogeneous network is a random collection of evenly distributed rods, whereas a heterogeneous network is comprised of ramified clusters with a fractal appearance. For volume fractions, Φ_p, less than $\Phi_p A \approx 0.7$, the network will need to be heterogeneous in order to percolate [12].

Computer simulations by Peng et al. first demonstrated that a percolating arrangement of rod-shaped particles could be formed during a demixing process [13]. The rods were preferentially wet by the minority phase and had highly anisotropic interactions; the resulting networks had enhanced mechanical and electrical properties [14]. Hore and Laradji modeled the behaviour of purely repulsive colloidal rods which also partitioned into a single domain during phase separation [15]. They found that the phase separation process could be dramatically slowed; although arrest in a percolating structure was not observed for the compositions studied on the timescale of the simulations. Experimentally, it has been possible to create a percolating arrangement of nano-rods in a phase-separating polymer mixture. For sufficiently high concentrations of 'sticky' nanoparticles, Li et al. found that a continuous percolating domain of rods formed during phase separation [16]. More recently, Xavier and Bose studied the behaviour of multi-walled carbon nanotubes in a phase separating polymer mixture [17]. At the low concentrations employed, the kinetics were strongly modified but the system underwent complete separation.

The purpose of this Communication is to demonstrate that phase separation (see Figure 1a,b) can be used to organize anisotropic particles which preferentially partition into one of the phases. To do this, we carry out experiments using dispersions of colloidal rods (see Figure 1c) in a partially miscible host solvent. We find that a percolating network forms from a small quantity of rods provided that the volume fraction, Φ_v, of rods in the water-rich phase exceeds 66%. At significantly lower Φ_v, the rods form isolated clusters, following phase separation, that rapidly sediment to the base of the vial. If Φ_v is only slightly below the 66% threshold, then the system becomes exquisitely sensitive to the quench route. Finally, slowly but decisively destabilizing the percolating network leads to a sudden transition from a collapsing network to hindered phase separation. We show that the different outcomes can be precisely controlled.

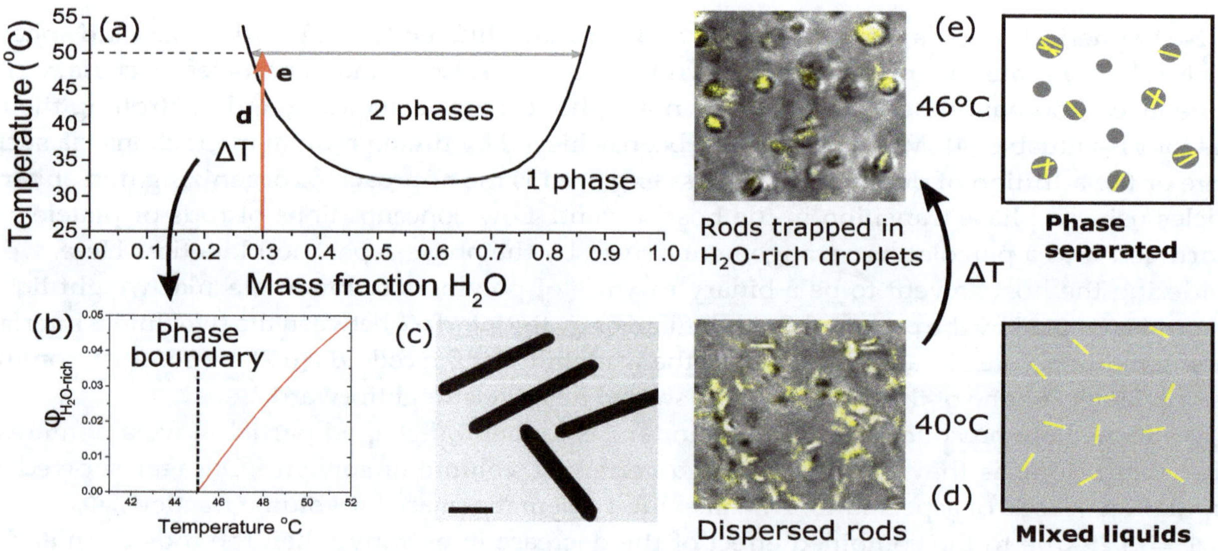

Figure 1. Showing how phase separating water and 1-propoxy-2-propanol can be used to create different phase volumes by controlling the depth of the temperature quench. **(a)** Our partially miscible solvents exhibit a lower critical solution temperature; **(b)** the temperature step into the demixed region can be used to control the volume of the minority phase, here the water-rich phase; **(c)** We use colloids shaped like sphero-cylinders, scale bar 1 µm; **(d,e)** The colloids, shown in yellow, are dispersed at low temperature below the binodal line; see red d in **(a)**, where the liquids are mixed. On warming the samples, the liquids demix and the colloids are confined within the minority phase; see red e in **(a)**.

2. Results and Discussion

First we describe experiments using the lowest volume fractions of rods, Figure 1c. The sample is quenched from room temperature to 46 °C by submerging it in a thermostated water bath. The choice of liquids and particles is outlined in Materials and Methods, below. Inside the vial [18], the sample changes temperature at a rate of 1 °C/s. This shallow quench leads to liquid–liquid phase separation and the formation of 1 vol. % of water-rich phase (see Figure 2a). If 0.1 vol. % of rods are added to the sample at the beginning of the experiment, then the liquids separate by nucleation and growth and the rods are observed in the droplets after the transition (see Figure 1d,e).

The silica surfaces are coated with silanol groups and a layer of physically bound water; it is no surprise that they are hydrophilic [19]. The water-rich phase is the more dense and so the rod-filled droplets eventually collect at the base of the vial. Qualitatively, the same thing is observed when 0.5 vol. % of rods are used. Immediately following the quench, the sample appears macroscopically homogeneous; however, within 2 h, all of the clusters of particles have sedimented to the base of the vial. This sample composition gives a volume fraction of $\Phi_v \approx 33\%$ in the water-rich phase, Figure 2b. It is perhaps surprising that we observe little that is special due to the shape of the particles in this regime given that the 0.5 vol. % sample should be approaching the isotropic to nematic transition during the phase separation process ($\Phi_{ons} = 34\%$).

Maintaining the same quench depth (i.e., to 46 °C) but raising the volume fraction of rods to 2 vol. % yields entirely different behaviour. Superficially, following the quench, the sample again appears macroscopically homogeneous. Now, however, on tipping it becomes immediately obvious that the sample has a solid-like character, Figure 2a,c. Evidently, the combination of phase separation and the 2 vol.% of colloidal particles ($\Phi_v \approx 67\%$) are conveying a yield stress to the sample.

Microscopic observations, Figure 2c, reveal a fine network of colloidal rods which percolate across the sample. There are no obvious droplets of water-rich phase; instead, the solvent appears to tightly envelop the particles. This is very similar to the case of gels held together by capillary bridges [20,21]. This phenomena has also been studied using rods and fully immiscible liquids [22]; no network formation was found in that case. With our rods and partially miscible fluids, we find that the system's attempt to lower the liquid–liquid interfacial area leads to significant effective attractions between the rods. It is well known that such attractive interactions greatly enhance the likelihood of percolation [23].

Networks formed from colloidal rods and nanotubes have been categorized as homogeneous or heterogeneous depending on whether the components are evenly distributed [12,23]. The micrographs show a 'spikey' network with some strands much longer than the individual rods, Figure 2c; qualitatively, this is consistent with the idea of a heterogeneous network. Quantitatively, we expect that our rods will only be able to percolate at volume fractions below $\Phi_p \approx 7\%$ via the formation of a heterogeneous network [12].

Because the combined volume of the water-rich phase and particles make up 3 vol. % of the sample, this is consistent. Within the water-rich phase, the volume fraction of rods is larger than that expected for the highest density amorphous packing, $\Phi_a \approx 50\%$. Hence, we anticipate that the liquid domain has pulled the particles together in a disordered arrangement of a liquid envelope combined with capillary bridges, Figure 2a,b. The latter are responsible for giving the network its strength.

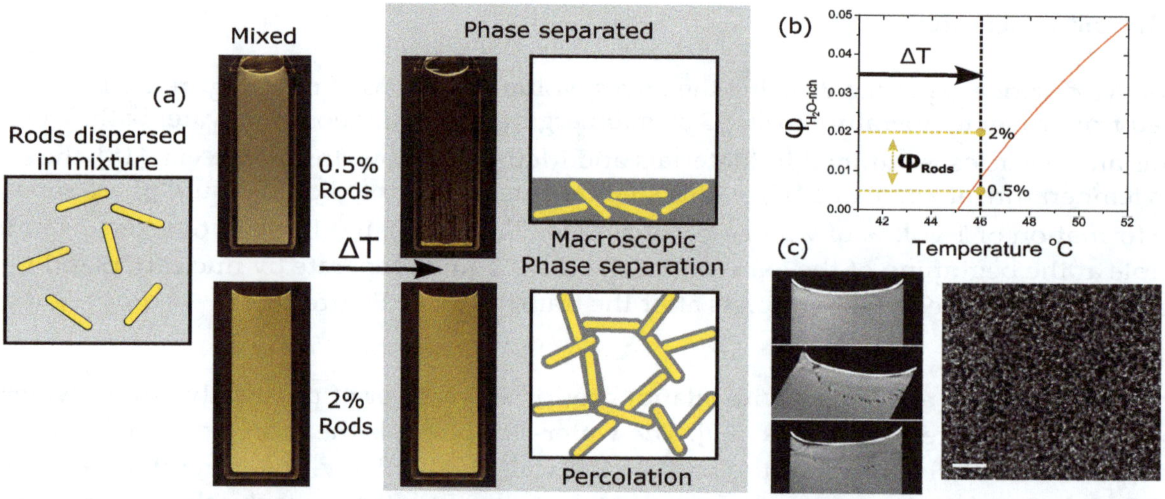

Figure 2. Showing how different concentrations of rods lead to radically different behaviour. (**a**) Images and cartoons of samples with rod volume fractions which are smaller and significantly larger than the volume of the minority phase; (**b**) The variation of the minority phase volume as a function of temperature compared to the volume of rods in the two cases of interest; (**c**) Image of a sample with 2% rods as it is tipped. There are indications that flow is resisted. The confocal micrograph shows the internal network formed by the rods, scale bar 50 μm.

We now increase the depth of the quench, finishing at a temperature of 48 °C, while keeping the volume fraction of rods at 2 vol. % and we show how the behaviour depends on the timing of the temperature changes. Increasing the final temperature increases the volume of the water-rich phase, here to 2 vol. %. Macroscopically, the sample is initially homogeneous; however, within an hour, it has sedimented slightly and it is clearly non-uniform, Figure 3b. Microscopically, the network now looks quite different to the one formed by the same quantity of particles in a quench to 46 °C. The strands of the network are thicker and there are significant gaps between different sections of the network, Figure 3e,f. For comparison, we now take a sample of identical composition to 48 °C by a different route. We first quench the sample to 46 °C to form a stable network and then it is warmed in the bath to 48 °C. The stable network, Figure 3a left panel, macroscopically sags and collapses noticeably under its own weight, Figure 3a right panel. The changes on the microscopic scale are much less significant compared to the alternative heating route: some bright droplets are evident and some narrow network threads may have broken, Figure 3d. Evidently, reducing the volume fraction of rods within the water-rich phase to $\Phi_v \approx 50\%$ has prevented the composite from forming a stable network without first sedimenting.

The relationship between sedimentation and aggregation has been considered by Allain et al. [24]. For a sufficiently high concentration of attractive particles, they demonstrate that a percolating network will form without settling due to gravity, i.e., aggregation will beat sedimentation. This is reminiscent of the situation for 2 vol. % of rods quenched to 46 °C. For a lower concentration of particles, the relative importance of sedimentation and aggregation is more finely balanced. They show that, initially, clusters of attractive particles will form that will not percolate; instead, the clusters will sediment and eventually form a percolating network of clusters which will not occupy the sample to the top [24]. This seems to accurately capture the behaviour of 2 vol. % of rods quenched directly to 48 °C. The significantly different appearance of the network is consistent with pre-formed clusters of rods having subsequently connected to span the sample. The boundary between network formation and cluster formation has been probed in detail for spherical particles and fully immiscible liquids by Heidlebaugh et al. [25]. In our experiments, these two scenarios have been observed for identical concentrations of particles but different volumes of the water-rich phase. The very different behaviour of these two systems is driven by the fact that a quench to 48 °C creates a larger volume water-rich phase which allows the rods to reorganize. The rods first make contact and then align in bundles and it is these bundles that are tightly enough enveloped in the water-rich phase to be 'sticky'. The bundles cluster, sediment and eventually form the network of clusters. This creates an effective change in the particle concentration. The change in thickness of the network strands is visible in Figure 3e,f.

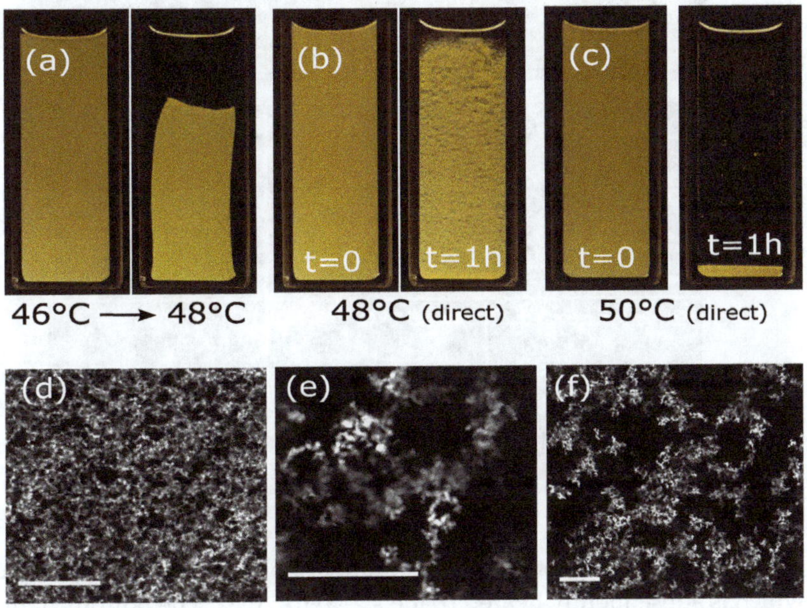

Figure 3. (**a**) Sample of 2 vol. % rods (left panel) first immersed in a water bath at 46 °C, and subsequently (right panel) slowly heated in the water bath to 48 °C; (**b**) Sample of 2 vol. % rods immersed in a water bath at, 48 °C, (left panel) immediately, and (right panel) one hour after immersion; (**c**) Sample of 2 vol. % rods immersed in a water bath at 50 °C, (left panel) immediately, and (right panel) one hour after immersion; (**d–f**) Confocal laser scanning microscopy images taken at 48 °C of samples shown in (**a**) confocal image (**d**) and (**b**) confocal images (**e**,**f**). Note changes in the 100 μm scale bars.

Finally, we demonstrate that a sharp transition occurs when there is a very significant rise in the water-rich phase volume, Figure 4. Here, an initial stable network has been created by quenching 2 vol. % rods into the demixed regime. This network is then steadily destabilized by gentle warming. In Figure 4a, the self-supporting network is seen to steadily collapse as the volume of the water-rich phase is increased. Ultimately, only a water-rich phase, densely packed with rods, remains at the base of the vial. Microscopically, we begin to see small local changes to the network as it is warmed, Figure 4b panels 2–3. Small bright droplets become increasingly prevalent and the mesh size of the network increases noticeably. Approaching $\Phi_v \approx 30\%$ there is a sudden change where the network is replaced by a large fluid domain. From this point onwards, the two liquid/particle system now resembles a conventional phase separation, Figure 4b panel 6. As predicted using computer simulations [15], the coarsening of the domain pattern is extremely slow due to the presence of the rods.

The sudden change from network to rod-filled liquid domain, Figure 4, is a consequence of the change in the rod–rod interactions. For networks formed from $\Phi_v > 66\%$, there are no rearrangements; the structure is self-supporting and robust. It appears that the water-rich phase envelope around the rods is responsible for strong, short-range attractions which are akin to the effect of a primary minimum in the interaction potential. We believe that the junctions between rods have some resemblance to capillary bridges [20,21]; although, our system is quite different because the rods are fully coated with the water-rich phase. Once the water-rich phase volume increases, the apparent strong attractions no longer control the system and a liquid domain filled with rods then emerges. The flow and accompanying slow coarsening only begin once the dominance of capillarity has ended.

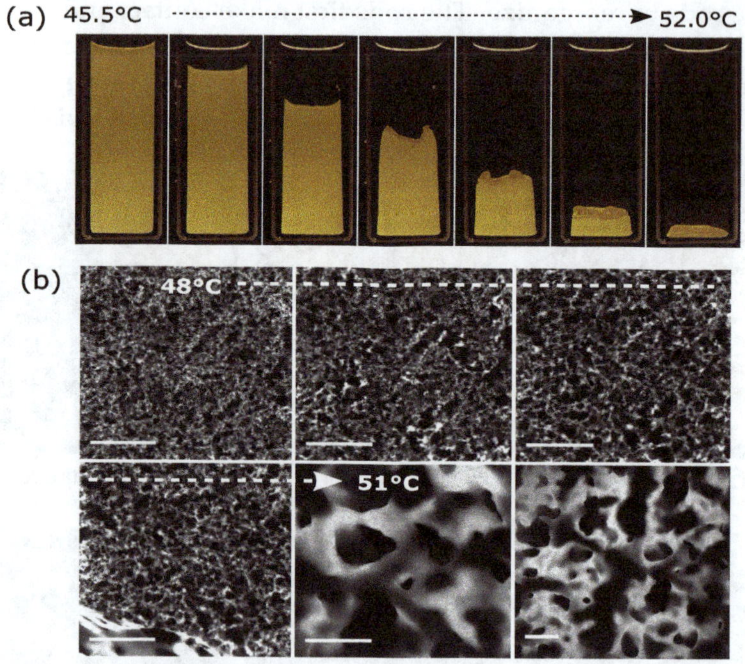

Figure 4. (**a**) A sample (2 vol. % rods) quenched to 45.5 °C, $\Phi_v \approx 80\%$, was then slowly heated to 52 °C, $\Phi_v \approx 30\%$; (**b**) A time series of confocal micrographs showing (panel 1) the structure of a similar sample at 48 °C, which is subsequently heated (panels 2–5) to 51 °C. The bottom right panel is a lower magnification image of the structure in panel 5. All scale bars 100 μm.

3. Materials and Methods

The binary fluid system used here is a mixture of water and 1-propoxy-2-propanol (PGPE), Figure 1a [26]. The rods are prepared by a multi-step route [27] that yields hollow cylinders of length 3.5 μm, aspect ratio $A = 9.7$ with $\sigma_A = 16\%$. To begin with, the rods are dispersed in the single-fluid phase at room temperature, typically with a volume fraction in the range 0.5–2%. Away from the phase boundary, the rods remain well dispersed demonstrating that they repel one another in the single-fluid phase. Over a period of 24 h, the rods sediment to the base of the vial; our experiments are usually carried out within 2 h unless otherwise mentioned above and hence are not greatly affected. It is important to note that the behaviour is very different close to the phase boundary. We have previously studied this in some detail [28] and we avoid this region here. In the experiments described above, all samples contain 30 wt % water in PGPE; we study the behaviour of the rods after quenching to different depths in the two-fluid phase via a change in temperature. Using the Lever rule, the change in temperature can be converted into a volume fraction of water-rich phase (see Figure 1b). This is the phase into which the rods always partition.

4. Conclusions

To conclude, we have shown that a well-controlled volume of the water-rich phase, created via demixing, can direct the organization of colloidal rods to form different structures and domains. While the water-rich phase in our experiments never occupies more than a few percent of the total sample volume, the concentration of rods within this phase can become very high. This creates effective attractive interactions between the rods leading to the formation of clusters, networks, networks of clusters and phase separating domains as the concentration is varied. Hence we have demonstrated that phase separation can be used to organize anisotropic particles which preferentially partition into one of the domains. Being able to trigger clustering, network formation, etc., via a sudden change in temperature and the relative insensitivity to the precise wetting properties of the particles are the strengths of this approach.

Acknowledgments: This work was supported by the Marie Curie Initial Training Network COMPLOIDs No. 234810 and BBSRC grant BB/M027597/1.

Author Contributions: Niek Hijnen and Paul S. Clegg conceived and designed the experiments; Niek Hijnen synthesized the particles, carried out the experiments and analyzed the data; Paul S. Clegg wrote the paper.

References

1. Glotzer, S.C.; Solomon, M.J. Anisotropy of building blocks and their assembly into complex structures. *Nat. Mater.* **2007**, *6*, 557–562.
2. Dugyala, V.R.; Daware, S.V.; Basavaraj, M.G. Shape anisotropic colloids: Synthesis, packing behavior, evaporation driven assembly, and their application in emulsion stabilization. *Soft Matter* **2013**, *9*, 6711–6725.
3. Li, T.; Brandani, G.; Marenduzzo, D.; Clegg, P.S. Colloidal Spherocylinders at an Interface: Flipper Dynamics and Bilayer Formation. *Phys. Rev. Lett.* **2017**, *119*, 018001.
4. Yao, S.; Zhu, Y. Nanomaterial-enabled stretchable conductors: Strategies, materials and devices. *Adv. Mater.* **2015**, *27*, 1480–1511.
5. Vigolo, B.; Coulon, C.; Maugey, M.; Zakri, C.; Poulin, P. An Experimental Approach to the Percolation of Sticky Nanotubes. *Science* **2005**, *309*, 920–923.
6. Imperiali, L.; Clasen, C.; Fransaer, J.; Macosko, C.W.; Vermant, J. A simple route towards graphene oxide frameworks. *Mater. Horizons* **2014**, *1*, 139–145.
7. Hijnen, N.; Cai, D.; Clegg, P.S. Bijels stabilized using rod-like particles. *Soft Matter* **2015**, *11*, 4351–4355.
8. Onsager, L. The effects of shape on the interaction of colloidal particles. *Ann. N. Y. Acad. Sci.* **1949**, *51*, 627–659.
9. Bolhuis, P.; Frenkel, D. Tracing the phase boundaries of hard spherocylinders. *J. Chem. Phys.* **1997**, *106*, 666–687.
10. Williams, S.R.; Philipse, A.P. Random packings of spheres and spherocylinders simulated by mechanical contraction. *Phys. Rev. E Stat. Nonlinear Soft Matter Phys.* **2003**, *67*, 051301.
11. Chaikin, P.M.; Donev, A.; Man, W.; Stillinger, F.H.; Torquato, S. Some observations on the random packing of hard ellipsoids. *Ind. Eng. Chem. Res.* **2006**, *45*, 6960–6965.
12. Philipse, A.P.; Wierenga, A.M. On the Density and Structure Formation in Gels and Clusters of Colloidal Rods and Fibers. *Langmuir* **1998**, *14*, 49–54.
13. Peng, G.; Qui, F.; Ginzberg, V.V.; Jasnow, D.; Balazs, A.C. Forming Supramolecular Networks from Nanoscale Rods in Binary, Phase-Separating Mixtures. *Science* **2000**, *288*, 1802–1804.
14. Buxton, G.A.; Balazs, A.C. Predicting the mechanical and electrical properties of nanocomposites formed from polymer blends and nanorods. *Mol. Simul.* **2004**, *30*, 249–257.
15. Hore, M.J.A.; Laradji, M. Prospects of nanorods as an emulsifying agent of immiscible blends. *J. Chem. Phys.* **2008**, *128*, 054901.
16. Li, L.; Miesch, C.; Sudeep, P.K.; Balazs, A.C.; Emrick, T.; Russell, T.P.; Hayward, R.C. Kinetically Trapped Co-continuous Polymer Morphologies through Intraphase Gelation of Nanoparticles. *Nano Lett.* **2011**, *11*, 1997–2003.
17. Xavier, P.; Bose, S. Mapping the intriguing transient morphologies and the demixing behavior in PS/PVME blends in the presence of rod-like nanoparticles. *Phys. Chem. Chem. Phys.* **2015**, *17*, 14972–14985.
18. Hijnen, N. Colloidal Rods and Spheres in Partially Miscible Binary Liquids. Ph.D. Thesis, University of Edinburgh, Edinburgh, UK, 2013.
19. White, K.A.; Schofield, A.B.; Wormald, P.; Tavacoli, J.W.; Binks, B.P.; Clegg, P.S. Inversion of particle-stabilized emulsions of partially miscible liquids by mild drying of modified silica particles. *J. Colloid Interface Sci.* **2011**, *359*, 126–135.
20. Koos, E.; Willenbacher, N. Capillary forces in suspension rheology. *Science* **2011**, *331*, 897–900.
21. Koos, E. Capillary suspensions: Particle networks formed through the capillary force. *Curr. Opin. Colloid Interface Sci.* **2014**, *19*, 575–584.
22. Maurath, J.; Bitsch, B.; Schwegler, Y.; Willenbacher, N. Influence of particle shape on the rheological behavior of three-phase non-brownian suspensions. *Colloids Surf. A Physicochem. Eng. Asp.* **2016**, *497*, 316–326.
23. Kyrylyuk, A.V.; van der Schoot, P. Continuum percolation of carbon nanotubes in polymeric and colloidal media. *Proc. Natl. Acad. Sci. USA* **2008**, *105*, 8221–8226.

24. Allain, C.; Cloitre, M.; Wafra, M. Aggregation and Sedimentation in Colloidal Suspensions. *Phys. Rev. Lett.* **1995**, *74*, 1478–1481.
25. Heidlebaugh, S.J.; Domenech, T.; Iasella, S.V.; Velankar, S.S. Aggregation and Separation in Ternary Particle/Oil/Water Systems with Fully Wettable Particles. *Langmuir* **2014**, *30*, 63–74.
26. Bauduin, P.; Wattebled, L.; Schrödle, S.; Touraud, D.; Kunz, W. Temperature dependence of industrial propylene glycol alkyl ether/water mixtures. *J. Mol. Liquids* **2004**, *115*, 23–28.
27. Hijnen, N.; Clegg, P.S. Simple Synthesis of Versatile Akaganéite-Silica Core-Shell Rods. *Chem. Mater.* **2012**, *24*, 3449–3457.
28. Hijnen, N.; Clegg, P.S. Colloidal aggregation in mixtures of partially miscible liquids by shear-induced capillary bridges. *Langmuir* **2014**, *30*, 5763–5770.

9

Physicochemical Properties and the Gelation Process of Supramolecular Hydrogels

Abdalla H. Karoyo and Lee D. Wilson *

Department of Chemistry, University of Saskatchewan, 110 Science Place, Saskatoon, SK S7N 5C9, Canada; abk726@mail.usask.ca

* Correspondence: lee.wilson@usask.ca

Academic Editor: Clemens K. Weiss

Abstract: Supramolecular polysaccharide-based hydrogels have attracted considerable research interest recently due to their high structural functionality, low toxicity, and potential applications in foods, cosmetics, catalysis, drug delivery, tissue engineering and the environment. Modulation of the stability of hydrogels is of paramount importance, especially in the case of stimuli-responsive systems. This review will update the recent progress related to the rational design of supramolecular hydrogels with the objective of understanding the gelation process and improving their physical gelation properties for tailored applications. Emphasis will be given to supramolecular host–guest systems with reference to conventional gels in describing general aspects of gel formation. A brief account of the structural characterization of various supramolecular hydrogels is also provided in order to gain a better understanding of the design of such materials relevant to the nature of the intermolecular interactions, thermodynamic properties of the gelation process, and the critical concentration values of the precursors and the solvent components. This mini-review contributes to greater knowledge of the rational design of supramolecular hydrogels with tailored applications in diverse fields ranging from the environment to biomedicine.

Keywords: gel; sol; aggregation; cyclodextrin; hydration

---

1. Introduction

Polymer gels are generally defined as 3D networks swollen by a large amount of water [1]. In particular, polysaccharide-based hydrogels are important due to their diverse chemical structure and rich functionality [2,3]. In general, gels play a vital role in the biomedical field (e.g., contact lenses and ocular implants, wound dressing, tissue regeneration and engineering) [4–8], in the environment [9–12], food processing [13] and personal care products such as cosmetics and disposable diapers [14]. Most recently, polysaccharide-based hydrogels have found application in agriculture as controlled-release devices for fertilizers and agrochemicals [15]. Additional applications include the use of self-healing pH sensitive superabsorbent polymers to self-seal cracks in concrete [16,17]. A number of responsive hydrogel systems based on natural and synthetically modified polysaccharides have been reported. For example, significant contributions on responsive systems are reported by the research groups of Rinaudo [18–21], Saito [22], and others [5,6,23–25]. Polysaccharides are abundant and readily available from renewable sources such as plants and algae, and various microbial organisms [26]. Such polysaccharides have a large variety of compositional and structural properties, making them facile to produce and versatile for gel formation as compared with synthetic polymers. Selected examples of natural polysaccharides for the preparation of stimuli-responsive hydrogels are listed in Table 1.

In contrast to conventional gel formation, the combined use of polymer chains along with stable and selective supramolecular cross-links offer versatile constructs that afford facile modification of structural parameters of the polymer backbone that include the strength and dynamics of cross-linking interactions, and responsiveness to multiple stimuli [27]. Supramolecular hydrogels (or aqua gels) are hydrophilic materials which undergo self-assembly to form 3D continuous networks of macromolecules, where water resides within the interstitial domains of the polymer network [28,29]. Supramolecular hydrogels are a relatively new class of soft and responsive materials of great research interest owing to the broad application of such systems in areas that range from tissue engineering and carrier systems [30] to environmental remediation [31]. The formation of supramolecular materials through host–guest interactions is a powerful method to create non-conventional stimuli-responsive hydrogels. This relates to the host–guest interactions present which can be modulated to fine tune the stability and responsiveness of the resulting gel system based on the choice of macromolecular scaffold. Numerous studies have been reported on supramolecular hydrogels that show reversible response to environmental stimuli; however, the responsive behaviour of many of these materials often depend on the inherent properties of the building blocks rather than the resulting supramolecular interactions. Examples of inherent responsive hydrogel systems include the temperature-induced *rod-to-coil* transition of poly(*N*-isopropyl acrylamide) (PNIPAM) [32–34] and oligo(ethylene glycol)s (OEGs) [35,36], the pH induced-protonation of poly(vinyl pyridine) [37,38], and the photosensitive behaviour of azobenzenes [4,39,40].

Table 1. Selected examples of polysaccharide biopolymers reported for the preparation of hydrogels.

Origin	Gelator/Precursor	Responsive Feature	References
Plant cell walls, wood, seeds, & roots	Pectins, cellulose, galacto-/gluco-mannans	Chemical species (arsenic), pH, & temperature	[21,25,26,41]
Seaweeds	Carrageenans, alginates, agar	Light, & temperature	[5,42–45]
Animals, organisms, bacteria	Hyaluronan, chitosan, chondroitins, xanthan, succinoglycan, gelatin, gellan	Temperature, & pH	[19,46–50]
Sugars	Cyclodextrins, galactose, glucose	Redox, light, temperature, & chemical species	[25,51–54]

Host–guest carbohydrate-based hydrogels that employ macrocyclic building blocks such as cyclodextrins (CDs) as porogens [48,52–57] are unique for various reasons: (i) tunable physicochemical properties (e.g., mechanical stability, viscosity, etc.); (ii) specificity and effectiveness of the host–guest molecular recognition that lead to stable hydrogel structures; and (iii) wide applications of host–guest hydrogels in various fields such as the environment, biomedicine, delivery systems, and food technology. In general, the formation of supramolecular-based hydrogels occur *via* two processes; either through self-assembly of monomer units to form aggregates that gelate in aqueous solvents, or derived from polymer units and/or host–guest interactions *via* multi-component inclusion systems that contain macromolecular scaffolds [19,51]. The latter is used to form hydrogels based on CD macromolecules as the key building block. Unlike chemical gels which involve covalent bonding (e.g., divinyl sulfone cross-linked gellan gels [58]), physical gel formation occurs *via* multiple, weak non-covalent interactions (e.g., H-bonding, π–π and van der Waals interactions, and hydrophobic effects) [51,59].

Despite recent research advances in physical supramolecular hydrogels, the rational design of such systems is sparsely reported. This knowledge gap poses a significant challenge for the use of these materials for specific applications. Specifically, the optimization of the strength of supramolecular hydrogels is important, especially in the case of stimuli-responsive hydrogels. For example, temperature sensitive hydrogels can encapsulate drugs during the gelation process and the duration or rate of their release at the site of action will depend on the stability of the hydrogels [60]. Thus, the tunability and responsive nature of such materials depend on several factors, such as the solute–solute and solute–solvent interactions, for the gelation process. Yui et al. [61,62] and Zhao et al. [63] have reported an approach for improving the stability of hydrogels by tuning the hydrophile–lipophile balance (HLB) of the precursor materials as a way of stabilizing the macromolecular assembly. The stability of the supramolecular assembly can be tuned by varying the nature and relative feed ratios of the host/guest system or the precursor materials [23]. Furthermore, the role of the aqueous solvent on the hydrogel stability is significant [20], according to the HLB of the system. This mini-review presents a coverage of the literature in the past five years concerning the physicochemical properties and structural variables that can be tuned to improve the formation and stability of hydrogels for tailored applications. Examples of spectroscopic (e.g., NMR, XRD, and FT-IR) and microscopic (e.g., TEM and SEM) studies related to the rheological and structural information of the hydrogel assembly will be presented in order to gain a greater understanding of the rational design of these materials. In particular, the nature of the intermolecular interactions, the thermodynamic properties of the gelation process, and the critical concentration of the precursors and the solvent will be reviewed. Because of their low toxicity, and potential applications in foods, cosmetics, drug delivery, tissue engineering and catalysis, attention will be directed to polysaccharide-based hydrogels with a special emphasis on CD-based host–guest hydrogel systems.

2. The Structure of Supramolecular Hydrogels

2.1. Design Strategy of Hydrogels: Mechanism of Gelation

Many gels are formed by simply heating a gelator or mixture of gelators in aqueous, organic or a co-solvent system to form a solution, followed by cooling. To gain a greater understanding of the mechanism of gel formation, the system may be categorized by the primary (1°), secondary (2°), and tertiary (3°) structure [64]. The 1° structure is determined by the molecular level recognition (e.g., host–guest interactions) that is largely influenced by nonspecific hydrophobic interactions. The 2° and 3° structures are determined by the molecular associations of individual polymer chains and their subsequent aggregation to form gels, respectively. In the case of host–guest systems, the formation of poly-pseudorotaxanes (PPRs) by threading a polymer chain into a series of CD cavities is one of the most common known self-assembly motifs in supramolecular hydrogels [54,65]. Strong hydrogen bonds between the adjacent PPRs function as non-covalent (physical) cross-links that lead to microcrystalline aggregation, thus promoting physical gel formation. A challenge for polysaccharide biopolymer gels concerns the molecular mechanism of non-covalent cross-linking in stimuli-reversible gelation. Generally, this is due to the role of non-covalent interactions; namely, H-bonding, hydrophobic and electrostatic interactions that result as the 3D networks are formed [66,67]. Moreover, the gelation process of many biopolymers is preceded by a transition from a random coil state to an ordered helix conformation, where the subsequent aggregation of the helices forms an extended network [67,68], as shown in the relationship below.

$$Coil \rightarrow Helix \rightarrow Gel \quad (1)$$

The *coil–helix* (or *coil–globule*) transition in relationship (1) above is manifested as a volume phase transition (VPT) on a macroscopic level, where polymer gels can exist in either swollen or collapsed phases [35], as illustrated in Scheme 1.

VPT occurs between the phases in response to chemical and/or physical stimuli, such as temperature and pH, as a result of modification of the HLB of the system. For gelation to occur, the helix formation must lead to association and branching (or aggregation) of the polymer strands to form an infinite 3D network. Two possible mechanisms for association and branching occur as described in Figure 1. In Figure 1a, the association and branching may occur on the helical level, i.e., the formation of the helices and the aggregation of the polymer strands occur simultaneously. This is the classical mechanism of gelation which has been observed in many polysaccharide gels [47,68,69]. By comparison, Figure 1b represents a network formation at the super-helical level, where fully developed helices undergo self-assembly and aggregation to form a gel. Other more complex mechanisms which are beyond the scope of this mini-review have been proposed [70].

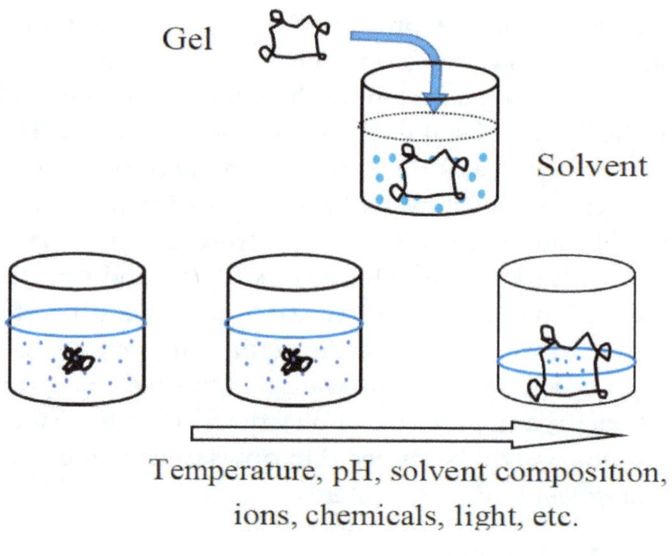

Scheme 1. Phase transition of a gel indicated as a reversible and discontinuous volume change in response to various stimuli. Redrawn from [67].

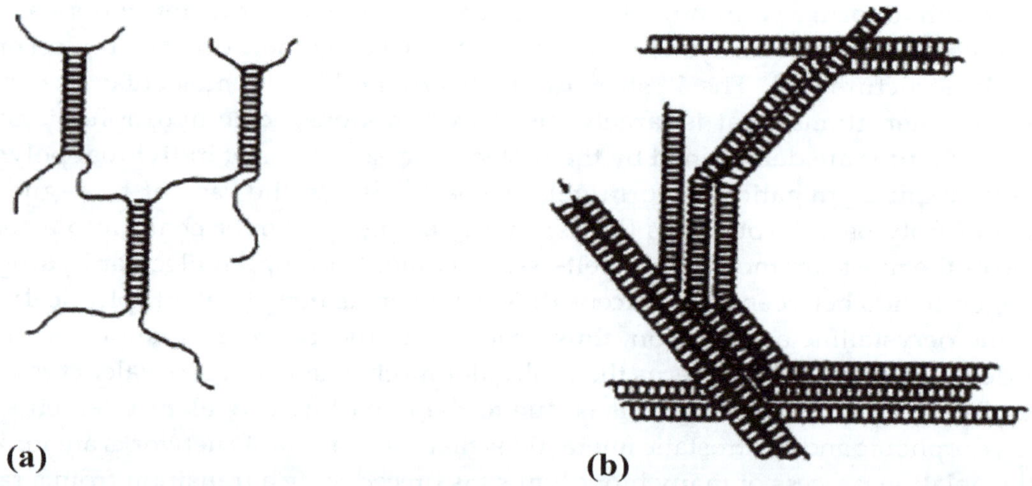

Figure 1. Schematic representations of possible mechanisms of network formation in helical gels: gelation on: (**a**) the helical level; and (**b**) super-helical level.

2.2. *Volume Phase Transition (VPT) in Polymer Gels*

Knowledge of the phase transitions in polymer gels is essential for establishing a greater understanding of the fundamentals underlying the associative interactions and molecular recognition in polymer networks. This effect occurs because the process of cross-linking of supramolecular assemblies in aqueous solution is known to result in VPTs as a function of external stimuli (cf. Scheme 1). The gel VPT was first predicted in 1968 by Dušek and Petterson, using the Flory–Huggins (FH) theory for polymer solutions [71]. A discontinuous volume change of a gel based on the analogy of the *coil–globule* transition of a polymer in a solution was reported, as depicted in Scheme 1. The thermodynamics of VPT of gels were described by Shibayama and Tanaka [72], where different competing interactions (e.g., Coulomb interactions between charged groups and counterions, van der Waals, hydrophobic, electrostatic, and hydrogen bonding interactions) control the gel size. For example, in the case of the volume phase transition of *N*-isopropylacrylamide (NIPA) gel [73], the enthalpic contribution due to hydrophobic interactions lead to gel collapse and entropic contributions due to the rubber elasticity that favour swelling of the gel. Therefore, gel size is determined by a simple additivity of the enthalpy/entropy contributions to the overall Gibbs energy of the process. The phenomenon of a gel phase transition resembles the gas-liquid phase transition. Thus, a discontinuous phase transition can be observed depending on the degree of ionizable groups and stiffness of the polymer chains that constitute the 3D network [18]. Theoretically, the swelling of a gel is determinable by minimizing the Gibbs energy per polymer segment with respect to polymer density or alternatively by modulating the pressure. The FH theory is an oversimplified mean-field theory which quantitatively describes the phase transition of a polymer network, where the Gibbs energy of a gel can generally be written as:

$$\Delta G = \Delta G_{rubber} + \Delta G_{counterion} + \Delta G_{mixing} \quad (2)$$

The terms on the right side of Equation (2) correspond to the Gibbs energy of rubber elasticity, ionization, and mixing, respectively. The terms for the total number of persistent (monomer) units and the polymer–solvent interaction energy, the number of ionized groups per chain, and a set of virial terms including the charge-charge repulsion, are contained within the generalized FH equation. Using the FH equation of state, Dušek and Patterson [71] concluded that a discontinuous volume change of a gel occurs when an external stress is imposed upon it. Generally, the VPT described by the oversimplified FH equation relate to polymer–solvent interactions and the tendency of the polymer chains to either repel (swell) or attract (contract) each other. Therefore, the role of the solvent in gel formation cannot be over emphasized. However, many studies have traditionally focused on solute–solute (intrinsic) interactions as the source of both binding enthalpy and recognition in the polymer self-assembly. In part, this relates to a limited understanding of the role of solvent effects in gel systems. The separation of the actual enthalpy of binding into an intrinsic (solute–solute) enthalpy (ΔH_i) and the enthalpies of solution for the bound (ΔH_b) and unbound (ΔH_u) species is shown by the Born–Haber cycle as presented in Scheme 2. The association of polymer chains in solution is therefore composed of solute- and solvent-associated processes.

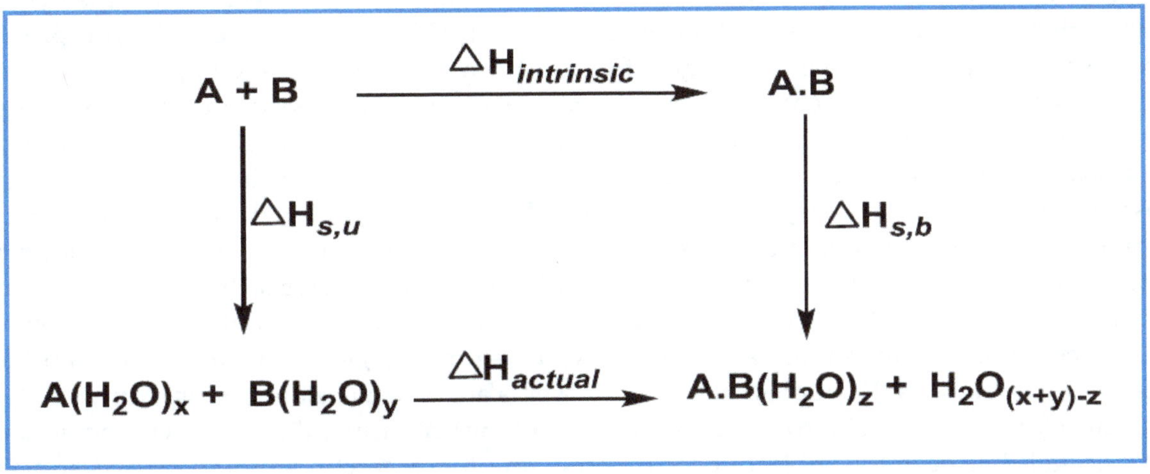

Scheme 2. Born–Haber cycle showing separation of the actual (measured) enthalpy (ΔH_{actual}) into intrinsic enthalpy ($\Delta H_{intinsic}$) and the enthalpy of solution values for the bound ($\Delta H_{s,b}$) and unbound species ($\Delta H_{s,u}$). The physical states of the reactants and products are not shown.

2.3. *Classification of Polysaccharide-Based Gels*

Polysaccharide-based physical hydrogels can be classified according to various criteria [8,74–76], such as the source of the precursors, polymer composition, polymer configuration, type of cross-linking, and so on. In this mini-review, attention is made on the polymer composition as a basis to classify polysaccharide hydrogels. The method of gel preparation leads to the formation of some important classes of hydrogels, namely, homo-polymer, co-polymer, and multi-polymer inter-penetrating networks (IPNs).

2.3.1. Single Component Homo-Polymer Gels

These types of gels are derived from a single species of monomer unit. Meena et al. [45] and Yoshida and Takahashi [49] reported χ-carrageenan and gellan as respective examples of polysaccharide-based single component thermo-reversible gels. Hydrogels reported by Jung et al. [51] represent single component polymer-based gels. Swellable functionalized starch and cellulose materials can also exist as 3D networks in water and may form hydrogels that are responsive to changes in temperature and pH [12]. Self-inclusion complexes of functionalized β-CD may exist as supramolecular polymers in water and undergo hydrogel formation [77]. The mechanism of the gelation in single component thermo-responsive hydrogels has been studied by X-ray diffraction and optical rotation studies. The process is based on the dissolution of the gelator material at high temperature, formation of double helices upon cooling, followed by aggregation of these helices [47], as described above. Stabilization of the double-helix is achieved by inter-chain H-bonding interactions. Gels based on aggregation of triple helices are also known [22]. Other examples of single component polymer gels were reported by Jejurikar et al. [42]. These gels were prepared from Ca^{2+} and Ba^{2+} cross-linked alginate materials. In the case of ionic cross-linked polymer gels, the structure that invokes the double helix is mimicked by cross-linking of specific polymers using multivalent counterion species such as Ca^{2+}. Thermal induced formation of single component polymer-based gels were reported using chitosan derivatives [46] where such systems form due to the presence of hydrophobic interactions. The ability of such systems to form gels depends on the density and length of the hydrophobic side chains. Phase separation is related to hydrophobic chain segregation as the main mechanism of gel formation [46].

2.3.2. Two Component Co-Polymer Gels

Co-polymer gels are composed of two types of interacting systems where one type is hydrophilic in nature. Host–guest complexes containing CDs (e.g., β-CD with cholesterol [78]) represent the simplest examples of co-polymer hydrogels formed from two polymers/components (where one or both are saccharides). Two component polymer gels can be formed from: (1) two non-gelling; (2) one gelling and one non-gelling; or (3) two gelling polymers. The mechanism of gel formation in this class of materials involves specific interactions (mostly H-bonding and/or hydrophobic, electrostatic, and π–π interactions) that give rise to gels whose properties depend on the structure of each polymer, as well as the relative concentration of each component [21]. In general, the structural characterization of the resulting gel due to the combination of two polysaccharides is dependent on variable factors. However, the structure will be governed partly by the kinetics of phase separation and the viscosity of the materials. An example of category (1) gels is a mixture of xanthan and galacto- or glucomannans. Examples of category (2) gels are dextran (non-gelling)/amylose (gelling) and galactomannan (non-gelling)/χ-carrageenan (gelling) mixtures [79]. The characterization of category (2) gels has been the subject of controversy among researchers. Some reports suggest that interpenetrating networks (IPNs; Section 2.3.3) of the two polymers are formed, while others suggest a phase separated entrapment of the non-gelling polymer within the gel network. Category (3) gels, e.g., agarose/χ-carrageenan mixture, form gels based mainly on IPNs. Evidence of IPNs relate to their formation by independent gelation of each polysaccharide, where a network of one polymer is formed followed by that of a second polymer in an independent fashion.

Pseudo-copolymer systems based on multi-component host–guest polysaccharide-based gels have been extensively studied [1]. For example, Yui and coworkers [61,62] have prepared thermo-reversible supramolecular hydrogels using dextran or chitosan as hydrophilic backbones and short PEG or PPG side chains for inclusion complexation with CDs. The mechanism of gelation for these types of hydrogels is based on the phase separation of the hydrated backbone (e.g., dextrans or chitosan) and the hydrophobic-driven aggregation of inclusion complexes via physical cross-linking (cf. Scheme 3).

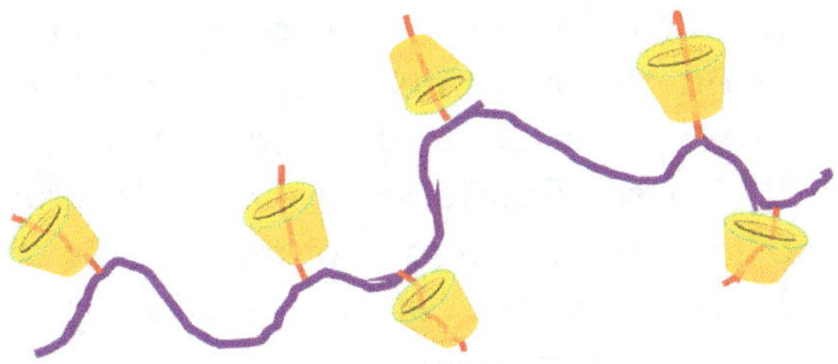

Scheme 3. Hypothetical structure of a polymer network showing a chitosan backbone (purple line) with hydrophilic PEG pendants (red lines) and CD pendants (toroid).

2.3.3. Multi-Polymer Inter-Penetrating Networks (IPNs)

This category represents an important class of hydrogels combined of at least two polymers as a network assembly, where at least one component is independently synthesized and/or cross-linked in the presence of the other component without the formation of any covalent bonds [80]. Generally, IPNs are synthesized for the purpose of combining individual properties of two or more polymers. In many cases, new properties which are not observed in the single networks are observed in the prepared gel [43,81]. Various polysaccharides are used to prepare IPN-based gels, such as alginate, dextran, xanthan, chitosan, and guar gum [43,82].

2.4. Solvent–Gelator Interactions

Solvent interactions have a significant influence on the self-assembly of gels since the solvent constitutes about ~99% of the system by weight. Many biophysical processes in nature, such as enzyme–substrate interactions, the self-assembly of bilayers in biomembranes, surfactant aggregation, and kinetic solvent effects in water-rich solutions are predominantly governed by hydrophobic interactions [83]. Generally, the tendency of water molecules to avoid unfavourable entropic configurations with apolar solutes and constitutes a driving force for their aggregation. The self-assembly of gelators is a very important step in gel formation; however, the modulation of the HLB is equally important for ensuring that the 3D network imbibes the solvent within the polymer framework. Sparse studies [84] have reported the molecular level details of solvent effects in self-assembly and gelation processes. A variety of approaches by which the solvent effects can be quantified through physical parameters and equations have been reviewed in detail elsewhere [84,85]. A useful accounting method for specific solvent–solute interactions in gels involve the use of Kamlet–Taft parameters where α (hydrogen bond donor ability), β (hydrogen bond acceptor ability) and π^* (polarizability) are defined [86]. Various studies have demonstrated that the α parameter strongly determines whether a hydrogen bonding gelator will undergo self-assembly in a given solvent [87,88]. Polar protic solvents with good hydrogen bond donating ability with high α values such as water, methanol or formic acid can interact competitively with the gelator. This leads to promotion of solute–solvent interactions and thwarts the formation of a gelator–gelator 3D network, leading to macroscopically homogeneous solutions. The β parameter represents the ability of the solvent to accept hydrogen bonds. Thus, solvents with high β values (e.g., tetrahydrofuran and ethyl acetate) can affect gelator–gelator interactions, resulting in a disruption of the self-assembly process, along with a lowering of the thermal stability of the gel network to a variable extent. In the case of the polarizability of a solvent (π^*), the self-assembly and gelation can be modulated by the ability of the solvent to interact with the peripheral (surface or backbone) groups of the gelator, thus affecting the stability of the gel. Various structural factors that modulate the HLB need to be considered in order to modify solute–solvent interactions as a methodology for controlling gel formation (cf. Section 2.5).

2.5. Characterization of Hydrogels

Hydrogels and their stability are characterized by several methods including spectroscopy (e.g., nuclear magnetic resonance (NMR), Fourier transform infrared (FT-IR), ultra-violet visible

(UV-vis), induced circular dichroism (ICD), and fluorescence), microscopy (e.g., scanning electron microscopy (SEM) and transmission electron microscopy (TEM)), diffraction (e.g., SAXS, SANS, and XRD), rheology (e.g., ball drop method and viscosity), calorimetry (e.g., DSC and ITC), and computational methods [89]. These techniques provide structural characterization of gels in terms of the nature of the intermolecular interactions, solute–solvent interactions, thermodynamics of the gelation process and the critical concentration of the gelators and the solvents. Selected examples of interest are briefly presented in this mini-review.

2.5.1. Spectroscopy Techniques

Various techniques such NMR, UV-Vis, FT-IR, and Raman spectroscopy have been used to characterize the structure of hydrogels [15,24,29,54,84]. Solution- and solid-state NMR techniques at high resolution give useful structural information about the formation of H-bonds during gelation. The chemical shift changes of a specific nuclei during a *gel-sol* transition can be tracked using temperature dependent NMR measurements, where signals are broadened in the gel state [7]. Furthermore, chemical shift changes of a gel material can be used to probe local changes in a microenvironment such as aggregation [61,62]. ^{13}C solid-state NMR can be used to study the 2° structure of gels giving an insight into their molecular arrangements [7]. UV-vis and FT-IR/Raman spectroscopy can be used to probe π–π interactions, and H-bonding, respectively. Although ICD may not directly support gel formation, it can be used as a supplementary technique to show evidence of double helix formation [90].

2.5.2. Rheology

Rheology can be used to determine the mechanical properties of supramolecular hydrogels [90]. Gel structure has been characterized using such techniques as ball drop, inverted tube and modulus (viscosity) methods [91]. Despite its simplicity, the inverted-tube method has been used successfully to determine the stability of hydrogels. Figure 2 shows various solutions (PEG, PEG-α-CD, Ada-PEG, and Ada-PEG-α-CD) with different gel formation abilities as characterized by the inverted tube method. Note that the differences in structure of the mixtures shown in Figure 2 can be related to the HLB concept (cf. Scheme 3). Modulus methods have also been widely used to determine the strength of hydrogels [65,90], cross-over point of storage (G') over loss moduli (G'') and can be used to measure gelation kinetics of physical and chemical gel systems. Hence, by monitoring G' and G'', the viscoelasticity and the gelation time of hydrogels can be determined through variations of such factors as concentration and time. Stable gels are typically characterized by higher values of G' over G''.

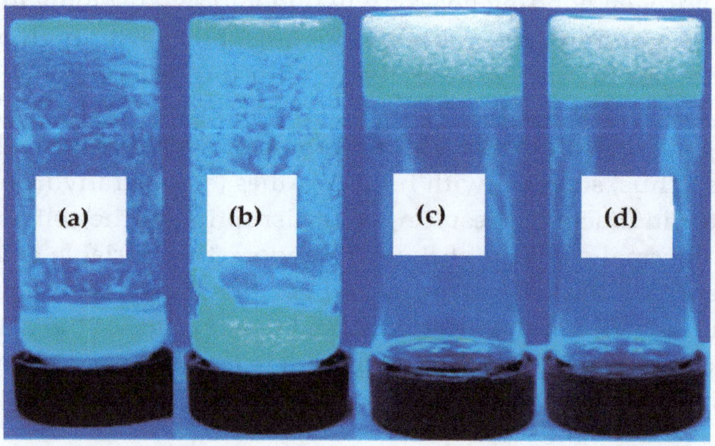

Figure 2. Optical photos of the complexes of: (**A**) mPEG1.1K; and (**B**) mPEG2K with α-CD; and invertible supramolecular hydrogels formed by: (**C**) Ada-PEG1.1K; and (**D**) Ada-PEG2K and α-CD.

2.5.3. Diffraction Techniques

Diffraction techniques such as SAXS (small-angle X-ray scattering), SANS (small-angle neutron scattering) and XRD (X-ray diffraction) have been successfully applied in hydrogel characterization to elucidate their nanoscale structure and to provide insight on the molecular order of the system [92]. A typical example of hydrogel characterization by XRD was presented by Guo and co-workers (cf. Figure 9 in Reference [65]) for the adamantane (Ada)-polyethylene glycol (PEG)-α-CD hybrid supramolecular structure, where sharp XRD diffraction peaks ~$2\theta = 19.8°$ represent the extended channel structure of α-CD. The channel structure corresponds to the formation of α-CD-PEG inclusion complexes, which provides the driving force for the gel formation in this system, consistent with the mechanism of gelation in pseudo-copolymer multicomponent systems as depicted in Scheme 3 above.

2.5.4. Microscopy Methods

Various microscopy methods such as Total Emission Microscopy (TEM), Scanning Electron Microscopy (SEM), Atomic Force Microscopy (AFM) and Scanning Tunnelling Microscopy (STM) have been used to characterize the morphology and microstructure of hydrogels [6,29,51,84]. TEM and SEM imaging can be used to visualize how small belts and fibres can entangle (aggregate) to form a 3D network. Thus, microscopy of such systems with variable morphology provide insight on the formation of H-bonds, π–π stacking, entangled networks, and other self-assembled structures [51]. While TEM and SEM give insight on the morphology of aggregation, AFM and STM are high resolution techniques that can be used to study the conformation of a gel [93]. Moreover, SEM/TEM [94] and AFM [95] can provide valuable information regarding the pore size and pore size distribution of such 3D networks.

2.5.5. Modeling

Computational techniques have been used to model the structural motif of supramolecular hydrogel networks [90,96]. Birchall and coworkers [90] successfully generated the structural features of some amphiphilic systems (**1–6**) that contain aromatic (fluorene or naphthalene) moieties and sugar (galactosamine or glucosamine) residues using molecular modeling [90] (cf. Figure 3b). Four possible modes by which dimers of four amphiphilic monomer units (**1**,**2**; **2**,**2**; **3**,**2**; and **4**,**2**) may be formed were proposed; (i) "J-stacking", which involves a mixture of XH–π interactions between the aromatic residues (where X denotes a heteroatom); (ii) aromatic-aromatic π–π stacking (F2F); (iii) H-bonding between the sugar moieties (S2S); and (iv) solely XH–π interactions between the aromatic group and the sugar (F2S) (cf. Figure 3a).

Each of the four possible configurations were optimized for the lowest energy structures and the most stable dimers were estimated from the relative binding energy of each pair. Molecular modeling can be used to determine if two or more systems form hydrogels based on whether the structural models display favourable binding energy value and their topology may lead to aggregation. The most favourable configuration for gel formation should minimize competitive solvent interactions and promote aggregation of the individual units whilst allowing the solvent to be imbibed within the 3D network. The terminal position of the sugar moieties in the F2F configuration (cf. Figure 3a) may promote competitive H-bonding with the solvent which limits the ability of the units to aggregate into a gel structure.

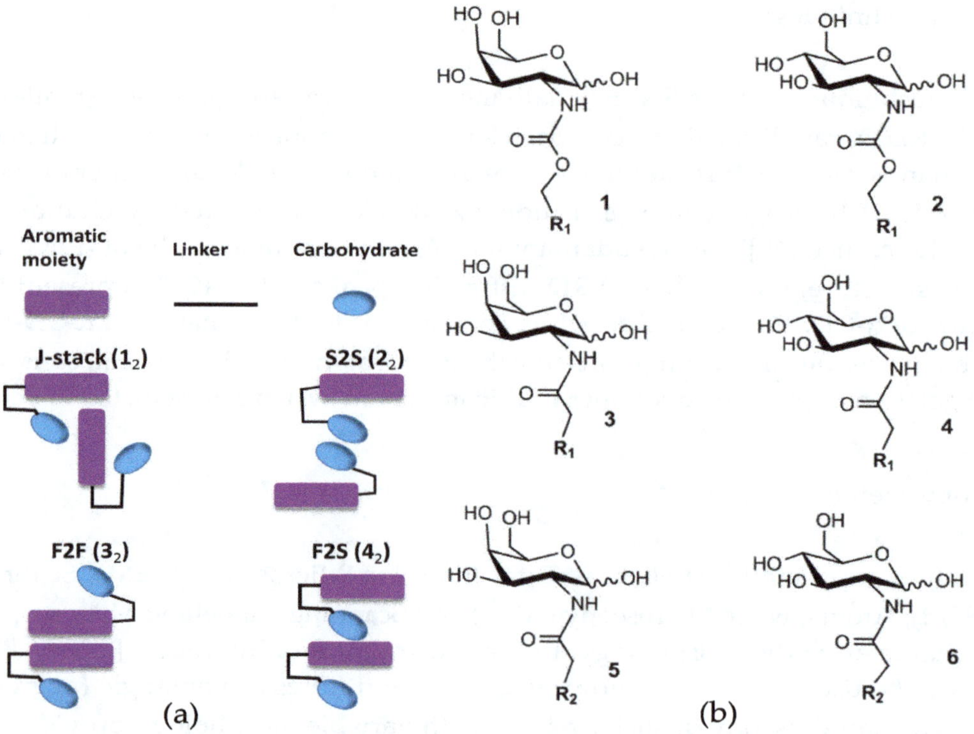

Figure 3. (**a**) Schematic representing dimer configurations of various amphiphilic systems (**1–6**); (**b**) Structures of the aromatic carbohydrate amphiphiles **1–6** containing different aromatic moieties (R_1 and R_2) and either a galactosamine or glucosamine residues. R_1 and R_2 represent fluorene and naphthalene residues, respectively. Redrawn from [90].

2.6. Improving the Stability and Performance of Hydrogels

In the foregoing sections, insight concerning the gelation process was revealed to gain a better understanding of how stimuli-responsive supramolecular hydrogels behave. One of the challenges of the design of supramolecular hydrogels is to modulate their physicochemical properties for optimum and specific applications. In the case of physical hydrogels where CD-based (host–guest) interactions are involved, the stability of the 3D network is variable and may significantly be weakened which limits the widespread application in areas of biomedical and environmental science. The destabilized structure of CD-based hydrogels is known owing to the weak non-covalent host–guest interactions along with the unfavourable entropy of binding of the restricted host and guest molecules within the polymer network. Selected physicochemical properties such as the stability of the 3D scaffold, the critical aggregation point, and the system viscosity can be manipulated to enhance the stability of hydrogels. Structural variations of this type have been achieved through design strategies that vary the nature/combination and feed ratios of the precursor materials, the use of amphiphiles or polymer inclusion complexes (PICs), and incorporation of nanoparticles (NPs), *vide infra*.

2.6.1. Use of Polymer Inclusion Complexes (PIC)

The strategy of polymer inclusion complexation (PIC) is known to enhance the stability of supramolecular hydrogels. PIC-based supramolecular hydrogels involving polymers that contain cyclodextrin (CD) as an ideal host due to its hydrophobic inclusion sites have been widely investigated [48,53,57,61–63]. Small molecules or linear polymers can serve as guest molecules for the CD hosts. For example, PEG has been widely used since its first report in 1990 [96] to form supramolecular linear necklace-like PPR nanostructures with CD. Yui et al. [61,62] obtained a relatively low sol–gel transition temperature by using PIC formation between short PEG chains-modified chitosan (or dextran) and α-CD were synthesized by a series of coupling reactions. The phase-separation of the crystalline domains formed by the host–guest interaction between the α-CD and PEG provide the basis for understanding the supramolecular association and dissociation, namely; a sol–gel transition. The gelation properties of PIC-based hydrogels can be further tuned by adjusting the PEG content due to its variable hydrophilicity profile. Similarly, the solution concentration and the mixing ratio of the host and guest system may be varied. The optimal gel formation is a function of the fraction of crystalline PIC micro-domains, where an ideal host–guest stoichiometry will give the most stable network. At non-stoichiometric ratios, little or no gelation will be observed because of limited physical cross-linker domains. Hence, a delicate compromise between the host–guest stoichiometry and a chemical balance between components with variable hydrophile–lipophile balance (HLB) of the PIC system must be met for optimum gelation to occur.

2.6.2. Use of Host–Guest Macromers (HGMs)

The use of host–guest macromers (HGMs) [48] differ from PICs since the latter forms PPRs, whereas; macromers (cf. Figure 4c) are formed by HGMs. Basically, the HGM approach involves the self-assembly of pre-functionalized host and guest molecules. A previous report [48] on a responsive hydrogel formed from hyaluronic acid biopolymer functionalized with adamantane (AD) to form the guest polymer (AD_xHA) and a polymerizable acrylate, (Ac)-functionalized β-CD (Ac-β-CD) as the host (cf. Figure 4). HGMs were reported to form via self-assembly between AD_xHA and Ac-β-CD driven by efficient host–guest interactions of the less bulky host system and the guest polymer. The stability of the HGM-based hydrogels is understood due to the interaction between the macrocyclic host and guest within the HGM system, and is more favourable than those of the bulky polymer units. The resulting hydrogel [48] was found to be robust with potential utility as a carrier device with extended release properties.

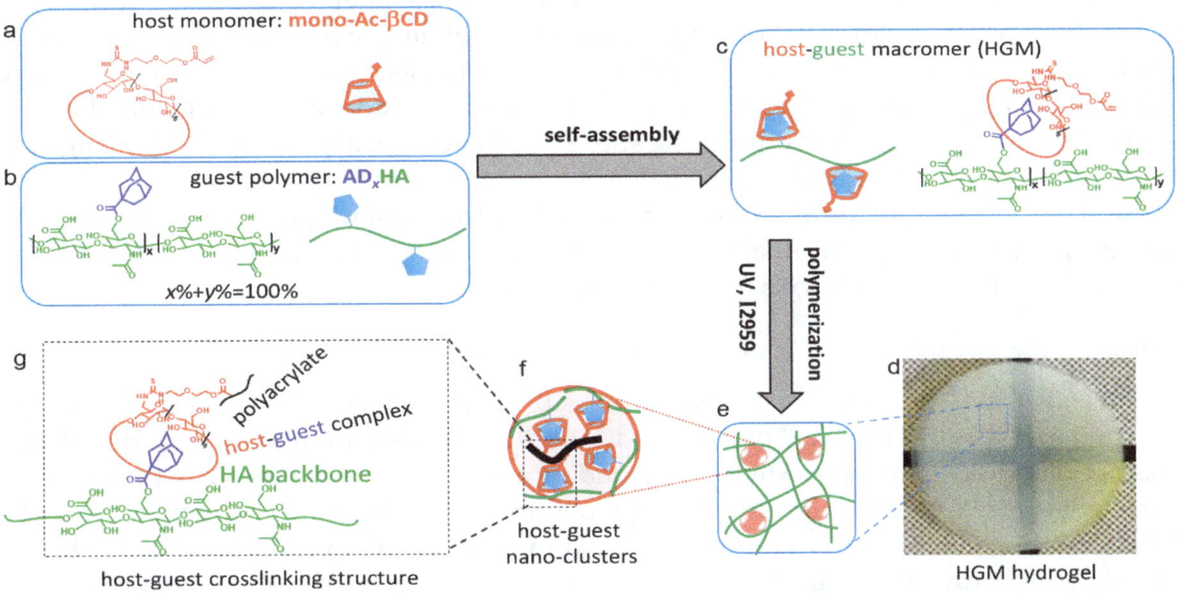

Figure 4. Chemical illustration of: **(a)** the host monomer (mono-Ac-βCD); **(b)** the guest polymer (AD_xHA); and **(c)** the host–guest macromer (HGM). **(d–g)** Representation of various hydrogel/host–guest structures.

2.6.3. Use of Amphiphiles

The use of amphiphiles was long proposed [51,65,90] to promote the gelation process of hydrogels, along with an enhancement of gel stability through the HLB phenomenon, as described above. Co-polymers bearing Ada pendants as those described by Wei et al. [48] and Koopmans et al. [57]

have been widely used to provide the hydrophobic end group requirement. A typical example of such association phenomena is the study reported by Guo et al. [65] where an aqueous solution of α-CD and LMW PEG undergoes precipitation over hydrogel formation (cf. Figure 3). This occurs when complexation occurs between α-CD and LMW PEG at both ends of the PEG, where the unbound PEG becomes too short to form a network. However, when the PEG was functionalized with the highly hydrophobic Ada, a gel was formed. The Ada group serves to (i) decrease the amount of threaded CD which provides sufficient unbound PEG; and (ii) provide additional physical cross-links via hydrophobic aggregation. On this basis, it can be concluded that for hydrogels which are composed of an amphiphilic block copolymer and a CD, the driving force for gelation is a combination of inclusion complexation between CD and PEG blocks, as well as the aggregation of the hydrophobic Ada blocks via favourable interactions illustrated in Scheme 3. Jung et al. [51] reported the first example of a hydrogel formation using a sugar-based amphiphile, where well-defined bilayer aggregates self-assemble via intermolecular hydrogen bonding, π–π stacking and hydrophobic interactions. The synergetic role of various interactions is essential for the successful design and stabilization of such hydrogel systems.

2.6.4. Use of Hybrid Hydrogels and Nano-Fillers

The incorporation of nanoparticles (NPs) has emerged as one of the latest strategies of modulating the mechanical strength and the viscosity of supramolecular hydrogels. Guo et al. [65] used modified β-CD-silica NPs (β-CD-SiO_2) to enhance the gelation of low molecular weight (LMW) PEG-α-CD hydrogels. Cooperative binding of the hydrogel and the NPs yield a strong network structure that leads to a nanoparticle-hybridized supramolecular hydrogel. The storage modulus (G') and the viscosity of the hybrid hydrogel containing ca. 9 wt % of the modified NPs were increased by ca. 4- and 10-fold relative to the native hydrogel. The effect of the incorporation of silver and ferrous NPs into the supramolecular hydrogel networks was reported to enhance their stability. Ma et al. [92,97] reported a PEG-PCL(poly-ε-caprolactone)-α-CD hydrogel hybridized with magnetic Fe_3O_4 NPs. The introduction of magnetic Fe_3O_4 NPs in PEG-PCL-α-CD hydrogel was concluded to speed up the gelation time and to improve the stability of the hydrogel nanocomposite. The interaction of PEG-PCL with the dispersed Fe_3O_4 provided favourable conditions for the subsequent complexation with α-CD. Wang et al. [98] reported hydrogel systems from PEO-PPO-PEO triblock copolymers, α-CD and an inorganic nanotube. The addition of the nanotube resulted in a suppressed hydrophobic aggregation of the middle PPO block lowering the viscosity of the final hydrogel. Thus, the introduction of NPs can be used as a way to modulate the stability of hydrogels for tailored applications.

2.7. Modulating the Viscosity: Influence on Host–Guest Complexation

Various studies have indicated that hydrogel stability generally increase as the gelator concentration increases [57,60,96]. This effect indicates that the formation of hydrogels with stronger networks occurs as a result of more extensive H-bonding and efficient π–π stacking of the hydrogelators. Supramolecular hydrogels investigated by Koopmans and Ritter [57] represent typical examples that reveal the effects of viscosity, concentration and pH in modulating hydrogel stability. Their report studied a series of hydrogel systems based on acrylamide-Ada polymers bearing variable spacer units, as host molecules, and epichlorohydrin-CD globular/linear copolymers as host molecules, respectively (cf. Scheme 3 in Reference [57]). The effects of host/guest polymer concentration, length/amount of hydrophobic Ada chains/groups, and the amount of cross-linker for the α-CD on the stability of the hydrogel were investigated. pH effects were reported to alter the viscosity of the hydrogel. The study by Koopmans and Ritter [57] demonstrate that the viscosity of the medium can affect the stability of a hydrogel network, which reaches a maximum value at specific host–guest stoichiometry, temperature and pH conditions reported therein. Thus, the stability of a hydrogel can be tuned to specific applications by varying the concentration of the host/guest system. Apart from the host/guest concentration, the hydrogel viscosity and its stability depend on other factors that can influence host–guest complexation. For example, the HLB effect (e.g., the length/number of the hydrophobic Ada moiety/groups), amount of cross-linker used for the host molecule (in the case of cross-linked hosts), the pH of the hydrogel medium, and the type of host polymer (*globular vs. linear*).

3. Conclusions

In this mini-review, a general outline describing the mechanism of gel formation in polysaccharide-based supramolecular hydrogels was presented. While the stabilization of helices is the generally proposed pathway of gel formation in single polymer networks, phase separation of polymer networks is supported by the overall mechanism of *gel* $\rightarrow$ *sol* transition in polymer networks. The Flory–Huggins relationship and the utility of the Born–Haber cycle provide insight on the molecular level cross-linking of polymers in aqueous solution. By comparison, the gel formation process involves solute- and solvent-associated steps. As well, several strategies were outlined that can be used to fine-tune the physicochemical properties of biopolymer networks. Many of these strategies relate to stabilizing the 3D structure of the polymer scaffold in such gels as follows: (1) controlling the HLB of the polymer system; (2) controlling the cross-linking of the co-polymers; (3) providing favourable conditions for host–guest complexation; and (4) controlling independent variables such as pH of the medium, along with the nature and concentration of the precursors. Generally, the process of gelation in biopolymers is poorly understood due to inadequate understanding of the role of solvation phenomena (thermodynamic, kinetic, and structural effects) in aqueous media. The role of solvation phenomena in gel formation processes is a suggested direction of future research that deserves further attention.

Acknowledgments: The authors gratefully acknowledge the University of Saskatchewan and Howard Wheater of the Global Institute for Water Security (GIWS) for supporting this research work through CERC Project Number GIWS Award 21927.

Author Contributions: Abdalla Karoyo wrote the first draft of this paper and all authors contributed extensively to subsequent editing of the manuscript.

Abbreviations

3D	Three dimensional
Ada or AD	Adamantane
CD	Cyclodextrin
DSC	Differential scanning calorimetry
FT-IR	Fourier transform infra-red
HGM	Host–guest macromer
HLB	Hydrophile lipophile balance
ICD	Induced circular dichroism
IPN	Interpenetrating network
LMW	Low molecular weight
NMR	Nuclear magnetic resonance
NP	Nanoparticle
PEG	Polyethylene glycol
PIC	Polymer inclusion network
PPR	Polypseudorotaxane
SANS	Small-angle neutron scattering
SAXS	Small-angle X-ray scattering
SEM	Scanning electron microscopy
TEM	Total emission microscopy
STM	Scanning Tunneling Microscopy
UV/Vis	Ultra-violet visible
XRD	X-ray diffraction

References

1. *Polymer Gels and Networks*; Osada, Y.; Khokhlov, A.R. (Eds.) Marcel Dekker, Inc.: New York, NY, USA, 2002.
2. Morimoto, N.; Winnik, F.M.; Akiyoshi, K. Botryoidal Assembly of Cholesteryl–Pullulan/Poly(N-isopropylacrylamide) Nanogels. *Langmuir* **2007**, *23*, 217–223. [CrossRef] [PubMed]
3. Akiyoshi, K.; Deguchi, S.; Moriguchi, N.; Yamaguchi, S.; Sunamoto, J. Self-aggregates of hydrophobized polysaccharides in water. Formation and characteristics of nanoparticles. *Macromolecules* **1993**, *26*, 3062–3068. [CrossRef]
4. De Las Heras Alarcon, C.; Pennadam, S.; Alexander, C. Stimuli responsive polymers for biomedical applications. *Chem. Soc. Rev.* **2005**, *34*, 276–285. [CrossRef] [PubMed]
5. Giammanco, G.E.; Carrion, B.; Coleman, R.M.; Ostrowski, A.D. Photoresponsive Polysaccharide-Based Hydrogels with Tunable Mechanical Properties for Cartilage Tissue Engineering. *ACS Appl. Mater. Interfaces* **2016**, *8*, 14423–14429. [CrossRef] [PubMed]
6. Lü, S.; Gao, C.; Xu, X.; Bai, X.; Duan, H.; Gao, N.; Feng, C.; Xiong, Y.; Liu, M. Injectable and Self-Healing Carbohydrate-Based Hydrogel for Cell Encapsulation. *ACS Appl. Mater. Interfaces* **2015**, *7*, 13029–13037. [CrossRef] [PubMed]
7. Yan, C.; Pochan, D.J. Rheological properties of peptide-based hydrogels for biomedical and other applications. *Chem. Soc. Rev.* **2010**, *39*, 3528–3540. [CrossRef] [PubMed]
8. Shetye, S.P.; Godbole, A.; Bhilegaokar, S.; Gajare, P. Hydrogels: Introduction, Preparation, Characterization and Applications. *Hum. J.* **2015**, *1*, 47–71.
9. Paulino, A.T.; Belfiore, L.A.; Kubota, L.T.; Muniz, E.C.; Tambourgi, E.B. Efficiency of hydrogels based on natural polysaccharides in the removal of Cd2+ ions from aqueous solutions. *Chem. Eng. J.* **2011**, *168*, 68–76. [CrossRef]
10. Guilherme, M.R.; Reis, A.V.; Paulino, A.T.; Fajardo, A.R.; Muniz, E.C.; Tambourgi, E.B. Superabsorbent hydrogel based on modified polysaccharide for removal of Pb2+ and Cu2+ from water with excellent performance. *J. Appl. Polym. Sci.* **2007**, *105*, 2903–2909. [CrossRef]
11. Copello, G.J.; Mebert, A.M.; Raineri, M.; Pesenti, M.P.; Diaz, L.E. Removal of dyes from water using chitosan hydrogel/SiO2 and chitin hydrogel/SiO2 hybrid materials obtained by the sol–gel method. *J. Hazard. Mater.* **2011**, *186*, 932–939. [CrossRef] [PubMed]
12. Udoetok, I.; Wilson, L.; Headley, J. Quaternized Cellulose Hydrogels as Sorbent Materials and Pickering Emulsion Stabilizing Agents. *Materials (Basel)* **2016**, *9*, 645. [CrossRef]
13. *Encapsulation Technologies and Delivery Systems for Food Ingrediaent and Nutraceuticals*; Garti, N.; McClements, D.J. (Eds.) Woodhead Publishing Ltd.: Cambridge, UK, 2012.
14. Parente, M.E.; Ochoa, A.A.; Ares, G.; Russo, F.; Jimenez-Kairuz, A. Bioadhesive Hydrogels for Cosmetic Applications. *Int. J. Cosmet. Sci.* **2015**, *37*, 511–518. [CrossRef] [PubMed]
15. Guilherme, M.R.; Aouada, F.A.; Fajardo, A.R.; Martins, A.F.; Paulino, A.T.; Davi, M.F.T.; Rubira, A.F.; Muniz, E.C. Superabsorbent hydrogels based on polysaccharides for application in agriculture as soil conditioner and nutrient carrier: A review. *Eur. Polym. J.* **2015**, *72*, 365–385. [CrossRef]
16. Mignon, A.; Snoeck, D.; Schaubroeck, D.; Luickx, N.; Dubruel, P.; van Vlierberghe, S.; de Belie, N. pH-Responsive superabsorbent polymers: A pathway to self-healing of mortar. *React. Funct. Polym.* **2015**, *93*, 68–76. [CrossRef]
17. Mignon, A.; Graulus, G.J.; Snoeck, D.; Martins, J.; de Belie, N.; Dubruel, P.; van Vlierberghe, S. pH-Sensitive superabsorbent polymers: A potential candidate material for self-healing concrete. *J. Mater. Sci.* **2014**, *50*, 970–979. [CrossRef]
18. Hyon, S.; Cha, W.; Ikada, Y. Polymer Bulletin 9. *Polym. Bull.* **1987**, *29*, 119–126.
19. Milas, M.; Rinaud, M. Gellan gum, a bacterial gelling polymer. In *Novel Macromolecules in Food Systems*; Doxastakis, G., Kiosseogluou, V., Eds.; Elsevier: Amsterdam, The Netherlands, 2000; pp. 239–263.
20. Rinaudo, M. Advances in Characterization of Polysaccharides in Aqueous Solution and Gel State. In *Polysaccharides: Structural Diversity and Functional Versitility*; Dumitriu, S., Ed.; Marcel Decker: New York, NY, USA, 2004; p. 237.
21. Rinaudo, M. Gelation of Polysaccharides. *J. Intell. Mater. Syst. Struct.* **1993**, *4*, 210–215. [CrossRef]

22. Saitô, H.; Ohki, T.; Takasuka, N.; Sasaki, T. A 13C-N.M.R.-Spectral study of a gel-forming, branched (1→3)-β-D-glucan, (lentinan) from lentinus edodes, and its acid-degraded fractions. Structure, and dependence of conformation on the molecular weight. *Carbohyd. Res.* **1977**, *58*, 293–305. [CrossRef]
23. Pasqui, D.; de Cagna, M.; Barbucci, R. Polysaccharide-based hydrogels: The key role of water in affecting mechanical properties. *Polymers (Basel)* **2012**, *4*, 1517–1534. [CrossRef]
24. Crescenzi, V.; Paradossi, G.; Desideri, P.; Dentini, M.; Cavalieri, F.; Amici, E.; Lisi, R. New hydrogels based on carbohydrate and on carbohydrate-synthetic polymer networks. *Polym. Gels Netw.* **1997**, *5*, 225–239. [CrossRef]
25. Himmelein, S.; Lewe, V.; Stuart, M.C.A.; Ravoo, B.J. A carbohydrate-based hydrogel containing vesicles as responsive non-covalent cross-linkers. *Chem. Sci.* **2014**, *5*, 1054–1058. [CrossRef]
26. Coviello, T.; Matricardi, P.; Marianecci, C.; Alhaique, F. Polysaccharide hydrogels for modified release formulations. *J. Control. Release* **2007**, *119*, 5–24. [CrossRef] [PubMed]
27. Appel, E.A.; del Barrio, J.; Loh, X.J.; Scherman, O.A. Supramolecular polymeric hydrogels. *Chem. Soc. Rev.* **2012**, *41*, 6195–6214. [CrossRef] [PubMed]
28. Suzaki, Y.; Taira, T.; Osakada, K. Physical gels based on supramolecular gelators, including host–guest complexes and pseudorotaxanes. *J. Mater. Chem.* **2011**, *21*, 930–938. [CrossRef]
29. Sukul, P.K.; Malik, S. Supramolecular hydrogels of adenine: Morphological, structural and rheological investigations. *Soft Matter* **2011**, *7*, 4234–4241. [CrossRef]
30. Peppas, N.A.; Hilt, J.Z.; Khademhosseini, A.; Langer, R. Hydrogels in biology and medicine: From molecular principles to bionanotechnology. *Adv. Mater.* **2006**, *18*, 1345–1360. [CrossRef]
31. Kiyonaka, S.; Sugiyasu, K.; Shinkai, S.; Hamachi, I. First thermally responsive supramolecular polymer based on glycosylated amino acid. *J. Am. Chem. Soc.* **2002**, *124*, 10954–10955. [CrossRef] [PubMed]
32. Ge, Z.; Xu, J.; Hu, J.; Zhang, Y.; Liu, S. Synthesis and supramolecular self-assembly of stimuli-responsive water-soluble Janus-type heteroarm star copolymers. *Soft Matter* **2009**, *5*, 3932–3939. [CrossRef]
33. Ren, L.; Liu, T.; Guo, J.; Guo, S.; Wang, X.; Wang, W. A smart pH responsive graphene/polyacrylamide complex via noncovalent interaction. *Nanotechnology* **2010**, *21*, 335701–335706. [CrossRef] [PubMed]
34. Klaikherd, A.; Nagamani, C.; Thayumanavan, S. Multi-Stimuli Sensitive Amphilic Block Copolymer Assemblies. *J. Am. Chem. Soc.* **2009**, *131*, 4830–4838. [CrossRef] [PubMed]
35. Li, Y.; Guo, H.; Zheng, J.; Gan, J.; Wu, K.; Lu, M. Thermoresponsive and self-assembly behaviors of poly(oligo(ethylene glycol) methacrylate) based cyclodextrin cored star polymer and pseudo-graft polymer. *Colloids Surfaces A Physicochem. Eng. Asp.* **2015**, *471*, 178–189. [CrossRef]
36. Hu, Z.; Cai, T.; Chi, C. Thermoresponsive oligo(ethylene glycol)-methacrylate-based polymers and microgels. *Soft Matter* **2010**, *6*, 2115–2123. [CrossRef]
37. Liu, S.; Jiang, M.; Liang, H.; Wu, C. Intermacromolecular complexes due to specific interactions. 13. Formation of micelle-like structure from hydrogen-bonding graft-like complexes in selective solvents. *Polymer (Guildf)* **2000**, *41*, 8697–8702. [CrossRef]
38. Gohy, J.F.; Varshney, S.K.; Jérôme, R. Water-soluble complexes formed by poly(2-vinylpyridinium)-block-poly(ethylene oxide) and poly(sodium methacrylate)-block-poly(ethylene oxide) copolymers. *Macromolecules* **2001**, *34*, 3361–3366. [CrossRef]
39. Kim, J., II; Kim, D.Y.; Kwon, D.Y.; Kang, H.J.; Kim, J.H.; Min, B.H.; Kim, M.S. An injectable biodegradable temperature-responsive gel with an adjustable persistence window. *Biomaterials* **2012**, *33*, 2823–2834. [CrossRef] [PubMed]
40. Yan, J.T.; Li, W.; Zhang, X.Q.; Liu, K.; Wu, P.Y.; Zhang, A.F. Thermoresponsive cyclodextrins with switchable inclusion abilities. *J. Mater. Chem.* **2012**, *22*, 17424–17428. [CrossRef]
41. Udoetok, I.A.; Dimmick, R.M.; Wilson, L.D.; Headley, J.V. Adsorption properties of cross-linked cellulose-epichlorohydrin polymers in aqueous solution. *Carbohydr. Polym.* **2016**, *136*, 329–340. [CrossRef] [PubMed]
42. Jejurikar, A.; Lawrie, G.; Martin, D.; Grøndahl, L. A novel strategy for preparing mechanically robust ionically cross-linked alginate hydrogels. *Biomed. Mater.* **2011**, *6*, 025010–025021. [CrossRef] [PubMed]
43. Kulkarni, A.R.; Soppimath, K.S.; Aminabhavi, T.M.; Rudzinski, W.E. In-vitro release kinetics of cefadroxil-loaded sodium alginate interpenetrating network beads. *Eur. J. Pharm. Biopharm.* **2001**, *51*, 127–133. [CrossRef]

44. Seoud, M.A.; Maachi, R. Biodegradation of Naphthalene by Free and Alginate Entrapped Pseudomonas sp. *Z. Naturforsch. C* **2003**, *58*, 726–731. [CrossRef] [PubMed]
45. Meena, R.; Lehnen, R.; Schmitt, U.; Saake, B. Effect of oat spelt and beech xylan on the gelling properties of kappa-carrageenan hydrogels. *Carbohydr. Polym.* **2011**, *85*, 529–540. [CrossRef]
46. Holme, K.R.; Hall, L.D. Chitosan Derivatives Bearing C10-Alkyl Glycoside Branches: A Temperature-Induced Gelling Polysaccharide. *Macromolecules* **1991**, *24*, 3828–3833. [CrossRef]
47. Djabourov, M.; Leblond, J.; Papon, P. Gelation of acqueous gelatin solutions. II. Rheology of the sol–gel transition. *J. Phys. Fr.* **1988**, *49*, 333–343. [CrossRef]
48. Wei, K.; Zhu, M.; Sun, Y.; Xu, J.; Feng, Q.; Lin, S.; Wu, T.; Xu, J.; Tian, F.; Xia, J.; et al. Robust Biopolymeric Supramolecular "Host-Guest Macromer" Hydrogels Reinforced by in Situ Formed Multivalent Nanoclusters for Cartilage Regeneration. *Macromolecules* **2016**, *49*, 866–875. [CrossRef]
49. Yoshida, H.; Takahashi, M. Structural-Change of Gellan Hydrogel Induced by Annealing. *Food Hydrocoll.* **1993**, *7*, 387–395. [CrossRef]
50. Hosseinzadeh, H. Full-Polysaccharide Superabsorbent Hydrogels Based on Carrageenan and Sodium Alginate. *Middle-East J. Sci. Res.* **2012**, *12*, 1521–1527.
51. Jung, J.H.; John, G.; Masuda, M.; Yoshida, K.; Shinkai, S.; Shimizu, T. Self-assembly of a sugar-based gelator in water: Its remarkable diversity in gelation ability and aggregate structure. *Langmuir* **2001**, *17*, 7229–7232. [CrossRef]
52. Nakahata, M.; Takashima, Y.; Yamaguchi, H.; Harada, A. Redox-responsive self-healing materials formed from host–guest polymers. *Nat. Commun.* **2011**, *2*, 511. [CrossRef] [PubMed]
53. Wu, Y.; Guo, B.; Ma, P.X. Injectable electroactive hydrogels formed via host-guest interactions. *ACS Macro Lett.* **2014**, *3*, 1145–1150. [CrossRef]
54. Yu, J.; Ha, W.; Sun, J.; Shi, Y. Supramolecular Hybrid Hydrogel Based on Host–Guest Interaction and Its Application in Drug Delivery. *ACS Appl. Mater. Interfaces* **2014**, *6*, 19544–19551. [CrossRef] [PubMed]
55. Liao, X.; Chen, G.; Liu, X.; Chen, W.; Chen, F.; Jiang, M. Photoresponsive pseudopolyrotaxane hydrogels based on competition of host-guest interactions. *Angew. Chemie Int. Ed.* **2010**, *49*, 4409–4413. [CrossRef] [PubMed]
56. Amiel, C.; Sebille, B. New Associating polymer systems Involving Water Soluble β-Cyclodextrin Polymers. *J. Incl. Phenom.* **1996**, *25*, 61–67. [CrossRef]
57. Koopmans, C.; Ritter, H. Formation of physical hydrogels via host-guest interactions of β-cyclodextrin polymers and copolymers bearing adamantyl groups. *Macromolecules* **2008**, *41*, 7416–7422. [CrossRef]
58. Annaka, M.; Ogata, Y.; Nakahira, T. Swelling Behavior of Covalently Cross-Linked Gellan Gels. *J. Phys. Chem. B* **2000**, *104*, 6755–6760. [CrossRef]
59. Chen, G.; Jiang, M. Cyclodextrin-based inclusion complexation bridging supramolecular chemistry and macromolecular self-assembly. *Chem. Soc. Rev.* **2011**, *40*, 2254–2266. [CrossRef] [PubMed]
60. Zhang, X.; Huang, J.; Chang, P.R.; Li, J.; Chen, Y.; Wang, D.; Yu, J.; Chen, J. Structure and properties of polysaccharide nanocrystal-doped supramolecular hydrogels based on cyclodextrin inclusion. *Polymer* **2010**, *51*, 4398–4407. [CrossRef]
61. Choi, H.S.; Kontani, K.; Huh, K.M.; Sasaki, S.; Ooya, T.; Lee, W.K.; Yui, N. Rapid Induction of Thermoreversible Hydrogel Formation Based on Poly(propylene glycol)-Grafted Dextran Inclusion Complexes. *Macromol. Biosci.* **2002**, *2*, 298–303. [CrossRef]
62. Huh, K.M.; Ooya, T.; Lee, W.K.; Sasaki, S.; Kwon, I.C.; Jeong, S.Y.; Yui, N. Supramolecular-structured hydrogels showing a reversible phase transition by inclusion complexation between poly(ethylene glycol) grafted dextran and α-cyclodextrin. *Macromolecules* **2001**, *34*, 8657–8662. [CrossRef]
63. Zhao, S.; Lee, J.; Xu, W. Supramolecular hydrogels formed from biodegradable ternary COS-g-PCL-b-MPEG copolymer with α-cyclodextrin and their drug release. *Carbohydr. Res.* **2009**, *344*, 2201–2208. [CrossRef] [PubMed]
64. Maity, G.C. Supramolecular Hydrogels. *J. Phys. Sci.* **2008**, *12*, 173–186.
65. Guo, M.; Jiang, M.; Pispas, S.; Yu, W.; Zhou, C. Supramolecular Hydrogels Made of End-Functionalized Low-Molecular-Weight PEG and α-Cyclodextrin and Their Hybridization with SiO_2 Nanoparticles through Host–Guest Interaction. *Macromolecules* **2008**, *41*, 9744–9749. [CrossRef]
66. Tanaka, F. Thermoreversible gelation strongly coupled to coil-to-helix transition of polymers. *Colloids Surf. B Biointerfaces* **2004**, *38*, 111–114. [CrossRef] [PubMed]

67. Tanaka, T. Phase transitions of gels. *Nippon Gomu Kyokaishi* **1991**, *64*, 219–231. [CrossRef]
68. Viebke, C.; Piculell, L.; Nilssont, S. On the Mechanism of Gelation of Helix-Forming Biopolymers. *Macromolecules* **1994**, *27*, 4160–4166. [CrossRef]
69. Shukla, P. Thermodynamics and kinetics of gelation in the poly(γ-benzyl α,l-glutamate)—Benzyl alcohol system. *Polymer (Guildf)* **1992**, *33*, 365–372. [CrossRef]
70. Moris, E.R.; Rees, D.A.; Robinson, G. Cation-specific aggregation of carrageenan helices: Domain model of polymer gel structure. *J. Mol. Biol.* **1980**, *138*, 349–362. [CrossRef]
71. Dusek, K. My fifty years with polymer gels and networks and beyond. *Polym. Bull.* **2007**, *58*, 321–338. [CrossRef]
72. Shibayama, M.; Tanaka, T. Volume phase transition and related phenomena of polymer gels. *Adv. Polym. Sci.* **1993**, *109*, 1–62.
73. Tanaka, T.; Filmore, D.; Sun, S.T.; Nishio, I.; Swislow, G.; Shah, A. Phase Transition in Ionic Gels. *Rev. Lett.* **1980**, *45*, 1636–1639. [CrossRef]
74. Das, N. Preparation methods and properties of hydrogel: A review. *Int. J. Pharm. Pharm. Sci.* **2013**, *5*, 112–117.
75. Ahmed, E.M. Hydrogel: Preparation, characterization, and applications: A review. *J. Adv. Res.* **2015**, *6*, 105–121. [CrossRef] [PubMed]
76. Aminabhavi, T.M.; Deshmukh, A.M. Polysaccharide-Based Hydrogels as Biomaterials. In *Polymeric Hydrogels as Smart Biomaterials*; Kalia, S., Ed.; Springer: Basel, Switzerland, 2016; pp. 45–71.
77. Deng, W.; Yamaguchi, H.; Takashima, Y.; Harada, A. A chemical-responsive supramolecular hydrogel from modified cyclodextrins. *Angew. Chemie Int. Ed.* **2007**, *46*, 5144–5147. [CrossRef] [PubMed]
78. Van de Manakker, F.; van der Pot, M.; Vermonden, T.; van Nostrum, C.F.; Hennink, W.E. Self-Assembling Hydrogels Based on β-Cyclodextrin/Cholesterol Inclusion Complexes. *Macromolecules* **2008**, *41*, 1766–1773. [CrossRef]
79. Kalichevsky, M.T.; Orford, P.D.; Ring, S.G. The incompatibility of concentrated aqueous solutions of dextran and amylose and its effect on amylose gelation. *Carbohydr. Polym.* **1986**, *6*, 145–154. [CrossRef]
80. Shivashankar, M.; Mandal, B.K. A review on interpenetrating polymer network. *Int. J. Pharm. Pharm. Sci.* **2012**, *4*, 1–7.
81. Myung, D.; Waters, D.J.; Wiseman, M.E. Al Progress in the development of interpenetrating network hydrogels. *Polym. Adv. Technol.* **2008**, *19*, 4109. [CrossRef] [PubMed]
82. Al-Kahtani, A.A.; Sherigara, B.S. Controlled release of theophylline through semi-interpenetrating network microspheres of chitosan-(dextran-g-acrylamide). *J. Mater. Sci. Mater. Med.* **2009**, *20*, 1437–1445. [CrossRef] [PubMed]
83. Blokzijl, B.W.; Engberts, J.B.F.N. Hydrophobic Effects. Opinions and Facts. *Angew. Chem. Int. Ed. Engl.* **1993**, *32*, 1545–1579. [CrossRef]
84. Edwards, W.; Lagadec, C.A.; Smith, D.K. Solvent-gelator interactions-using empirical solvent parameters to better understand the self-assembly of gel-phase materials. *Soft Matter* **2011**, *7*, 110–117. [CrossRef]
85. Reichardt, C.; Welton, T. *Solvents and Solvents Effects in Organic Chemistry*, 4th ed.; Wiley-VCH Verlag GMbH&Co. KGaA: Weinheim, Germany, 2011.
86. Kamlet, M.J.; Abboud, J.L.M.; Abraham, M.H.; Taft, R.W. Linear solvation energy relationships. 23. A comprehensive collection of the solvatochromic parameters, .pi.*, alpha., and beta., and some methods for simplifying the generalized solvatochromic equation. *J. Org. Chem.* **1983**, *48*, 2877–2887. [CrossRef]
87. Lagadec, C.; Smith, D.K. Synthetically accessible, tunable, low-molecular-weight oligopeptide organogelators. *Chem. Commun.* **2011**, *47*, 340–342. [CrossRef] [PubMed]
88. Hirst, A.R.; Smith, D.K. Solvent effects on supramolecular gel-phase materials: Two-component dendritic gel. *Langmuir* **2004**, *20*, 10851–10857. [CrossRef] [PubMed]
89. *Handbook of Characterization of Sol-Gel Science and Technology: Processing, Characterization and Applications*; Almeida, R.M. (Ed.) Kuwer Academic Publishers: Boston, MA, USA, 2005.
90. Birchall, L.S.; Roy, S.; Jayawarna, V.; Hughes, M.; Irvine, E.; Okorogheye, G.T.; Saudi, N.; de Santis, E.; Tuttle, T.; Edwards, A.A.; et al. Exploiting CH-π interactions in supramolecular hydrogels of aromatic carbohydrate amphiphiles. *Chem. Sci.* **2011**, *2*, 1349. [CrossRef]
91. Steed, J.W.; Atwood, J.L. *Supramolecular Chemistry*, 2nd ed.; John Wiley & Sons, Ltd.: West Sussex, UK, 2009.

92. Ma, D.; Xie, X.; Zhang, L.-M. Effect of Molecular Weight and Temperature on Physical Aging of Thin Glassy Poly(2,6-dimethyl-1,4-phenylene oxide) Films. *J. Polym. Sci. B Polym. Phys.* **2007**, *45*, 1390–1398.
93. Decho, A.W. Imaging an alginate polymer gel matrix using atomic force microscopy. *Carbohydr. Res.* **1999**, *315*, 330–333. [CrossRef]
94. Yin, L.; Fei, L.; Cui, F.; Tang, C.; Yin, C. Superporous hydrogels containing poly(acrylic acid-co-acrylamide)/O-carboxymethyl chitosan interpenetrating polymer networks. *Biomaterials* **2007**, *28*, 1258–1266. [CrossRef] [PubMed]
95. Pernodet, N.; Tinland, B.; Sadron, I.C.; Pasteur, C.L. Pore size of agarose gels by atomic force microscopy. *Electrophoresis* **1997**, *18*, 55–58. [CrossRef] [PubMed]
96. Yu, G.; Yan, X.; Han, C.; Huang, F. Characterization of supramolecular gels. *Chem. Soc. Rev.* **2013**, *42*, 6697–6722. [CrossRef] [PubMed]
97. Ma, D.; Zhang, L.M. Fabrication and modulation of magnetically supramolecular hydrogels. *J. Phys. Chem. B* **2008**, *112*, 6315–6321. [CrossRef] [PubMed]
98. Wang, W.; Wang, H.; Ren, C.; Wang, J.; Tan, M.; Shen, J.; Yang, Z.; Wang, P.G.; Wang, L. A saccharide-based supramolecular hydrogel for cell culture. *Carbohydr. Res.* **2011**, *346*, 1013–1017. [CrossRef] [PubMed]

10

Temperature-Triggered Colloidal Gelation through Well-Defined Grafted Polymeric Surfaces

Jan Maarten van Doorn, Joris Sprakel and Thomas E. Kodger *

Physical Chemistry and Soft Matter, Wageningen University & Research, Stippeneng 4, 6708 WE, Wageningen, The Netherlands; janmaarten.vandoorn@wur.nl (J.M.v.D.); joris.sprakel@wur.nl (J.S.)
* Correspondence: thomas.kodger@wur.nl

Academic Editor: Clemens K. Weiss

Abstract: Sufficiently strong interparticle attractions can lead to aggregation of a colloidal suspension and, at high enough volume fractions, form a mechanically rigid percolating network known as a colloidal gel. We synthesize a model thermo-responsive colloidal system for systematically studying the effect of surface properties, grafting density and chain length, on the particle dynamics within colloidal gels. After inducing an attraction between particles by heating, aggregates undergo thermal fluctuation which we observe and analyze microscopically; the magnitude of the variance in bond angle is larger for lower grafting densities. Macroscopically, a clear increase of the linear mechanical behavior of the gels on both the grafting density and chain length arises, as measured by rheology, which is inversely proportional to the magnitude of local bond angle fluctuations. This colloidal system will allow for further elucidation of the microscopic origins to the complex macroscopic mechanical behavior of colloidal gels including bending modes within the network.

Keywords: colloid gel; grafting density; temperature-triggered gelation; surface initiated atom transfer radical polymerization

---

1. Introduction

Colloidal particles are of significant importance to various fields in science and engineering and to consumer products, such as foods and paints. Upon inducing sufficiently strong attraction to a colloidal suspension, colloidal particles will aggregate and form a mechanically rigid percolating network above a critical volume fraction [1]. These structures, known as colloidal gels, can be regarded as a model for soft heterogeneous solids. Differing from polymeric gels, the bonds between particles in colloidal gels have a non-permanent nature enabling bonds to reform and individual particles to rearrange due to mechanical deformation or thermal fluctuations [2–4]. These rearrangements mainly govern the mechanical behavior of these soft solids and are of paramount importance to understanding the mechanics of soft heterogeneous solids [5,6].

Many efforts studying the particle dynamics within colloidal gels focus on the attraction strength as a control parameter. Systematic investigations on colloidal gels typically employ a depletion attraction [7], where both the range and depth of interaction may be tuned. However, apart from longitudinal fluctuations such as detaching and attaching, particles can also exhibit transverse modes of rearrangement such as sliding [3]. Where the first mode is mainly influenced by the inter-particle potential, the details of the other modes are difficult to unravel, but thought to be governed by the surface properties of the particles such as their friction coefficients [8]. The implications of such parameters on the assembly of colloidal systems may be profound, and are only briefly discussed in the literature; this is partly due to the fact that there does not yet exist an experimental means to investigate their effects.

In this paper, we synthesize a colloidal model system that is suitable for systematically studying the effect of particle surface properties such as grafting density and chain length, on the dynamics within colloidal gels. We control the grafting density and chain length by using surface initiated atom transfer radical polymerization (ATRP): The grafting density is tuned by copolymerizing a known volume of an ATRP initiator-monomer during particle formation and the chain length is tuned by adding a sacrificial initiator to the bulk solution during the ATRP reaction. We grow a temperature sensitive polymer, poly(*N*-isopropylacrylamide), from the particle surface to alter the interparticle potential dynamically. After inducing an attraction between particles by heating, a clear dependence on the magnitude of local bond angle fluctuations and linear mechanical behavior of the gel arises from both the grafting density and chain length. Lastly, we disperse these particles in a refractive index matching aqueous solution allowing for 3D confocal imaging during gelation.

2. Results and Discussion

The origin of the interparticle attraction between colloidal particles can be varied; common examples are depletion [7], electrostatic [9], or van der Waals; however, these sources of attraction cannot be easily triggered. Here, we induce inter-particle attraction using a temperature sensitive surface grafted polymer, poly(*N*-isopropylacrylamide) (pNIPAM) [10]. This polymer has a Lower Critical Solution Temperature (LCST) in water around 32 °C. Above this temperature, the polymers expel water and demixes from the aqueous solution which induces interparticle attraction [11]. When the temperature is lowered below the LCST, the polymer solubility is enhanced, resulting in good solvent for $T \ll$ LCST, and the interparticle potential becomes sterically repulsive. The precise value of the LCST is sensitive to the composition of the solvent [12] and as a result, we design our system to be stable in water. One of the challenges with studying concentrated particle suspensions is that the refractive index, n, mismatch between the water, $n = 1.333$, and the material of which the colloids are formed impedes experimental optical techniques due to multiple light scattering. To overcome this challenge, we synthesize monodispersed particles from poly(2,2,2-trifluoroethyl methacrylate) (ptFEMA) which has a relatively low refractive index of $n = 1.42$ which is suitable for refractive index matching. By forming particles with diameters between 0.5 μm and 3 μm, they are large enough to be easily visualized by optical microscopy and also small enough to undergo thermal fluctuations; here, we synthesize 1.00 μm diameter particles. Additionally, these particles are co-polymerized with 2-(2-bromoisobutyryloxy) ethyl methacrylate (BIEA) which acts as a monomer during particle synthesis and as an initiator for atom transfer radical polymer (ATRP) [13]. Due to its two sided functionality, this molecule is called an *inimer* [14]. Varying the co-polymerization volume percentage from 0.1% to 3.0% of *inimer* during particle synthesis enables tuning of the grafting density on the particle surface. Additionally, ATRP allows for precise control over the length of these grafted polymers [15]; resulting in the independent ability to tune both the length and density of polymer present on the particle surface as depicted in Figure 1.

During a typical ATRP reaction, the degree of polymerization is controlled by the molar ratio of the initiator to monomer. However, the precise molar value of surface available *inimer* molecules is difficult to determine. This leaves choosing the appropriate amount of monomer to establish a desired ratio challenging. To nevertheless control the length of the grafted polymers, we add a conventional ATRP initiator with identical ATRP initiation rate to the grafting reaction. This yields a free linear polymer with the same degree of polymerization (DP) as the polymers which are simultaneously grown from the surface [16]. Gel permeation chromatography (GPC) analysis of the linear polymer results in a clear dependence in the chain length for the desired DP, while still retaining a fairly monodisperse distribution as seen in Figure 2.

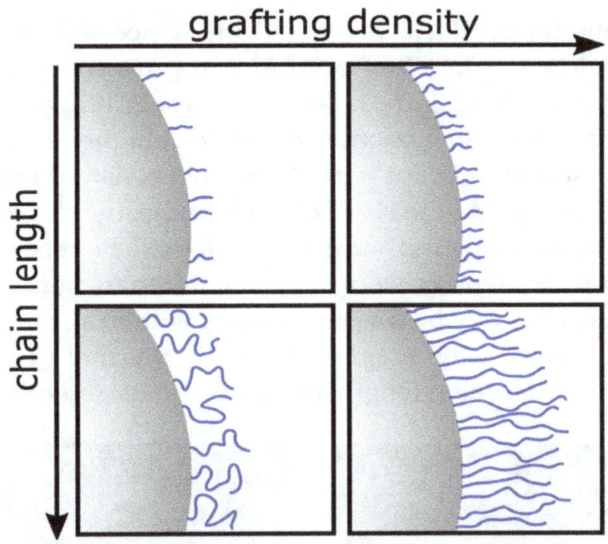

Figure 1. Controlled grafting density and chain length using surface initiated atom transfer radical polymer (ATRP) of poly(*N*-isopropylacrylamide) (pNIPAM).

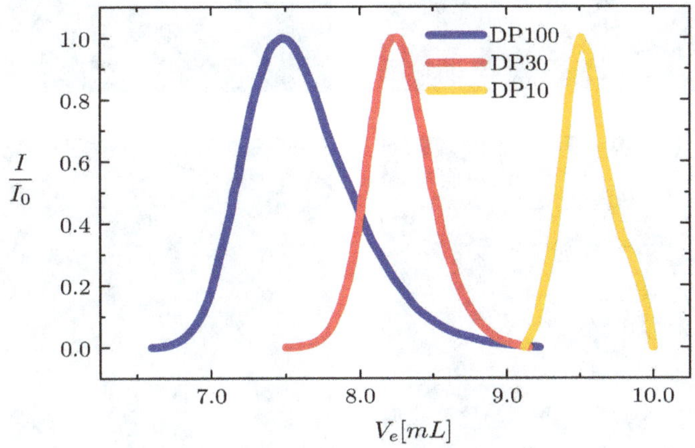

Figure 2. Gel permeation chromatography (GPC) elution profiles for polymers with different degrees of polymerization; with the elution volume of the polymers being inversely proportional to their respective degree of polymerisation. DP = 10, Mn = 2.9×10^2 g/mol, Mw = 3.1×10^2 g/mol, PDI = 1.10; DP = 30, Mn = 3.1×10^3 g/mol, Mw = 3.6×10^3 g/mol, PDI = 1.2; DP = 100, Mn = 9.3×10^3 g/mol, Mw = 1.5×10^4 g/mol, PDI = 1.6.

To obtain a temperature triggerable interaction, a pNIPAM surface modification is insufficient; electrostatic repulsion between particles must also be tuned. A controlled concentration of salt, 30 mM NaCl, is added to screen electrostatic repulsions to approximately the length scale of the shortest surface polymers; the calculated Debye screening length is $\kappa^{-1} = 1.7$ nm. It must be noted that at higher [NaCl], the LCST of pNIPAM decreases below room temperature [12] and additionally electrostatic repulsion is insufficient to prevent aggregation by van der Waals forces between particles; the precise salt concentration is crucial to obtain a temperature sensitive interaction potential via the pNIPAM grafted surfaces.

To study the structure and dynamics of aggregated surface modified particles, we employ bright-field microscopy. A two-dimensional array of colloidal particles is formed by simply letting the relatively dense ptFEMA colloids sediment onto the capillary wall. To prevent particles adhering to the capillary walls, the capillaries are coated with a polyeletrolyte multilayer which has been shown to eliminate wall interactions for pNIPAM layers [17,18]. Once sedimented, the sample is heated to a temperature slightly below the LCST of pNIPAM in pure water, the particles begin to form two-dimensional aggregates as seen in Figure 3. For the lowest grafting density, only a few aggregates are found at this temperature and volume fraction, ϕ, while at higher grafting density, large extended aggregates are visible. Correspondingly, for particles with a constant grafting density but differing chain length, the effects are similar: At the short chain lengths, the degree of aggregation is limited while at longer chain lengths, very few individual particles exist as seen in Figure 4. Aggregates of particles with the highest grafting density seem to be smaller than aggregates composed of particles with lower grafting densities. This may be due to particles with lower grafting densities rearranging more easily. Within each aggregate, the magnitude of the thermal fluctuations between particles appears to be directly related to the chain length and grafting density of the surface polymer.

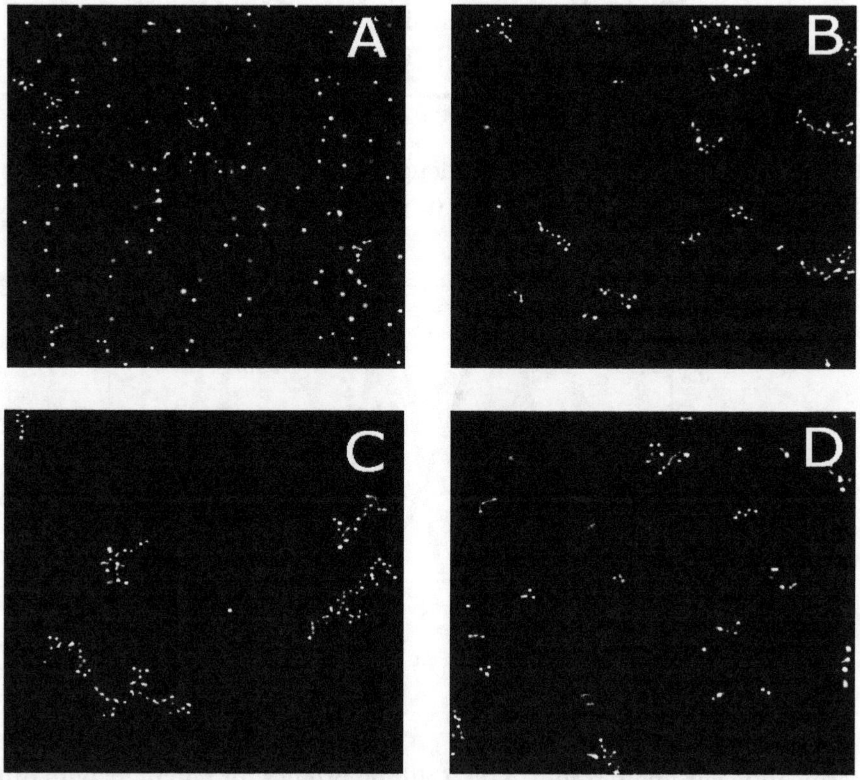

Figure 3. Optical microscopy images for different grafting densities at 32 °C in 30 mM NaCl for DP = 100. (**A**) 0.1%; (**B**) 0.3%; (**C**) 1.0%; and (**D**) 3.0%.

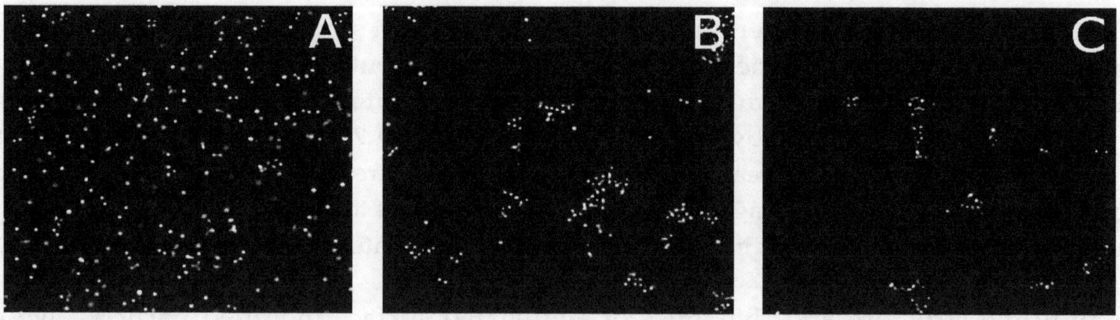

Figure 4. Optical microscopy images for different chain lengths at 32 °C in 10 mM NaCl for a grafting density of 3.0%. (**A**) DP = 10; (**B**) DP = 30; and (**C**) DP = 100.

By measuring the angle between neighboring particles over time, we are able to directly quantify the amplitude of the bond angle fluctuations as a proxy for the friction coefficient. Centers of neighboring particle are first located and tracked over time; after which the angle, $\theta(t)$, is calculated as seen in Figure 5 inset. The fluctuations about the mean angle, $\theta(t) = \theta(t) - \langle\theta(t)\rangle$, are shown for two grafting densities in Figure 5 (see Supplementary Movies). At lower grafting density, angular fluctuations are large. Conversely, at a higher grafting density, the angular fluctuations are minimized. A smaller amplitude in θ corresponds to more hindrances in thermally activated motion between particles occurring which points at a higher friction between the particle surfaces [18]. Polymer brushes, with their high grafting densities, have repeatedly been found to be low friction interfaces, seemingly contradictory to the above observations [19–21]. However, temperature sensitive polymer brushes tethered to a substrate have been shown to switch from low to high friction above the LCST of pNIPAM which supports the different amplitudes of θ seen in Figure 5 [22,23]. Therefore, increasing grafting density also increases the friction between particles; the consequences of this increased friction may be profound. We hypothesize that colloidal gels with lower friction coefficients and therefore more flexible bonds are capable of relaxing applied stresses and would result in a lower elastic modulus.

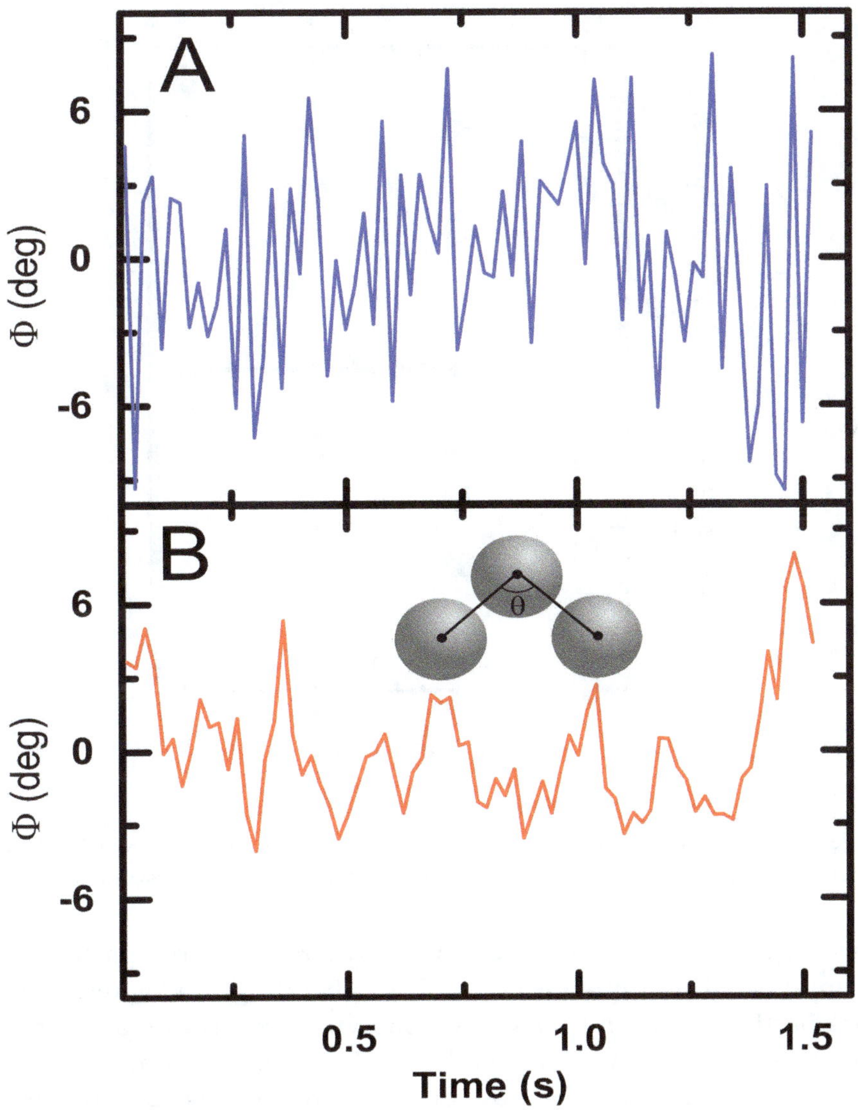

Figure 5. Bond angle fluctuations for samples of DP = 100 with 0.3% (**A**) and 3% (**B**) grafting density. The variance of the fluctuations are 6.7 $(\text{deg})^2$ and 19.2 $(\text{deg})^2$ respectively. Inset; schematic representation of bond angle calculation between neighboring particles.

To directly investigate whether more flexible bonds lead to a lower elastic modulus, we use bulk rheology. At a higher volume fraction, $\phi = 0.28 \pm 0.02$, the particle dispersions form elastic 3D colloidal gels upon heating above the LCST. We compare the mechanical behavior of colloidal gels with differing grafting densities and chain length of the surface pNIPAM polymer. Though the precise volume fraction of the dispersion is not known, the resulting differences of linear mechanical response in these gels are larger than the variance caused by the uncertainty in ϕ as seen in Figure 6. The elastic modulus of colloidal gels has been shown to scale as, $G' = (\kappa_0/a)(\phi - \phi_c)^p$ where κ_0 is the two-particle spring constant, a is the particle size, p is a scaling exponent which depends on the nature of the network deformation, and ϕ_c is the critical volume fraction which is typically $\phi_c \leq 0.08$ [7,24]. Here, $\phi = 0.28 >> \phi_c$, therefore, the uncertainty in ϕ cannot account for the large variation in the elastic moduli seen in Figure 6; it must arise from changes in κ_0. For the highest grafting densities, the elastic and viscous responses of the gels converge for all chain lengths. By contrast, at lower grafting densities the gels are significantly weaker by nearly two orders of magnitude for the longest chain length; this drop in elasticity corresponds well with the larger magnitude in θ observed microscopically as seen in Figure 5.

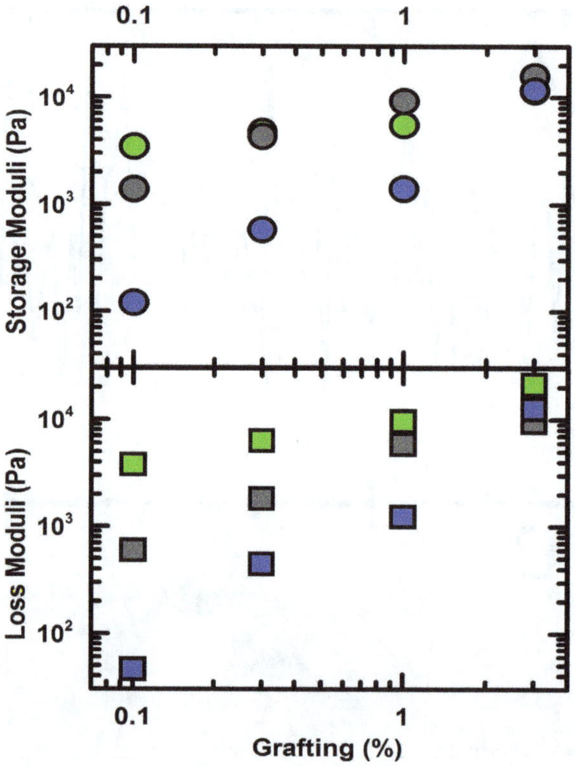

Figure 6. Storage and loss moduli after heating dispersions at $\phi = 0.28$ with 30 mM NaCl to 45 °C with green for DP = 10, gray for DP = 30, blue for DP = 100. All moduli are measured at 1 Hz and $\gamma < 0.03$.

Colloidal networks resist mechanical deformation by stretching interparticle bonds and bending particle strands composed of multiple particles. These bending modes result in angular changes between individual particles, $\Delta\theta$, and have been shown to contribute significantly to the elastic response of colloidal networks [7]. Therefore, hindering these bending modes can directly increase the elastic response which is seen in Figure 6. How precisely the microscopic changes, grafting density and chain length, manifest as differences in the macroscopic rheology including yielding is beyond the scope of this work and has been the subject of extensive simulation studies [25–27].

Finally, these model ptFEMA particles may be fluorescently labeled and dispersed in a refractive index matching solution of 50 wt % sucrose with 10 mM NaCl. By refractive index matching the particles to the suspending solution, light scattering is minimized and by combining fluorescent labeling, 3D confocal microscopy images may be captured. As the thermo-responsive nature of the polymer brush is retained in the sucrose solution, the dispersion may still be heated from a colloidal liquid into a colloidal gel while being imaged deep into the sample, ≈75 μm as shown in Figure 7. From the individual particle locations, the radial distribution function, $g(r/a)$, was calculated and shown in Figure 7C; the $g(r/a)$ clearly show a transition from a liquid dispersion of particles to a colloidal gel by heating. This ability to dynamically induce gelation by heating this particle dispersion with its controlled pNIPAM surface polymer is similar to previous work where the authors quantified the kinetics and structure of pNIPAM grafted nanoparticles [28]. In these dynamic light scattering studies, only the fractal dimension was determined as individual particle kinetics are not available. From this work, a detailed kinetic aggregation framework was proposed to connect the local particle-level dynamics to the macroscopic rheology, effectively describing many experimental rheology results on colloidal gels [29]. The model system proposed here will allow for a detailed study of this kinetics framework to different gelation processes as well as allowing direct observation of microscopic sliding dynamics between particles after gelation in three dimensions using confocal microscopy.

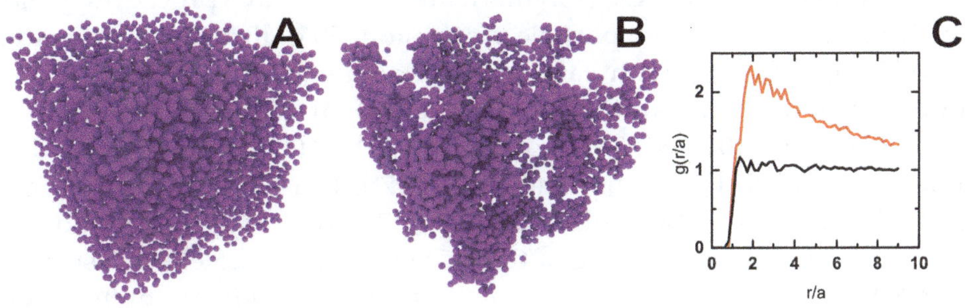

Figure 7. Computer-reconstructed visualizations of a sample with particle coordinates obtained from three-dimensional confocal microscopy data. The field of view is 67 μm × 67 μm × 75 μm. (**A**) a liquid dispersion of particles, $\phi \approx 0.15$, at 25 °C in 50 wt % Sucrose with 10 mM NaCl; (**B**) a colloidal gel of the same dispersion at 50 °C; (**C**) calculated radial distribution functions normalized for particle size, a, for gel (red, **A**) and liquid dispersion (black, **B**).

3. Conclusions

We have developed a thermally responsive colloidal system with controlled grafting density and chain length of a pNIPAM polymer on the particle surface. Upon heating, such dispersions form a colloidal gel. Both the microscopic bond angle fluctuations and macroscopic elastic moduli exhibit a clear dependence on both grafting density and chain length. The unique combination of complete transparency, tunable particle surface properties and temperature-triggerable interactions paves the way to the study of gelation kinetics in three-dimensions with high resolution.

4. Materials and Methods

All materials were purchased from TCI Europe (Zwijndrecht, Belgium) and used as received unless otherwise noted. *N*-Isopropyl acrylamide (NIPAM) monomer was recrystallized from *n*-hexane prior to use. Additionally, the *inimer* monomer, 2-(2-bromoisobutyryloxy) ethyl methacrylate (BIEA), was synthesized as previously reported [13,14].

4.1. Particle Synthesis

We synthesized poly(2,2,2-trifluoroethyl methacrylate) (ptFEMA) colloidal particles co-polymerized with 2-(2-bromoisobutyryloxy) ethyl methacrylate using free radical dispersion polymerization [13]. To a 500 mL round bottom flask is added 30 mL water, 270 mL methanol, 25 mL

2,2,2-trifluoroethyl methacrylate, 250 mg 2,2-azobis(2-methylpropionitrile), 250 mg 3-sulfopropyl methacrylate potassium salt (Sigma-Aldrich, St. Louis, MI, USA) and 25 μL of BIEA (0.1 vol % to monomer). The flask is placed under reflux conditions in a silicone oil bath preheated to 80 °C and allowed to polymerize for 4 h. The resulting particles have a hydrodynamic diameter, $a = 1.00$ μm with a polydispersity index, PDI = $(\sigma(a)/\langle a \rangle) = 0.01$ as determined by Dynamic Light Scattering (DLS). The reaction is repeated with 75 μL, 250 μL and 750 μL of BIEA to arrive at 0.3 vol %, 1.0 vol % and 3.0 vol % of *inimer* respective to monomer, with no measurable change to particle diameter or polydispersity.

4.2. Surface Initiated ATRP

Particle dispersions were washed three times by centrifugation at 250× *g* into a 1 wt % solution of L23 surfactant (Sigma-Aldrich) to a final volume of 200 mL. To graft polymers from the particle surface, 50 g dimethylformamide, 2 g NIPAM (1.7×10^{-2} moles), 0.47 mL tris[2-(dimethylamino)ethyl]amine (1.7×10^{-3} moles), 0.253 mL ethyl α-bromoisobutyrate (1.7×10^{-3} moles) were added to a 250 mL round bottom flask. The solution was bubbled with nitrogen for 15 min, after which 0.168 g of Cu(I)Cl (1.7×10^{-3} moles) was added to initiate the polymerization. The above procedure yielded a DP = 10 as shown in Figure 2. For DP = 30, 50 mL particle dispersion, 50 g dimethylformamide, 2 g NIPAM (1.7×10^{-2} moles), 0.156 mL tris[2-(dimethylamino)ethyl]amine (0.56×10^{-3} moles), 0.084 mL ethyl α-bromoisobutyrate (0.56×10^{-3} moles) were added to a 250 mL round bottom flask, bubbled, and initiated with 0.056 g of Cu(I)Cl (0.56×10^{-3} moles). For DP = 100, 50 mL particle dispersion, 50 g dimethylformamide, 2 g NIPAM (1.7×10^{-2} moles), 0.047 mL tris[2-(dimethylamino)ethyl]amine (1.7×10^{-4} moles), 0.025 mL ethyl α-bromoisobutyrate (1.7×10^{-4} moles) were added to a 250 mL round bottom flask and initiated with 0.017 g of Cu(I)Cl (1.7×10^{-4} moles). These procedures were repeated for each BIEA volume ratio, 0.1%, 0.3 vol %, 1.0 vol % and 3.0 vol %, to yield a total of 12 different particle dispersions each with a unique grafting density and chain length. After surface modification, the dispersions were centrifuged and the supernatant collected and purified before GPC measurements. The sedimented particles were redispersed in 20 mL of demineralized water and each particle dispersion was dialyzed for 10 days again in deionized water to remove Cu(I)Cl and the surfactant L23. The hydrodynamic diameters of the particles after surface modification have been characterized by DLS using a second-order cumulants fit to the correlation functions. The results show an increasing trend only for the particles with the highest surface grafting density, 3%, from $a = 1020 \pm 68$ nm for the bare particles to DP = 10, $a = 996 \pm 63$ nm; DP = 30, $a = 1044 \pm 31$ nm; and DP = 100, $a = 1112 \pm 40$ nm.

The supernatant was heated to 80 °C overnight to remove water and then precipitated in diethyl ether, dissolved in chloroform, and precipitated again, a total of three times. The precipitate was dried and dissolved in water and mixed bed resins (AG501-X8, Bio-Rad, Hercules, CA, USA) were added to remove copper salts. The resins were filtered away and the now clean pNIPAM polymer was freeze dried. GPC measurements were performed on 5 mg/mL samples in a solution of tetrahydrofuran with 5 vol % triethylamine at a flow rate of 1 mL/min at 35 °C on an Agilent Technologies 1200, PLgel 5 μm Mixed-D column [30]. The column was calibrated prior to use with linear polystyrene dissolved in the above solvent.

4.3. Fluorescent Labeling

A single dispersion, 1 vol % BIEA with DP = 100, was fluorescently labeled. A miniemulsion was prepared by tip sonication, containing 0.2 mL toluene, 5 mg boron-dipyrromethene 543 dye (Excition, Inc., West Chester, OH, USA), and 4 mL 1 wt % solution of L23 surfactant. To this miniemulsion, 1.5 mL of particle dispersion at $\phi = 0.30$ was added. This dispersion was mixed for 3 days to allow the particles to swell and take up the dye. Subsequently, dry nitrogen was blown over the top of the dispersion to remove toluene and kinetically trap the dye inside the particles. This fluorescently labeled dispersion was dialyzed against deionized water to remove L23. Sucrose was then added as a powder and dissolved to a final concentration of 50 wt % which resulted in a refractive index matched dispersion.

4.4. Microscopy

Bright field and confocal microscopy experiments were performed in capillaries of 40 mm $\times$ 4 mm $\times$ 0.2 mm inner dimensions coated with polyelectrolyte multilayers. Capillaries were first plasma treated, then submerged into a 1 M NaCl solution with 1 wt % poly(diallydimethyl ammonium) chloride (Mw $\approx 5 \times 10^5$ g/mol, Sigma-Aldirch), then washed extensively with deionized water, then submerged in a 1 M NaCl solution with 1 wt % poly(styrene sulfonate) (Mw $\approx 2 \times 10^5$ g/mol, Sigma-Aldrich) and finally washed extensively with deionized water. This layer-by-layer treatment was repeated three times for a total of six layers. A dilute suspension of each particle dispersion, $\phi = 0.001$, was prepared by diluting with either a 10 mM or 30 mM NaCl solution, loaded into a coated capillary, allowed to sediment over 1 h and finally heated to the desired temperature using a home-built objective and capillary heater. Samples were allowed to equilibrate for 10 min at each temperature before imaging. Images were then captured using a Nikon microscopy (Nikon Instruments, Amsterdam, The Netherlands) with a 60$\times$ water immersion objective at 50 fps using a Fastec HiSpec1 camera (Fastec Imaging Corporation, San Diego, CA, USA). Confocal microscopy 3D images were captured using a Zeiss LSM5 Pascal (Carl Zeiss AG, Oberkochen, Germany) with 488 nm excitation and 100$\times$ oil immersion objective. The refractive index matched dispersion in 50 wt % sucrose with 10 mM NaCl was first imaged at room temperature then quickly heated to 50 °C. Particle centers were located using standard locating software [31] using Matlab.

4.5. Rheology

For rheology measurements, each dialyzed dispersion was allowed to sediment over several days and the supernatant removed until the dispersion obtained a high volume fraction, $\phi > 0.30$. Each dispersion's volume fraction was measured by drying a known mass of dispersion, $\approx$1.00 g, in an 80 °C oven overnight; this method exhibited repeatability within 6% of the mean. To this measured dispersion, a small volume of water and 2.0 M NaCl was added to obtain $\phi = 0.28$ in 100 mM NaCl for each dispersion which was measured using an Anton Paar MCR501 rheometer (Anton Paar, Graz, Austria) with a 50 mm diameter cone-plate geometry. A solution of tetradecane was added around the geometry to minimize evaporation. The dispersion was heated to 45 °C in 10 min and allowed to gel further over 1 h, then measured at 1 Hz with an applied strain from $\gamma = 0.001$ to $\gamma = 1.00$ and an average value taken within the linear regime, typically $\gamma < 0.03$.

Acknowledgments: This work of Jan Maarten van Doorn is part of the Industrial Partnership Programme Hybrid Soft Materials that is carried out under an agreement between Unilever Research and Development B.V. and The Netherlands Organisation for Scientific Research (NWO). Thomas E. Kodger acknowledges support by a VENI personal grant also from NWO.

Author Contributions: Jan Maarten van Doorn, Joris Sprakel and Thomas E. Kodger conceived and designed the experiments and wrote the paper; Jan Maarten van Doorn and Thomas E. Kodger performed the experiments and analyzed the data.

Abbreviations

The following abbreviations are used in this manuscript:

pNIPAM	poly(*N*-isopropyl acrylamide)
ptFEMA	poly(2,2,2-trifluoroethyl methacrylate)
BIEA	2-(2-bromoisobutyryloxy) ethyl methacrylate
ATRP	Atom Transfer Radical Polymerization
GPC	Gel Permeation Chromatography
DP	Degree of Polymerisation

References

1. Trappe, V.; Prasad, V.; Cipelletti, L.; Segre, P.; Weitz, D.A. Jamming phase diagram for attractive particles. *Nature* **2001**, *411*, 772–775.
2. Buscall, R.; Choudhury, T.H.; Faers, M.A.; Goodwin, J.W.; Luckham, P.A.; Partridge, S.J. Towards rationalising collapse times for the delayed sedimentation of weakly-aggregated colloidal gels. *Soft Matter* **2009**, *5*, 1345–1349.
3. Rajaram, B.; Mohraz, A. Microstructural response of dilute colloidal gels to nonlinear shear deformation. *Soft Matter* **2010**, *6*, 2246–2259.
4. Sprakel, J.; Lindstrom, S.B.; Kodger, T.E.; Weitz, D.A. Stress Enhancement in the Delayed Yielding of Colloidal Gels. *Phys. Rev. Lett.* **2011**, *106*, 248303.
5. Coniglio, A.; de Arcangelis, L.; Gado, E.D.; Fierro, A.; Sator, N. Percolation, gelation and dynamical behaviour in colloids. *J. Phys. Condens. Matter* **2004**, *16*, 4831–4839.
6. Cipelletti, L.; Manley, S.; Ball, R.C.; Weitz, D.A. Universal Aging Features in the Restructuring of Fractal Colloidal Gels. *Phys. Rev. Lett.* **2000**, *84*, 2275–2278.
7. Prasad, V.; Trappe, V.; Dinsmore, A.D.; Segre, P.N.; Cipelletti, L.; Weitz, D.A. Rideal Lecture. *Faraday Discuss.* **2002**, *123*, 1–12.
8. Condre, J.M.; Ligoure, C.; Cipelletti, L. The role of solid friction in the sedimentation of strongly attractive colloidal gels. *J. Stat. Mech. Theory Exp.* **2007**, *2007*, 02010.
9. Russell, E.; Sprakel, J.; Kodger, T.; Weitz, D. Colloidal Gelation of Oppositely Charged Particles. *Soft Matter* **2012**, *8*, 8697.
10. Masci, G.; Giacomelli, L.; Crescenzi, V. Atom Transfer Radical Polymerization of *N*-Isopropylacrylamide. *Macromol. Rapid Commun.* **2004**, *25*, 559–564.
11. Pelton, R. Poly(*N*-isopropylacrylamide) (PNIPAM) is never hydrophobic. *J. Colloid Interface Sci.* **2010**, *348*, 673–674.
12. Zhang, Y.; Furyk, S.; Bergbreiter, D.E.; Cremer, P.S. Specific ion effects on the water solubility of macromolecules: PNIPAM and the Hofmeister series. *J. Am. Chem. Soc.* **2005**, *127*, 14505–14510.
13. Kodger, T.E.; Guerra, R.E.; Sprakel, J. Precise colloids with tunable interactions for confocal microscopy. *Sci. Rep.* **2015**, *5*, 14635.
14. Matyjaszewski, K.; Gaynor, S.G.; Kulfan, A.; Podwika, M. Preparation of Hyperbranched Polyacrylates by Atom Transfer Radical Polymerization. 1. Acrylic AB* Monomers in Atom Transfer Radical Polymerizations. *Macromolecules* **1997**, *30*, 5192–5194.
15. Jeyaprakash, J.D.; Samuel, S.; Dhamodharan, R.; Rühe, J. Polymer Brushes via ATRP: Role of Activator and Deactivator in the Surface-Initiated ATRP of Styrene on Planar Substrates. *Macromol. Rapid Commun.* **2002**, *23*, 277–281.
16. Von Werna, T.W.; Germack, D.S.; Hagberg, E.C.; Sheares, V.V.; Hawker, C.J.; Carter, K.R. A versatile method for tubing the chemistry and size of nanoscopic features by living free radical polymerization. *J. Am. Chem. Soc.* **2003**, *125*, 3831.
17. Zheng, H.; Lee, I.; Rubner, M.F. Two Component Particle Arrays On Patterned Polyelectrolyte Multilayer Templates. *Adv. Mater.* **2002**, *14*, 569–572.
18. Kodger, T.E.; Sprakel, J. Thermosensitive Molecular, Colloidal, and Bulk Interactions Using a Simple Surfactant. *Adv. Funct. Mater.* **2012**, *23*, 475–482.
19. Klein, J.; Kumacheva, E.; Mahalu, D.; Perahia, D.; Fetters, L.J. Reduction of frictional forces between solid surfaces bearing polymer brushes. *Nature* **1994**, *370*, 634–636.
20. Raviv, U.; Giasson, S.; Kampf, N.; Gohy, J.F.; Jérôme, R.; Klein, J. Lubrication by charged polymers. *Nature* **2003**, *425*, 163–165.
21. Chen, M.; Briscoe, W.; Armes, S.; Klein, J. Lubrication by biomimetic surface layers at physiological pressures. *Science* **2009**, *323*, 1698–1700.
22. Jones, D.; Smith, J.; Huck, W.; Alexander, C. Variable Adhesion of Micropatterned Thermoresponsive Polymer Brushes: AFM Investigations of Poly(*N*-isopropylacrylamide) Brushes Prepared by Surface-Initiated Polymerizations. *Adv. Mater.* **2002**, *14*, 1130–1134.
23. Malham, I.B.; Bureau, L. Density Effects on Collapse, Compression, and Adhesion of Thermoresponsive Polymer Brushes. *Langmuir* **2010**, *26*, 4762–4768.
24. Kantor, Y.; Webman, I. Elastic Properties of Random Percolating Systems. *Phys. Rev. Lett.* **1984**, *52*, 1891–1894.

25. Bijsterbosch, B.H.; Bos, M.T.A.; Dickinson, E.; van Opheusden, J.H.J.; Walstra, P. Brownian dynamics simulation of particle gel formation: from argon to yoghurt. *Faraday Discuss.* **1995**, *101*, 51–64.
26. Colombo, J.; Widmer-Cooper, A.; del Gado, E. Microscopic picture of cooperative processes in restructuring gel networks. *Phys. Rev. Lett.* **2013**, *110*, 198301.
27. Colombo, J.; del Gado, E. Stress localization, stiffening, and yielding in a model colloidal gel. *J. Rheol.* **2014**, *58*, 1089–1116.
28. Zaccone, A.; Crassous, J.J.; Béri, B.; Ballauff, M. Quantifying the reversible association of thermosensitive nanoparticles. *Phys. Rev. Lett.* **2011**, *107*, 168303.
29. Zaccone, A.; Winter, H.; Siebenbürger, M.; Ballauff, M. Linking self-assembly, rheology, and gel transition in attractive colloids. *J. Rheol.* **2014**, *58*, 1219–1244.
30. Cetintas, M.; Kamperman, M. Self-assembly of PS-b-PNIPAM-b-PS block copolymer thin films via selective solvent annealing. *Polymer* **2016**, *107*, 387–397.
31. Gao, Y.; Kilfoil, M.L. Accurate detection and complete tracking of large populations of features in three dimensions. *Opt. Express* **2009**, *17*, 4685.

11

Transport Phenomena in Gel

Masayuki Tokita

Department of Physics, Faculty of Science, Kyushu University, 744 Motooka, Fukuoka, Fukuoka 812-8581, Japan; tokita@phys.kyushu-u.ac.jp

Academic Editor: Clemens K. Weiss

Abstract: Gel becomes an important class of soft materials since it can be seen in a wide variety of the chemical and the biological systems. The unique properties of gel arise from the structure, namely, the three-dimensional polymer network that is swollen by a huge amount of solvent. Despite the small volume fraction of the polymer network, which is usually only a few percent or less, gel shows the typical properties that belong to solids such as the elasticity. Gel is, therefore, regarded as a dilute solid because its elasticity is much smaller than that of typical solids. Because of the diluted structure, small molecules can pass along the open space of the polymer network. In addition to the viscous resistance of gel fluid, however, the substance experiences resistance due to the polymer network of gel during the transport process. It is, therefore, of importance to study the diffusion of the small molecules in gel as well as the flow of gel fluid itself through the polymer network of gel. It may be natural to assume that the effects of the resistance due to the polymer network of gel depends strongly on the network structure. Therefore, detailed study on the transport processes in and through gel may open a new insight into the relationship between the structure and the transport properties of gel. The two typical transport processes in and through gel, that is, the diffusion of small molecules due to the thermal fluctuations and the flow of gel fluid that is caused by the mechanical pressure gradient will be reviewed.

Keywords: diffusion; friction; scaling theory; colloid gel

---

1. Introduction

Gel has the characteristics and common structure that consists of the three-dimensional polymer network and gel fluid. Because of such a diluted structure, many molecules are transported through gel. Such a characteristic of gel is used in separation technologies, namely, gel electrophoresis, gel permeation chromatography, and so forth. Therefore, information on the transport phenomena in and across gel is of importance in designing a separation system. Moreover, knowledge on transport phenomena in gel has become quite important recently for understanding the volume phase transition of gel and the pattern formation phenomena of gel [1–3]. It has been reported that the volume of gel reversibly changes more than a thousandfold in response to the slight environmental change in the vicinity of the volume phase transition point [4]. The huge amount of gel fluid, therefore, flows in and out in accordance with the swelling and shrinking of gel. Kinetic study on the volume change of gel indicates that the characteristic time of the swelling of gel, τ, is determined by the ratio of the longitudinal modulus of gel, E, the friction between gel fluid and the polymer network of gel, f, and the typical size of gel, a, as $\tau = a^2 f / E$ [5]. In contrast, the shrining pattern formation phenomena in the shrinking process of gel in the poor solvent is very complicated since the first event that occurs in the beginning of the shrinking process is the diffusion of the poor solvent into gel. Then, the squeeze out of gel fluid came in to create the shrinking patterns. Both the diffusion and the friction compete with each other in pattern formation in the shrinking gels. As a result, various shrinking patterns appear in the shrinking gel [6–8]. In addition to this, one of the most striking expectations is deduced from the study of the critical dynamics of the volume phase transition of gel, namely, the friction of gel becomes smaller and it vanishes at the volume phase transition point [9]. The diffusion and the friction, therefore, play important roles in the volume phase transition of gel as well as the pattern formation in gel. The detailed studies on the transport of substances by the diffusion and the flow of gel fluid are still required for the full understanding of the volume phase transition of gel and the pattern formation in gel. Since the polymer network of gel is regarded as an assembly of the molecular mesh, the relationship between the transport properties of gel and the mesh size of the polymer network of gel should first be clarified. The diffusion of small molecules in gel and the frictional properties of gel have been studied by many researchers so far [10–13]. Although many studies have been conducted, most measurements are made under not ideal conditions. In this review, we will discuss the diffusion and the friction of gel in terms of the structure of gel by which the simple view for the transport phenomena in gel will be deduced.

2. Gel

Gel is classified into two classes by the structure of the cross-linking point. One is called chemical gel and the other is known as physical gel. In the case of chemical gel, the cross-linking point consists of a covalent bond. On the other hand, the cross-linking point of physical gel consists of the assembly of polymers that is formed by the short range interactions such as the van der Waals interaction, the hydrophobic interaction, the hydrogen bond, and the Coulomb interaction between charges. The network structure of chemical gel is not varied by the change of the external environment such as the temperature, because the bonding energy of the covalent bond is much higher than the thermal energy. Hence, the cross-linking point of the polymer network is unchanged even if the temperature is changed, say, from 0 to 100 °C. Gel is in an equilibrium swelling state [14]. The volume of gel changes in response to environmental changes such as the temperature in the case of chemical gel. If the appropriate chemical structure is given to the polymer network of gel, then gel shows the volume phase transition phenomena. On the other hand, the bonding energy due to short range interactions is almost the same order of magnitude as the thermal energy. Accordingly, most physical gels show the sol-gel transition phenomena when the environmental conditions are changed. The sol-gel transition and the volume phase transition are the characteristic transitions in gel system [15].

In this review, we describe the transport properties of the chemical gels of acrylamide (AAm) and *N*- (NIPA) cross-linked by *N,N*-methylenebisacrylamide (BIS). The linear polymer chains of AAm or NIPA are connected by BIS to form a three-dimensional polymer network. The network structure is, therefore, determined both by the concentrations of the linear chain component and the cross-linker. The cross-linking density of gel, which is defined by the mole fraction of the cross-linker to the total amount of the linear component and the cross-linker, is fixed at 0.01 or 0.02 in most studies. The total concentration of gel is changed under the fixed cross-linking density. The gels obtained under these conditions are uniform; hence, gel is transparent and elastic. On the other hand, the cross-linking density is changed at a constant total concentration, usually fixed at 700 mM. The gel prepared at higher concentrations of BIS are opaque and brittle suggesting the formation of a non-uniform structure.

The mesh size of the polymer network is the most important parameter that defines the structure of the polymer network of gel. The gels used here are synthesized by the random copolymerization of the linear chain component and the cross-linker. As a result of this, the network structure of gel becomes more or less random. The size of the polymer mesh is, therefore, expressed by an averaged parameter that is called the correlation length of gel, ξ. The correlation length of gel represents, in a good solvent, the distance between two nearest neighbor contact points of polymers. The correlation length of gel varies with the concentration of the linear chain component, the cross-linking density, and the environmental conditions such as the temperature.

For most parts of the study, poly(acrylamide) gel (PAAm gel) is chosen as the sample gel since this gel is widely used in separation and the purification technologies. Hence, the physical and the chemical properties of PAAm gel are well known. For instance, the concentration of gel can be changed widely from 0.02 to 0.5 g/mL because of a high affinity with water. This is a definite advantage for the present purpose of the study because the correlation length of the polymer network of gel can be changed widely. In addition to this, PAAm gel does not interact strongly with many water soluble molecules. This is another great advantage for the present study since it widens the choice of the probe molecules. On the other hand, poly(n-siopropylacrylamide) gel (PNIPA gel) is known as the thermo-sensitive gel. The gel shows the discontinuous volume change at about 34 °C in water. The correlation length of PNIPA gel can be changed reversibly as a function of the temperature. Therefore, the critical behaviors of the frictional property can be revealed by studying the frictional property of PNIPA gel in the vicinity of the volume phase transition point.

3. Probe Diffusion in Gel

Many advancements in nuclear magnetic resonance technology have been made at the end of the last century. One of the useful technologies in the study of transport phenomena is the pulsed field gradient nuclear magnetic resonance method (PFG-NMR). The diffusion coefficient of the specific molecules, which will be called the probe molecule hereafter, can be determined by using PFG-NMR invasively if the probe molecule is distinguishable in the NMR spectrum of the whole system. The PGF-NMR method is firstly applied to study the diffusion of the molecules in simple solution. It is then applied to complex systems such as the semi-dilute polymer solution systems, and finally to polymer gels [10,16,17]. Previous studies suggest that it is a requirement to study the diffusion process by changing both the concentration of gel and the size of the probe molecule systematically to obtain the entire aspects of the diffusion process in gel. The diffusion of the probe molecule in gel is, however, extensively affected by short range interactions between the probe molecule and the polymer network. The presence of short range interactions is unfavorable because we firstly focus our attention only on the effects of the polymer network on the diffusion process of the probe molecules. Since the short range interactions depend strongly on the chemical structures of the polymer network and the probe molecules, one can minimize the complex situations by choosing appropriate chemicals for gel and the probe molecules. Then, we can deduce the effects of only the polymer network of gel on the diffusion of the probe molecules.

In this study, the diffusion coefficient of the probe molecules are measured in PAAm gel of the total concentration from 0.02 to 0.5 g/mL at a cross-linking density of 0.02. Although any substance can be a candidate for the probe diffusion experiments in PAAm gel, there are still several requirements for the probe molecule for the present purpose of the study as listed below.

1. Highly soluble in water because water is used as a solvent.
2. Compact molecules to be assumed as a spherical molecule.
3. Molecular weight spreads at least one order of magnitude.
4. Chemical shift does not overlap with that of PAAm gel.
5. Absence of strong interaction with the polymer network of PAAm gel.

Taking into account these requirements, we finally choose water (solvent), ethanol, glycerin, poly(ethylene glycol), and sucrose as the probe molecules. The molecular weight of each probe molecule is, 18, 46, 92, 200, and 342, respectively.

The sample PAAm gel is synthesized in an NMR tube of 10 mm in diameter. The solvent is the mixture of water and the heavy water (H_2O:D_2O = 9:1) which contains the probe molecule at a concentration of 10 wt % is prepared firstly. Then, an appropriate amount of AAm, BIS, and ammonium persulfate (initiator) are dissolved into the above solvent. The pre-gel solution thus obtained is de-gassed for 20 min. The reaction is initiated by raising the temperature to 60 °C for 1 h. The diffusion experiments are made on a JEOL FX-60Q (JEOL, Tokyo, Japan), which is equipped with a field gradient apparatus NMPL-502 (JEOL, Tokyo, Japan), at a frequency of 60 MHz. The temperature is controlled at 30.0 ± 0.5 °C. Details of the experimental procedure are given in the previous report [18].

In Figure 1, the diffusion coefficient of the probe molecules in PAAm gels are plotted as a function of the concentration of gel.

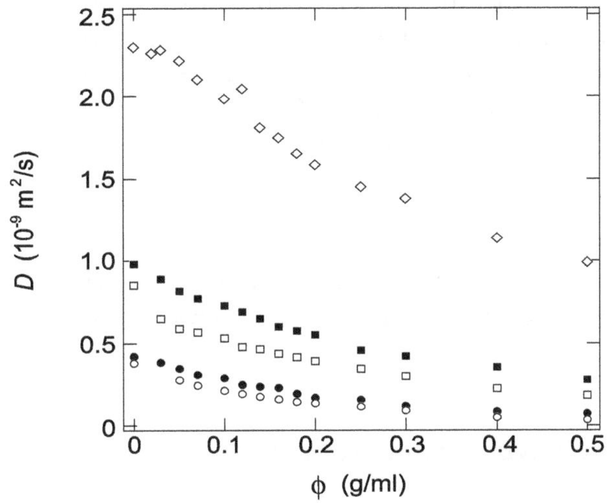

Figure 1. The dependence of the diffusion coefficient of probe molecules on the concentration of acrylamide (AAm) in gel. The probe molecules are water, diamonds, ethanol (filled squares), glycerin (open squares), poly(ethylene glycol) (filled circles), and sucrose (open circles), from top to bottom. The data points at the position of zero concentration represent the diffusion coefficient of each probe molecules in water that are measured by the same experimental setup. The temperature is fixed at 30.0 ± 0.5 °C.

The probe molecule, when dissolved in a simple fluid, thermally fluctuates in time and space. The thermal fluctuation and the hydrodynamic friction due to the surrounding fluid, ζ, determine the rate of diffusion. The diffusion coefficient, D_0, of the probe molecule of the hydrodynamic radius, R_h, in a simple fluid of viscosity, η, is expressed by the so-called Stokes–Einstein relationship as follows [19,20]:

$$D_0 = \frac{k_B T}{\zeta} = \frac{k_B T}{6\pi\eta R_h}. \tag{1}$$

Here, k_B and T represent the Boltzmann's constant and temperature. It is clear from the above equation that the diffusion coefficient of the probe molecule is essentially determined by the size of the probe molecule, R_h, when it is dissolved in a simple fluid at the constant temperature. However, it is found from Figure 1 that the diffusion coefficients of all probe molecules decrease monotonically with the concentration of AAm in the gel. The diffusion coefficient of the probe molecules decreases even at a very small concentration region of gel, *ca.*, less than 0.1 g/mL. The results obtained here, therefore, indicate that the diffusion of the probe molecule within the gel is far from the free diffusion in a simple fluid even in a tenuous gel. It is clear from Figure 1 that the diffusion coefficient of the probe molecule is a decreasing function of the molecular weight of the probe molecule, and it is also a decreasing function of the concentration of the gel. Furthermore, it may be clear that the concentration dependence of the diffusion coefficients of the probe molecules are similar to each other. It strongly suggests the presence of a universal function for the diffusion coefficient of the probe molecules in the gel. Since both the concentration of the gel and the size of the probe molecule is systematically changed in this study, such a function is written by these two parameters.

A similar phenomena has been also studied in semi-dilute polymer solutions and effort is devoted to finding the universal physical picture of the diffusion in these condensed polymer systems of the semi-dilute solution and gel [21,22]. Among others, one of the hopeful analyses of the diffusion processes in gel is the scaling approach [23]. In modern statistical theory of polymer systems, many physical quantities are expressed in terms of the scaling function, $f(x)$. The diffusion coefficient of the probe molecules in gel that is normalized by D_0 is also expected to be expressed by a scaling function as follows:

$$\frac{D}{D_0} = f(x). \tag{2}$$

Here, D and D_0 represent the diffusion coefficient of the probe molecule in the gel and that in the simple fluid, and x is the non-dimensional scaling variable. Since the diffusion coefficient is determined by the linear size of the probe molecule, R_h, one of the candidates for the scaling variable is a ratio of R_h and the correlation length. The correlation length of the polymer network of gel, ξ, is expressed as a function of the concentration of gel, ϕ, as follows [23]:

$$\xi \propto \phi^{-3/4}. \tag{3}$$

On the other hand, the size of the molecules with simple and compact structures is simply assumed as follows using the molecular weight of the probe molecule, M:

$$R_h \propto M^{1/3}. \tag{4}$$

We, therefore, assume the following scaling variable:

$$x = \frac{R_h}{\xi} \propto M^{1/3}\phi^{3/4}. \tag{5}$$

In other words,

$$\frac{D}{D_0} \propto f\left(M^{1/3}\phi^{3/4}\right). \tag{6}$$

The results shown in Figure 1 are, therefore, plotted as a function of the scaling variable, x, in Figure 2.

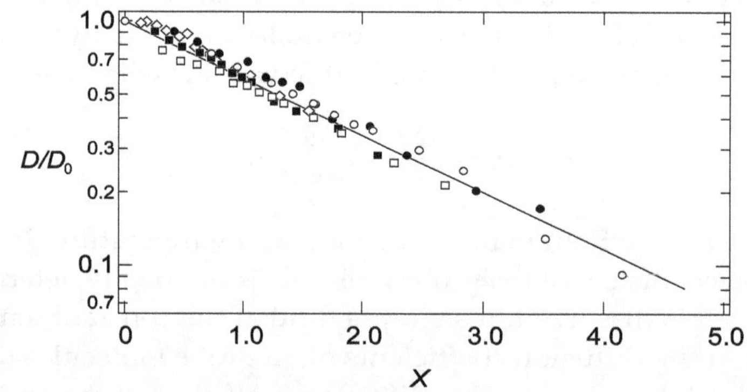

Figure 2. The normalized diffusion coefficient of the probe molecules in gel, D/D_0, is plotted as a function of the scaling variable, $x = M^{1/3}\phi^{3/4}$, in a semi-logarithmic manner. Symbols are the same as that of the Figure 1.

It is found that all data fall onto a single master curve. In addition to this, the results are well explained by a simple functional form as follows:

$$\frac{D}{D_0} = \exp\left(-\frac{R_h}{\xi}\right). \tag{7}$$

The results are in good agreement with the hydrodynamic calculations for the scaling function [21,22]. It is found that the diffusion coefficient of the probe molecules follows the simple scaling theory.

The results obtained here suggest strongly that the probe molecule that diffuses in the polymer network of PAAm gel mainly experiences resistance due to the polymer network. In addition, the effects of the polymer network on the diffusion of the probe molecules are determined by the ratio of the correlation length of the polymer network and the radius of the probe molecule. The results in Equation (7) indicate that the relative rate of the diffusion is equivalent if the ratio of the hydrodynamic radius and the correlation length of the polymer network, R_h/ξ, are equal. The smaller probe molecule diffuses faster within the gel. However, even the size of the probe molecule is larger, it diffuses in the gel at the similar rate as the smaller probe molecules if the mesh size of the gel is large enough. The rate of the diffusion in the gel is, in the simplest case, determined by the geometrical factor of the probe molecule and the polymer network of the gel. The parameter that determines the diffusion of the substance in gel is the ratio of the size of the probe molecule and the correlation length of gel, R_h/ξ. The results obtained here are schematically illustrated in Figure 3.

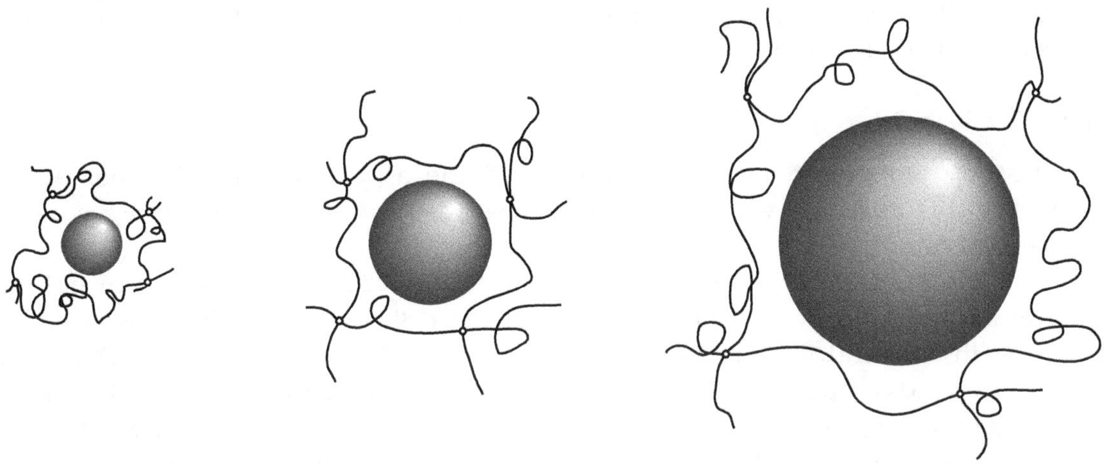

Figure 3. The schematic representation of Equation (7). The size of the probe molecules is different, but the ratio of the size and the correlation length, R_h/ξ, is similar in these three cases. The diffusion coefficient of the probe molecule that is normalized by the diffusion coefficient of the probe molecule in water, D/D_0, is almost the same in these cases.

4. Friction of Gel

Many efforts have been devoted to creating active systems such as artificial muscles after finding the volume phase transition of gel [24]. When gel collapses, a huge amount of gel fluid is squeezed out of gel. Hence, the acceleration of the drainage of gel fluid from gel is the key to success. In other words, an understanding of the friction between the polymer network of gel and gel fluid is required. The study of the kinetics of the swelling of gel indicates that the swelling of gel is determined by the collective diffusion coefficient of gel, D_c [25–27]:

$$D_c = \frac{E}{f}. \tag{8}$$

Here, E represents the longitudinal elastic modulus of the polymer network alone and f the friction coefficient between the polymer network of gel and gel fluid. Equation (8) indicates that the polymer network of gel moves under the drag force of the surrounding fluid.

The measurements of the friction between the polymer network of gel and gel fluid is, in principle, simple as schematically shown in Figure 4. However, only a little is known about the frictional properties of gels. The reason is simple—the friction coefficient of the polymer network is huge, while gel is fragile. Detailed study on the frictional properties of the PAAm gels are reported where the strategies to design the apparatus are also reported [28].

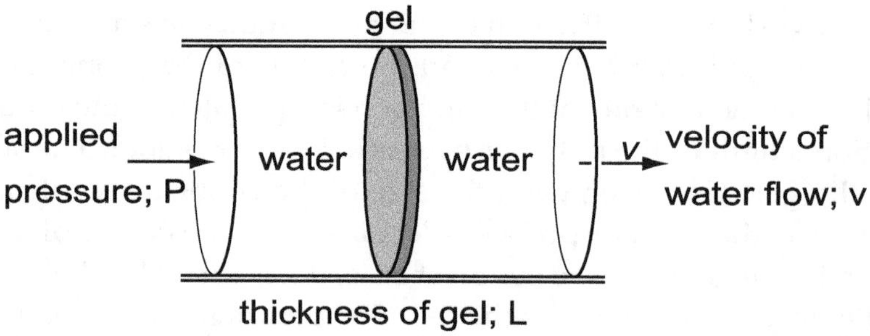

Figure 4. The schematic illustration of the principle of friction measurement. The gel of the thickness L is set in a column with the limb fixed. Then, small pressure, P, is applied to water, which covers the left side of the gel. The velocity, v, of water flowing out of gel is measured in the stationary state. The friction coefficient of gel is defined as $f = P/vL$.

4.1. Friction of Uniform Gel

The first study we should make is to obtain all the aspects of the frictional properties of uniform gel. For this purpose, we use PAAm gel because of its many advantages as described in the previous sections.

First of all, the temperature dependence of the friction coefficient of PAAm gel is studied. The results are given in Figure 5. The friction coefficient of PAAm gel decreases monotonously with the increase in temperature as shown in Figure 5. It may be natural to assume that the friction coefficient itself depends on the viscosity of the flowing fluid. Thus, the friction coefficient of PAAm gel is normalized by the viscosity of water and the results are also plotted in Figure 5. It is found that the friction coefficient of PAAm gel is independent of the temperature when the friction coefficient is normalized by the viscosity of water, $f(T)/\eta(T)$. The results obtained here indicate that the structure of the polymer network of PAAm gel is independent of the temperature. The results further indicate that the raw value of the friction coefficient includes the information on the viscosity of the flowing fluid. In the present case, the friction coefficient of PAAm gel is proportional to the viscosity of water, $f \propto \eta_{water}$.

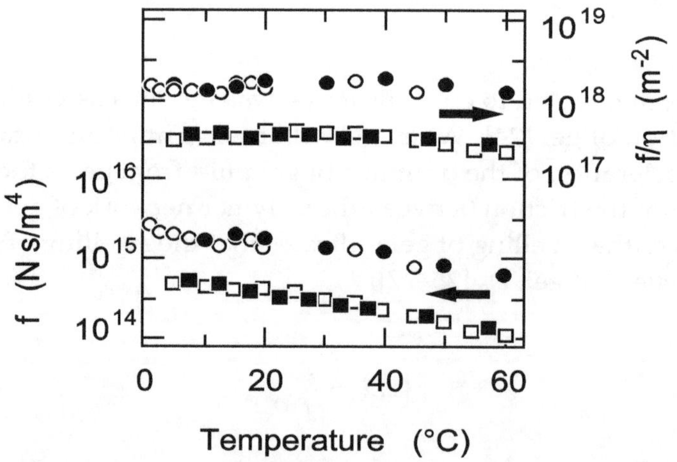

Figure 5. The temperature dependence of the friction coefficient of poly(acrylamide) (PAAm) gel. The concentration of gel is 5 % (squares) and 8% (circles). The friction coefficient that is normalized by the viscosity of water, $f(T)/\eta(T)$ is also shown in this figure. The solid symbols are used in the increasing of the temperature, and the open symbols are used in the decreasing of the temperature.

Secondly, we study the concentration dependence of the friction coefficient of PAAm gel. The raw value of the friction coefficient of PAAm gel is plotted as a function of the concentration of gel in a double logarithmic manner in Figure 6, since the measurements are made at a constant temperature of 20 °C, and hence, the viscosity of water does not show any singular behavior at this temperature.

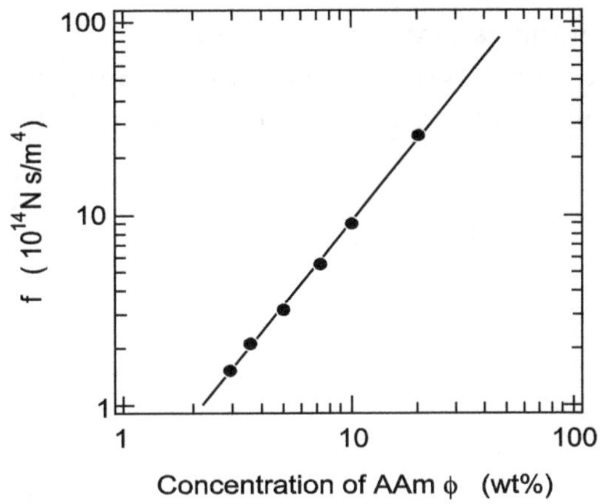

Figure 6. The concentration dependence of the friction coefficient of PAAm gel. The raw value of the friction coefficient is plotted as a function of the concentration of PAAm gel in a double logarithmic manner. The temperature is fixed at 20.0 ° C in these measurements. The straight line with the slope of 1.5 is the results of the least squares analysis.

The least squares analysis shows that the concentration dependence of the friction coefficient is well expressed by a power law relationship as,

$$f \propto \phi^{1.5}. \tag{9}$$

The scaling suggests that both the collective diffusion coefficient of gel and the elastic modulus of gel can be written as a function of the concentration as follows [23]:

$$D_c \propto \phi^{3/4}, \tag{10}$$

$$E \propto \phi^{9/4}. \tag{11}$$

Equation (8) with the above scaling relationships reveals that the friction coefficient of gel scales with the exponent of 6/4 for the power law relationship. The concentration dependence of the friction coefficient obtained here is in good agreement with the scaling theory. The results are also explained from the molecular structure of gel. When the pressure is applied to the system, water flows through the open space of the polymer network of gel. The gel is, therefore, regarded as the assembly of the pore. The rate of water flow is assumed to be proportional to the viscosity of the fluid, η, and is inversely proportional to the cross section of the pores. The cross section of the pores in gel is reasonably assumed to be proportional to the square of the correlation length of gel, ξ:

$$f \propto \frac{\eta}{\xi^2}. \tag{12}$$

It is suggested by the scaling theory that the correlation length of the polymer network depends on the concentration of gel as follows [23]:

$$\xi \propto \phi^{-3/4}. \tag{13}$$

The scaling again results as $f \propto \phi^{1.5}$. Equation (12) is rather useful when the friction coefficient of gel is discussed in terms of the structure of the gel.

Finally, the friction of PAAm gel is measured as a function of the cross-linking density. It is expected that the size of the pores in the polymer network of gel decreases with the cross-linking density of gel. In other words, the correlation length of the polymer network of gel decreases with the concentration of BIS. Therefore, it may be natural to assume that the friction coefficient of gel increases with the concentration of BIS according to Equation (12). However, the results, as shown in the Figure 7, clearly indicate that the friction of PAAm gel decreases with the concentration of BIS. In addition, it is also found that gel becomes translucent with the concentration of BIS and finally the gel becomes completely opaque when the concentration of BIS exceeds 3 % in mole fraction.

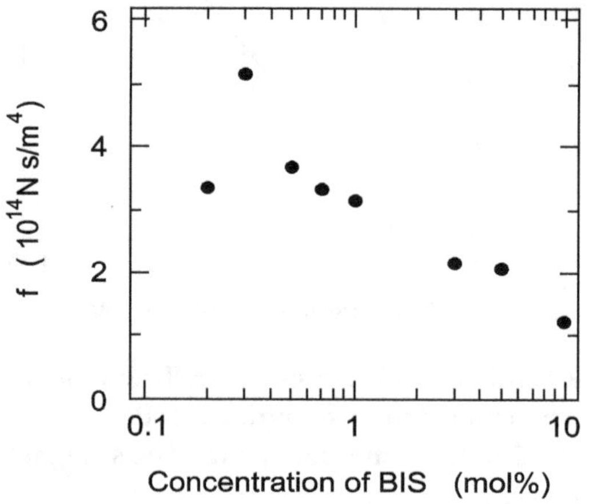

Figure 7. The cross-linker concentration dependence of the friction coefficient of PAAm gel. The total concentration of gel is fixed at 700 mM, and the mole fraction of the cross-linker is varied.

The increase in turbidity of the gel indicates the appearance of the density fluctuations in the polymer network of gel. The density fluctuations that appear in the present system are irreversible and independent of time, and hence, it indicates that the polymer network of gel itself is spatially non-uniform. In contrast, gel is transparent as long as the mole fraction of BIS is less than 3%. The friction of gel, however, clearly decreases with the concentration of BIS even in the concentration region where the transparent gels are obtained. These results thus strongly suggest that the seeds of the density fluctuations emerge in the polymer network of gel at a lower concentration region where gel looks completely transparent. The frictional property of gel, therefore, reflects the spatial homogeneity of gel in an extremely sensitive manner. However, we have to wait for the advancement of new technology that provides us with observations of the structure of the polymer network of the opaque gels in real space, such as the confocal laser scanning microscope (CLSM), for further detailed discussion of the phenomenon. We will return to this point again in the later section.

4.2. Friction of Non-Uniform Gel

Many studies have been devoted to clarifying the relationship between the chemical structure of the monomer unit and the biological phenomena [29,30]. Among others, the finding of the volume phase transition in poly(*N*-isopropylacrylamide) gel (PNIPA gel) is of importance since PNIPA gel shows the volume phase transition in response to the temperature change under water [31]. The transition in PNIPA gel is thought to be a result of the hydrophobic interaction between the *N*- isopropylacrylamide (NIPA) residues in the polymer chain. According to the fact that PNIPA gel shows the volume phase transition in water, the gel is widely used to study the aspects of the volume phase transition of the gel, for instance, such as the critical kinetics of the volume phase transition of gel [32]. One of the unsolved phenomena currently is the critical behavior of the friction of gel in the vicinity of the volume phase transition point. It is expected in the early stage of the studies of the volume phase transition of gel that the friction of gel becomes smaller in the vicinity of the volume phase transition point of gel [9]. Therefore, the critical behavior of the friction is studied in the vicinity of the volume phase transition temperature of PNIPA gel [33]. The results are given in the Figure 8. The results clearly indicate the drastic decrease of the friction coefficient of gel in the vicinity of the volume phase transition temperature. The friction coefficient of gel decreases more than three orders of magnitude as the temperature approaches the volume phase transition temperature. The change of the friction coefficient of gel is reversible against the temperature change. The results indicate that the reversible decrease of the friction coefficient of gel is caused by the emergence of the critical density fluctuations in the polymer network of gel. The characteristic time scale for the density fluctuations becomes longer in the vicinity of the phase transition point, and hence, the polymer network of gel looks like an assembly of the large pore of the radius that corresponds to the correlation length of the density fluctuations, ξ. Water flows through the open space that is created by the lower density regions of the polymer network, avoiding the denser region that blocks the flow. It is reasonably assumed that the correlation length of the density fluctuations diverge in the vicinity of the volume phase transition point. The friction of the polymer network of gel, therefore, becomes smaller and it practically vanishes at the phase transition point. The critical exponent for the divergence of the correlation length near the critical point is also estimated from these results. We, however, observed a relatively large discrepancy with the expected value. It is clear from the swelling curve of gel that is shown in Figure 8 that the isochore conditions employed here do not hit the exact critical point of gel. Therefore, the critical exponent we obtained here corresponds to the off-critical value. Although the critical slowing down is observed by the friction experiment of PNIPA gel, the critical behaviors of gel is still open to question.

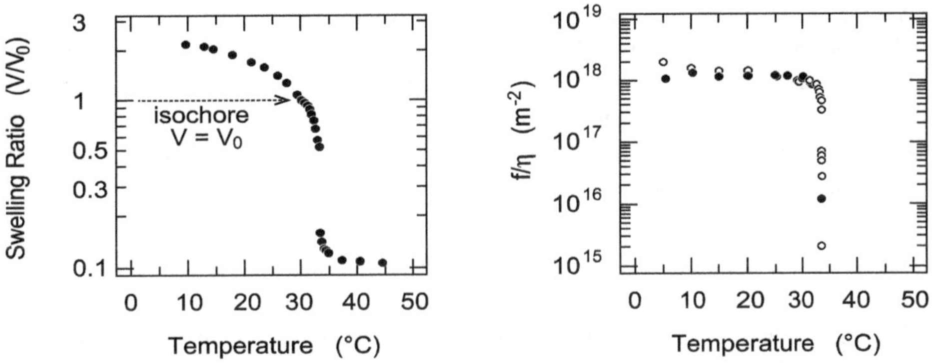

Figure 8. The temperature dependence of the swelling ratio of gel (left; V/V_0) and that of the friction coefficient of poly(*N*-siopropylacrylamide) (PNIPA) gel that is normalized by the viscosity of water (right; $f(T)/\eta(T)$). The friction is measured under a constant volume condition that is shown in the left figure as an arrow. Open circles are used in the increasing of the temperature and solid circles are used in the lowering of the temperature.

4.3. Friction of Colloid Gel

It is found that the friction of PAAm gel decreases with the cross-linking density in the previous section. The results suggest strongly that the non-uniform density fluctuations are frozen into the polymer network of gel whenever the cross-link is introduced. The structure of the polymer network of gel, therefore, may change if the concentration of BIS is increased further. The detailed study on the structure of the polymer network of gel, therefore, should be made firstly. Then, the relationship between the structure and the frictional properties of gel is discussed.

The formation of PAAm gel has been studied in detail by changing the concentration of AAm and BIS drastically. The results are expressed as a kind of phase diagram [34]. It has been found that PAAm gel becomes turbid at higher concentration regions of BIS. Therefore, the turbidity change of PAAm gel itself is a well known phenomenon. The structural study of opaque gels at these concentration regions has been also made by using spectroscopy techniques [35]. Although these studies strongly suggest the formation of the cluster of BIS in the opaque gel, the structure of the opaque PAAm gel in real space could not be obtained.

The idea of the confocal scanning microscopy has been reported in 50's [36]. It became a popular technique when the laser was used as the light source in the 1980s as the confocal laser scanning microscope (CLSM). Firstly, CLSM was used in the research area of biology, and eventually, it spread to the research area of materials science. In the area of gel science, the structure of PNIPA gel was firstly studied using CLSM. It is well known that PNIPA gel also becomes opaque if the reaction temperature is raised to 30 °C or more. The network structure of such opaque PNIPA gels is solved by CLSM imaging technique [37]. We also study the structure of the opaque PAAm gel [38]. In Figure 9, CLSM images of the opaque PAAm gel are shown.

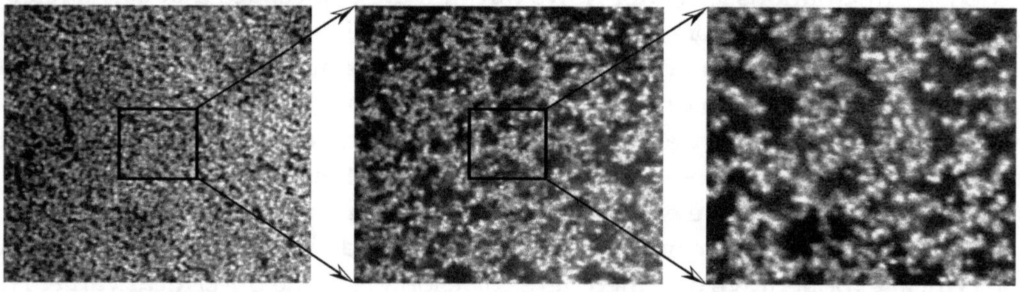

Figure 9. The confocal laser scanning microscope (CLSM) images of opaque PAAm gel. The magnification of the image is 10×, 40×, and 100× from left to right. The length of the side of these images is 189 μm, 48.8 μm, and 18.9 μm from left to right. The concentrations of AAm and *N,N*-methylenebisacrylamide (BIS) are 350 mM. The total concentration is 700 mM and the mole fraction of BIS is 50%. The squares in the image of 10× is the observation area of 40× observation. The square in the image of 40× is the observation are of 100× observation. These squares only show the relative size of the images and are shown only for a guide to the eye.

The total concentration of PAAm gel observed here is 700 mM and the mole fraction of the BIS is 50%. The images of the same gel are gained at various magnifications. The magnifications of the images are 10×, 40×, and 100×. The brighter portion in the images corresponds to the dense region of AAm and BIS, and the darker region is the dilute region of AAm and BIS. It is clearly shown that the opaque PAAm gel consists of the aggregate of the colloidal particles from these images. The detailed analysis of the confocal images of the opaque PAAm gel yields that the density of the particle is of the order of 1×10^3 kg/m^3. The density of the particle is almost the same as that of AAm and BIS in the solid state. Both AAm and BIS are densely polymerized into a sphere of the colloidal dimension. The particles thus formed in the opaque gels are, therefore, regarded as a hard sphere. In addition to this, the small angle neutron scattering from the opaque gel strongly suggest that gel is a fractal with the fractal dimension of $D_f \simeq 2.7$ [39–42]. The results suggest that the colloidal particles of AAm and BIS are formed and they aggregate by the process of the diffusion limited aggregation or the cluster-cluster aggregation process. We expect that water can easily flow through the darker portion in the colloidal gel since the mass is mostly localized in the colloidal particles, and, hence, the darker portion of the image serves as the open pore for the fluid flow. The friction experiments are made in the higher concentration region of BIS in gel. The results are given in Figure 10 [43].

It is found that the friction coefficient of gel decreases more than four orders of magnitude when the mole fraction of BIS is increased to 0.2 or more. The opacity of gel increases at the same concentration of BIS where the drastic decrease of the friction occurs. This suggests strongly that the structure of the polymer network of gel changes from the uniform polymer network to the colloidal aggregate. In accordance with the structural transition of the polymer network of gel, the friction of gel decreases more than four orders of magnitude. The friction of the opaque gel can be explained in terms of the structure of gel because the structural parameters can be deduced from the CLSM images. First of all, the density of the particle is high, and is almost the same as AAm and BIS in a solid state. Secondly, the particle is more or less spherical in shape. The particles are regarded as hard spheres. Water thus flows around the aggregates of hard spheres experiencing the hydrodynamic friction. Since the density of the sphere is almost the same as that of the solid, the aggregates only behave as the fixed obstacles for water flow. If only one solid particle of radius

R is suspended in the flowing fluid, the hydrodynamic friction due to the particle is expected to be $f_{obstacle} = 6\pi\eta R$. In the colloid gel, however, N particles are included in a unit volume of gel. The friction of the colloid gel thus becomes

$$f_{obstacle} = N6\pi\eta R. \tag{14}$$

Here, N represents the number of colloid particle in a unit volume of gel. Both N and R are calculated from the CLSM images of gel that are used as the samples of the friction experiments. Then, the friction of gel is calculated by using Equation (14). The calculated values show the same concentration dependence with the experimental results. The numerical values are, however, about one-half of the experimental values. One of the reasons for this discrepancy arises from the resolution limit of the CLSM that the focal plane of the CLSM has a finite thickness. In the case of the present system, the thickness of the focal plane is about 1 μm according to the manufacturer. In addition, the size of the particle in the present case is comparable with the thickness of the focal plane. Therefore, the overlap of the particles in the depth direction in the CLSM image is unavoidable. The number of the particle, N, that is simply calculated from the CLSM image therefore underestimates the actual number of the particle contained in the image. The friction of gel thus calculated becomes smaller than the actual values of the friction. The details are given in the previous reports [43,44].

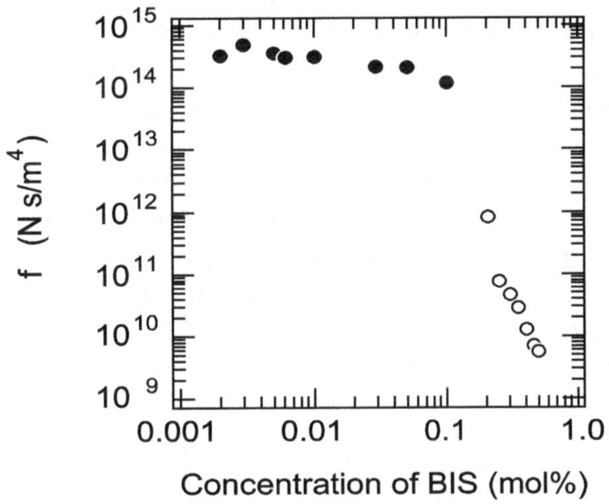

Figure 10. The cross-linker concentration dependence of the friction coefficient of PAAm gel at higher concentration region of BIS (open circles). The results shown in Figure 7 are also plotted in this figure (solid circles). The gels are prepared at a total concentration of 700 mM, and the mole fraction of the cross-linker is changed.

5. Conclusions

We review the probe diffusion of gel and the hydrodynamic friction of the polymer network. It is found that the diffusion coefficient of compact molecules in gel is well described by the scaling theory. The essential parameter that determines the diffusion coefficient is the ratio of the tow sizes, R/ξ, where R is the radius of the probe molecule and ξ the mesh size of the polymer network. In the study, most probe molecules are rather small and compact, and, hence, the scaling results are natural. In addition to this, it is also interesting that the diffusion coefficient of poly(ethylene glycol) follows the same scaling law since poly(ethylene glycol) is a linear polymer. In designing the diffusion experiments, we expected different diffusion mechanisms for poly(ethylene glycol), that is, the diffusion by the reptation [45,46]. The molecular weight of poly(ethylene glycol) used in the diffusion experiments is 200, which indicates that the polymer consists of only about five monomers. The results indicate that polymers of this size can be regarded as compact molecules. Hence, the diffusion experiments using the longer linear polymers is of interest for establishing the entire physical picture of the diffusion in gel. Since we have obtained the simplest view of the diffusion in gel, it may be of interest to study the system where the probe molecule interacts with the polymer network strongly through the typical interactions such as the hydrophobic interaction, the electrostatic interaction and so forth. Further detailed study on the diffusion in gel is required.

The results of the frictional properties of the polymer network of gel is also easily explainable by the scaling theory. The friction of the polymer network is essentially determined by the mesh size of gel. The measure of the mesh size of gel is the correlation length of the polymer network, ξ. In Equation (12), however, we only assume the proportionality between the friction coefficient and the correlation length, $f \propto \eta/\xi^2$. Thus, the numerical coefficient of this equation is not clear at present. It is well known that the Hagen–Poiseuille flow is one of well defined flow in hydrodynamics:

$$P = 8\eta L\frac{v}{r^2},$$

where P, L, r, and v are the hydrodynamic pressure, the length of the capillary, the radius of the capillary, and the velocity of the flowing fluid. Taking into account Equation (12), one obtains the proportional constant of eight for the friction of the capillary of radius r [47]. It seems that the proportional constant for Equation (12) depends on the structure of the system. Therefore, it will contribute in designing the flow through the porous medium if the numerical coefficient of Equation (12) is clarified. The frictional studies of the well defined gel or the combined experiments of the structure and the friction as well as the theoretical studies are required for understanding the flow processes in gel.

The diffusion of the molecule, which is dissolved in gel fluid, and the flow of gel fluid are independent phenomena from each other because the origin of the driving force for each process is not the same. Therefore, both transport processes can occur simultaneously in a system as is seen in the pattern formation in the shrinking gel. The well defined experimental study of such systems is of interest for obtaining information on the general transport phenomena in the soft material. For instance, it is expected that the crossover from the transport by flow to that by diffusion will occur at an extremely low pressure gradient. The detailed study of the crossover phenomena may open new insights into transport phenomena in biological systems.

Acknowledgments: The author thanks the Ministry of Education, Science, Sports, and Culture of Japan for the financial support (Grant-in-Aid fro Scientific Research, No. 06640523, No. 08640504, and No. 09440153). A part of this work is also supported by the International Joint Research Program of the New Energy and Industrial Technology Development Organization (NEDO).

Abbreviations

The following abbreviations are used in this manuscript:

PFG-NMR: Pulsed Field Gradient Nuclear Magnetic Resonance
AAm: Acrylamide
PAAm gel: Poly(acrylamide) gel
NIPA: *N*-isopropylacrylamide
PNIPA gel: Poly(*N*-isopropylacrylamide) gel
BIS: *N*, *N*-methylenebisacrylamide (cross-linker)
CLSM: Confocal laser scanning microscope

References

1. Tanaka, T. Collapse of gels and the critical endpoint. *Phys. Rev. Lett.* **1978**, *40*, 820–823.
2. Tanaka, T.; Sun, S.-T.; Hirokawa, Y.; Katayama, S.; Kucera, J.; Hirose, Y.; Amiya, T. Mechanical instability of gels at the phase transition. *Nature* **1987**, *325*, 796–798.
3. Matsuo E.S.; Tanaka, T. Patterns in shrinking gels. *Nature* **1992**, *358*, 482–485.
4. Tanaka, T. Gels. *Sci. Am.* **1981**, *244*, 124–136.

5. Tanaka, T.; Fillmore, D.J. Kinetics of swelling of gels. *J. Chem. Phys.* **1979**, *70*, 1214–1218.
6. Maskawa, J.; Takeuchi, T.; Maki, K.; Tsujii, K., Tanaka, T. Theory and numerical calculation of pattern formation in shrinking gels. *J. Chem. Phys.* **1999**, *110*, 10993–10999.
7. Tokita, M.; Suzuki, S.; Miyamoto, K.; Komai, T. Confocal laser scanning microscope imaging of a pattern in shrinking gel. *J. Phys. Soc. Jpn.* **1999**, *68*, 330–333.
8. Tokita, M.; Miyamoto, K.; Komai, T. Polymer network dynamics in shrinking patterns of gels. *J. Chem. Phys.* **2000**, *113*, 1647–1650.
9. Tanaka, T.; Ishiwata, S.; Ishimoto, C. Critical behavior of density fluctuations in gels. *Phys. Rev. Lett.* **1977**, *38*, 771–774.
10. Muhr, A.H.; Blanshard, J.M.V. Diffusion in gels. *Polymer* **1982**, *23*, 1012–1026.
11. Weiss, N.; van Villet, T.; Silberberg, A. Permeability of heterogeneous gels. *J. Polym. Sci. Polym. Phys. Ed.* **1979**, *17*, 2229–2240.
12. Hecht, A.M.; Geissler, E. Gel deswelling under reverse osmosis. *J. Chem. Phys.* **1980**, *73*, 4077–4080.
13. Geissler, E.; Hecht, A.M. Gel deswelling under revers osmosis. II. *J. Chem. Phys.* **1982**, *77*, 1548–1553.
14. Flory, P.J. *Principles of Polymer Chemistry*; Cornell University Press: Ithaca, NY, USA, 1953; Chapter XIII-3, pp. 576–589.
15. Papon, P.; Leblond, J.; Meijer, P.H.E. *The Physics of Phase Transition*; Springer-Verlag: Berlin/Heidelberg, Germany, 2002; Chapter 6, pp. 185–209.
16. Park, I.H.; Johnson, C.S., Jr.; Gabriel, D.A. Probediffusion in polyacrylamide gels as observed by means of holographic relaxation methods-search for a universal equation. *Macromolecules* **1990**, *23*, 1548–1553.
17. Gibbs, S.J.; Johnson, C.S., Jr. Pulsed field gradient NMR-study of probe motion in polyacrylamide gels. *Macromolecules* **1991**, *24*, 6110–6113.
18. Tokita, M.; Miyoshi, T.; Takegoshi, K.; Hikichi, K. Probe diffusion in gels. *Phys. Rev. E* **1996**, *53*, 1823–1827.
19. Crank, J. *The Mathematics of Diffusion*; Oxford University Press: New York, NY, USA, 1989.
20. Einstein, A. On the movement of small particles suspended in a stationary liquid demanded by the molecular-kinetic theory of heat. In *Investigations on the Theory of the Brownian Movement*; Fürth, R., Ed.; Dover Publications, Inc.: New York, NY, USA, 1956; pp. 1–18.
21. Langevin, D.; Rondelez, F. Sedimentation of large colloidal particles through semidilute polymer-solutions. *Polymer* **1978**, *53*, 875–882.
22. Cukier, R.I. Diffusion of Brownian spheres in semidilute polymer-solutions. *Macromolecules* **1984**, *17*, 252–255.
23. De Gennes, P.G. *Scaling Concepts in Polymer Physics*; Cornell University Press: Ithaca, NY, USA, 1979; pp. 128–162.
24. Tanaka, T.; Fillmore, J.D.; Sun, S.-T.; Nishio, I.; Swislow, G.; Shah, A. Phase transition in ionic gels. *Phys. Rev. Lett.* **1980**, *45*, 1636–1639.
25. Tanaka, T.; Hocker, L.O.; Benedek, G.B. Spectrum of light scattered from a viscoelastic gel. *J. Chem. Phys.* **1973**, *59*, 5151–5159.
26. Munch, J.P.; Candau, S.; Herz, J.; Hild, G. Inelastic light-scattering by gel modes in semi-dilute polymer solutions and permanent network at equilibrium swollen state. *J. Phys.* **1977**, *38*, 971–976.
27. Munch, J.P.; Lemarechal, P.; Candau, S.; Herz, J. Light-scattering spectroscopy of polydimethylsiloxane-toluene gel. *J. Phys.* **1977**, *38*, 1499–1509.
28. Tokita, M.; Tanaka, T. Friction coefficient of polymer networks of gels. *J. Chem. Phys.* **1991**, *95*, 4613–4619.
29. Ilmain, F.; Tanaka, T.; Kokufuta, E. Volume transition in a gel driven by hydrogen bonding. *Nature* **1991**, *349*, 400–401.
30. Annaka, M.; Tanaka, T. Multiple phases of polymer gels. *Nature* **1992**, *355*, 430–432.
31. Hirokawa, Y.; Tanaka, T. Volume phase transition in a nonionic gel. *J. Chem. Phys.* **1984**, *81*, 6379–6380.
32. Tanaka, T.; Sato, E.; Hirokawa, Y.; Hirotsu, S.; Peetermans, J. Critical kinetics of volume phase transition of gels. *Phys. Rev. Lett.* **1985**, *55*, 2455–2458.
33. Tokita, M.; Tanaka, T. Reversible decrease of gel-solvent friction. *Science* **1991**, *253*, 1121–1123.
34. Richards, E.G.; Temple, C.J. Some properties of polyacrylamide gels. *Nature* **1971**, *230*, 92–96.
35. Bansil, R.; Gupta, M.K. Effects of varying crosslinking density on polyacrylamide gels. *Ferroelectrics* **1980**, *30*, 63–71.
36. Minsky, M.L. Memoir on inventing the confocal scanning microscope. *Scanning* **1988**, *10*, 128–138.

37. Hirokawa, Y.; Jinnai, H.; Nishikawa, Y.; Okamoto, T.; Hashimoto, T. Direct observation of internal structures in poly(*N*-isoprpylacrylamide) chemical gels. *Macromolecues* **1999**, *32*, 7093–7099.
38. Doi, Y.; Tokita, M. Real space structure of opaque gel. *Langmuir* **2005**, *21*, 5285–5289.
39. Mandelbrot, B.B. *The Fractal Geometry of Nature*; W. H. Freeman and Company: New York, NY, USA, 1983; Chapters VIII and IX.
40. Weitz, D.A.; Huang, J.S. *Kinetics of Aggregation and Gelation*; Family, F., Landau, D.P., Eds.; Elsevier Science Publishing Company, Inc.: New York, NY, USA, 1984; pp. 19–28.
41. Aharony, A. *Fractals and Disordered Systems*; Bunde, A., Havlin, S., Eds.; Springer-Verlag: Berlin/Heidelberg, Germany, 1996; Chapter 4, pp. 177–199.
42. Mukai, S.; Miki, H.; Garamus, V.; Willmeit, R.; Tokita, M. Structural transition of non-ionic poly(acrylamide) gel. *Progr. Colloid Polym. Sci.* **2009**, *136*, 95–100.
43. Doi, Y.; Tokita, M. Friction coefficient and structural transition in a poly(acrylamide) gel. *Langmuir* **2005**, *21*, 9420–9425.
44. Tokita, M. Structure and frictional properties of colloid gel. *Polymers* **2014**, *6*, 651–666.
45. Doi, M.; Edwards, S.F. *The Theory of Polymer Dynamics*; Oxford University Press: New York, NY, USA, 1986.
46. Grosberg, A.Y.; Khokhlov, A.R.; *Giant Molecules—2nd Edition, Here, There, and Everywhere*; World Scientific Publishing Company, Co. Pte. Ltd.: Hackensack, NJ, USA, 2011; Chapter 12, pp. 239–259.
47. Tokita, M. Friction between polymer network of gels and solvent. *Adv. Polym. Sci.* **1993**, *110*, 27–47.

12

Carbamate-Based Bolaamphiphile as Low-Molecular-Weight Hydrogelators

Laurent Latxague, Alexandra Gaubert, David Maleville, Julie Baillet, Michael A. Ramin and Philippe Barthélémy *

ARNA laboratory, Univ. Bordeaux, ChemBioPharm, INSERM, U1212, CNRS UMR 5320, F-33000 Bordeaux, France; laurent.latxague@u-bordeaux.fr (L.L.); alexandra.gaubert@u-bordeaux.fr (A.G.); david.maleville@u-bordeaux.fr (D.M.); julie.baillet@outlook.fr (J.B.); michael.ramin@inserm.fr (M.A.R.)
* Correspondence: philippe.barthelemy@inserm.fr

Academic Editor: Clemens K. Weiss

Abstract: A new bolaamphiphile analog featuring carbamate moieties was synthesized in six steps starting from thymidine. The amphiphile structure exhibits nucleoside-sugar polar heads attached to a hydrophobic spacer via carbamate (urethane) functions. This molecular structure, which possesses additional H-bonding capabilities, induces the stabilization of low-molecular-weight gels (LMWGs) in water. The rheological studies revealed that the new bolaamphiphile **7** stabilizes thixotropic hydrogels with a high elastic modulus ($G' > 50$ kPa).

Keywords: nucleolipid; nucleoside; glycosyl; bolaamphiphiles; low-molecular-weight gels (LMWG); carbamate; hydrogels; supramolecular assemblies; thixotropy

---

1. Introduction

Supramolecular gels [1] are of interest in biomedicine for several biomedical applications including biological assays [2], wound healing [3], tissue engineering [4], or drug delivery [5]. The glycosyl-nucleoside-lipids (GNLs) [6–8] family belongs to amphiphilic structures composed of biological units such as sugars, lipids, and nucleic acids. Because of their intrinsic supramolecular properties, these amphiphiles have been studied as low-molecular-weight gelators (LMWGs). Among GNLs, bolaamphiphiles [9–14], which are composed of a hydrophobic chain covalently linked at both ends to hydrophilic head groups, have been recently reported as biocompatible LMWG-based hydrogels suitable for the culture of stem cells [15]. The molecular architectures previously reported exhibit lipid moieties covalently attached via ether functions (Figure 1a).

In order to improve the rheological properties of LMWG-based hydrogels, we modulated the bolaamphiphile architectures by inserting additional hydrogen bond functions. In this design, the hydrophobic segment is attached to the nucleobase via a triazole ring and a carbamate moiety at the 5′-position of thymidine.

Herein, we report the synthesis of a new carbamate-based bolaamphiphile, based on N-thymine glycosylated compounds as head groups and a lipid chain attached at the 5′-position of the nucleoside (Figure 1). The rheological studies revealed that such structures stabilize low-molecular-weight hydrogels at different concentrations including 1%, 3%, and 4%.

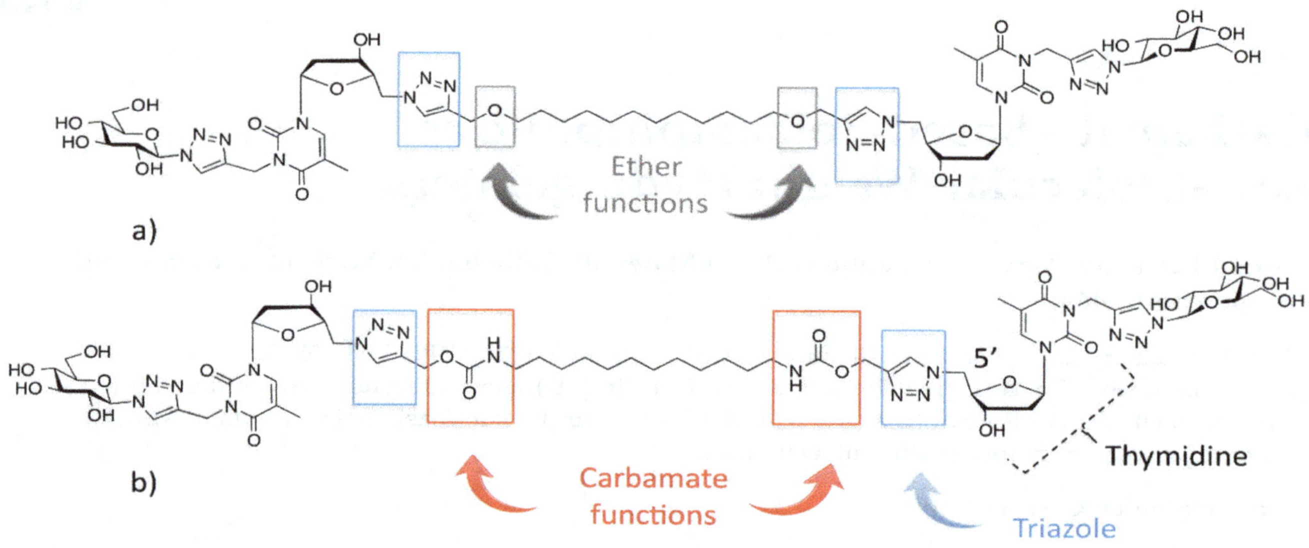

Figure 1. Schematic representations of bolaamphiphile structures. (**a**) Bolaamphiphile structure previously reported (see [15]). The hydrophobic segment is attached via ether functions. (**b**) Bolaamphiphile structures with carbamate functions described in this study.

2. Results and Discussion

2.1. Synthesis of the Carbamate-Based Bolaamphiphiles

In order to expend the current repertoire of the bolaamphiphiles, we selected the carbamate function to attach the hydrophobic central synthetic unit to the polar heads. The carbamate moiety, also called urethane, is a structural function used in many approved therapeutic compounds [16]. Its functionality, which is linked to amide-ester hybrid features, displays good chemical and biochemical stabilities. This function has been used in medicinal chemistry as a peptidic bond mimetic. Interestingly, the carbamate function can participate in hydrogen bonding through both the carboxyl and the NH motifs. Different hydrophilic and hydrophobic motifs can be connected to the carbamate O- or N-termini, or both. Hence, inserting this functionality into the GNL structures would allow us to modulate their supramolecular properties.

The synthetic strategy developed to access to the new carbamate bolaamphiphile is based on a key intermediate **5**: the 5′-desoxy-*N*-3-[1-((β-D-glucopyranoside)-1*H*-1,2,3-triazol-4-yl)methyl] azidothymidine (compound **5**, Scheme 1). The commercial starting material (thymidine) was first alkylated at the position 3′ by using propargyl bromide in dimethylformamide (DMF) in the presence of potassium carbonate to give compound **1** (Scheme 1). The propargyl moiety was then "clicked" with a peracetylated azidoglucose to provide intermediate **3** with a 60% yield (Scheme 1). The peracetylated azidoglucose **5** was then prepared due to a substitution reaction (SN_2) in the presence of triphenylphosphine (PPh_3), sodium azide (NaN_3), and tetrabromomethane (CBr_4) in DMF. This reaction provided the protected azidoglycoside **4** (Scheme 1, III), which was finally subjected to Zemplen's deacetylation to give the expected compound **5** (Scheme 1, IV). In parallel, the dicarbamate hydrophobic central synthetic unit **6** was prepared starting from commercial 1,12-diaminododecane. This diamine was reacted with the propargyl chloroformate to give the dicarbamate hydrophobic spacer **6** (Scheme 1, V). The latter was thus engaged in a click reaction with two equivalents of **5** to obtain the expected bolaamphiphile **7** with a 65% yield (Scheme 1, VI).

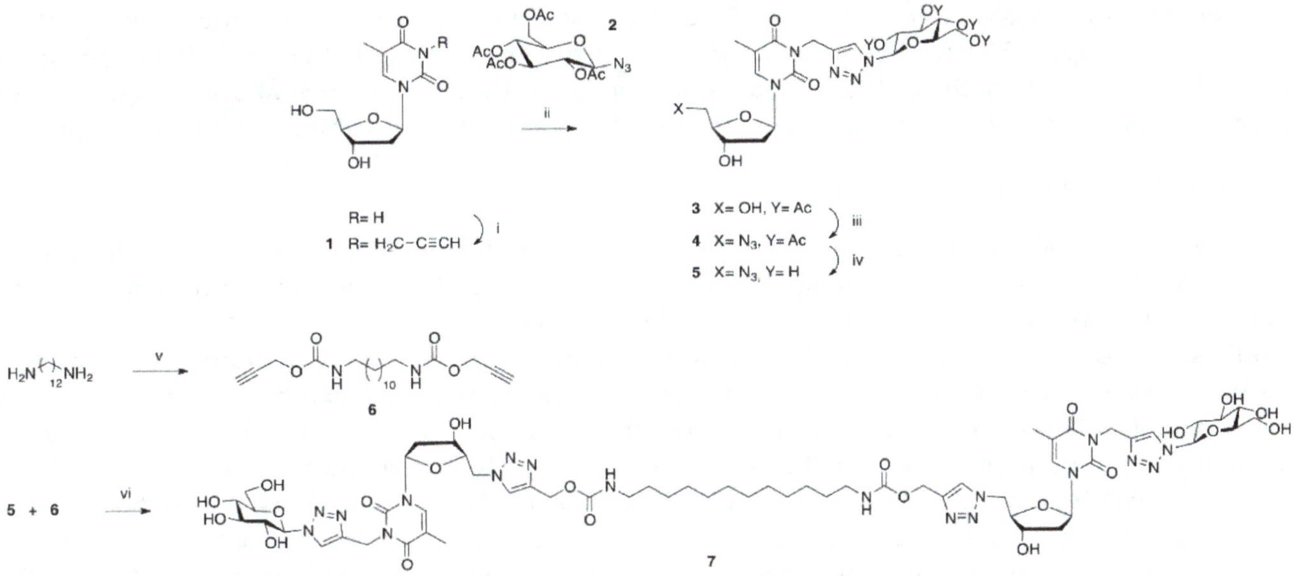

Scheme 1. General synthesis **(i)** K_2CO_3, propargyl bromide, tetra-n-butylammonium iodide (TBAI), dimethylformamide (DMF), r.t., 12 h (81%). **(ii) 2**, $^{t}BuOH/H_2O$ (1:1), $CuSO_4$, ascorbic acid, 75 °C, 7 h. **(iii)** PPh_3, NaN_3, CBr_4, DMF, r.t., 24 h. **(iv)** MeONa 1% in MeOH, r.t., 30 min. **(v)** Propargyl chloroformate, Et_3N, CH_2Cl_2, r.t., 24 h. **(vi)** Tetrahydrofuran (THF)/H_2O (1:1), $CuSO_4$, ascorbic acid, 65 °C, 4 h.

2.2. *Physicochemical Studies*

An important goal of the present study was to determine the impact of structural modifications (the insertion of carbamate) on the supramolecular properties. Thus, the physicochemical properties of compound **7** were studied, including (i) the stabilization of gel in water; (ii) the rheological properties; and (iii) the gel-sol transition temperature transition.

2.2.1. Gel Formation Assays

Bolaamphiphile carbamate **7** was dissolved in Milli-Q water to obtain gel concentrations of 1%, 3%, and 4% (*w*/*v*). To ensure complete dissolution of compounds, samples were heated at 60 °C and shaken at 1250 rpm for 30 min using a Thermomixer compact (Eppendorf, Hauppauge, NY, USA). Samples were cooled down at 4 °C for one day (gel at 3% and 4% (*w*/*v*)) or two days (gel at 1% (*w*/*v*)). The samples were allowed to return at room temperature. In order to evaluate the gel formation, samples were then turned upside-down; no flowing under its own weight indicated sample gelification. As shown in Figure 2, compound **7** enabled homogeneous hydrogel formation at low concentrations.

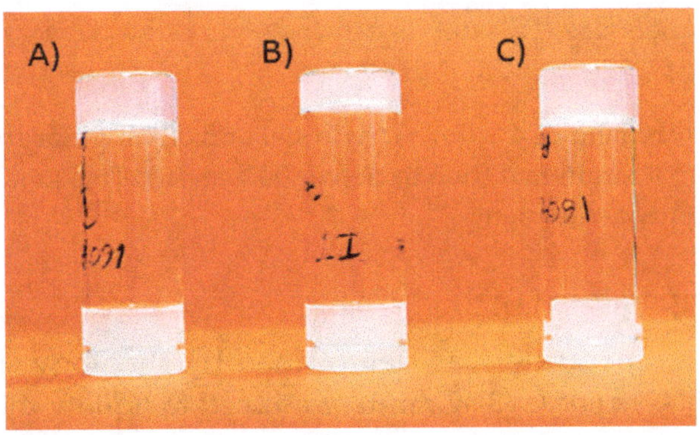

Figure 2. Hydrogel formation at 1% (**A**); 3% (**B**); and 4% (**C**) (*w*/*v*).

Gel formation was also assessed at 25 °C. However, using the same parameters (sonication, heating, and storage at RT), no homogeneous gel was observed. After incubation at 25 °C, gel formation occurred at the bottom of the test tube, and some non-gelified water remained above. Longer storage at room temperature (2 or 3 days up to one week) did not improve the stabilzation of the sample.

2.2.2. Rheological Studies

The mechanical properties of hydrogels stabilized by bolaamphiphile **7** were characterized by rheological measurements. The storage (G′) and loss (G″) moduli provided information about the visco-elastic properties of gels. The frequency dependence of G′ and G″ moduli for the hydrogel stabilized by compound **7** (4% w/v) indicates that G′ exceeded G″ below a strain of 0.5% and an angular frequency 1 Hz (data not shown). The hydrogel stabilized by **7** exhibits a dominant elastic character (solid-like) as expected for a supramolecular network. The G′ value is 57 kPa at an angular frequency of 1 rad/s for the hydrogel at 4% w/v. Importantly, compared with the ether analogue (Figure 1a), which possesses a G′ value is 30 kPa in similar rheological conditions, the elastic character of the hydrogel improved with the bolaamphiphile **7**. This behavior is likely due to the presence of carbamate functions in the amphiphile structure. The thixotropic property of the hydrogel **7** was also studied by rheology. The application of a high strain to the sample induces a melting of the gel; this phenomenon was studied by the evolution of the viscoelastic moduli (Figure 3). First, a low strain of 0.04% was applied to hydrogel **7** at 4% (w/v). In the gel state, G′ was higher than G″. Then, the hydrogel **7** was suddenly subjected to a higher strain of 15%. At this point, the G′ value was much lower, indicating that the gel liquefies under the stress. With a strain of 0.04% (as in the first step), the sample gradually returns to the gel state and recovers its original strength. The G′ value increases rapidly, and the sample completely recovers to its initial elastic modulus within 30 min.

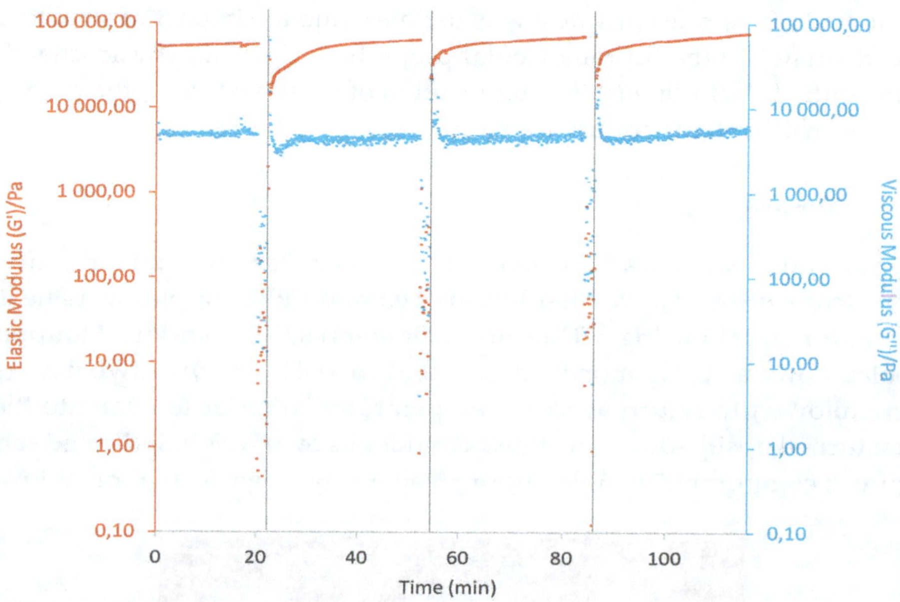

Figure 3. Step-strain measurement of 4% (w/v) hydrogel at fixed angular frequency of 1 Hz. The gel was swept from 0.04% to 15% strain and then back to 0.04% strain. This analysis was repeated at least three times for reproducibility assessment.

2.2.3. Gel–Sol Transition Temperature (Tgel)

The gel–sol transition temperature was determined by rheology (Kinexus® Pro+ rheometer). A temperature ramp from 25 °C to 60 °C (3 °C/min) was used to liquefy the carbamate-based hydrogel. The gel–sol temperature was recorded at the transition between gel and liquid states. An oscillatory stress of 3 Pa and a constant frequency (6.283 rad·s^{-1}) were applied during the assay. Graphically, the gel–sol temperature is characterized by the intersection of the two viscoelastic moduli G' and G''. The melting point was observed at 36.3 °C (Figure 4), which is close to 37 °C. This hydrogel could be of interest to the areas for which biomaterials must melt slowly close to 37 °C, such as in the transdermal delivery of drugs.

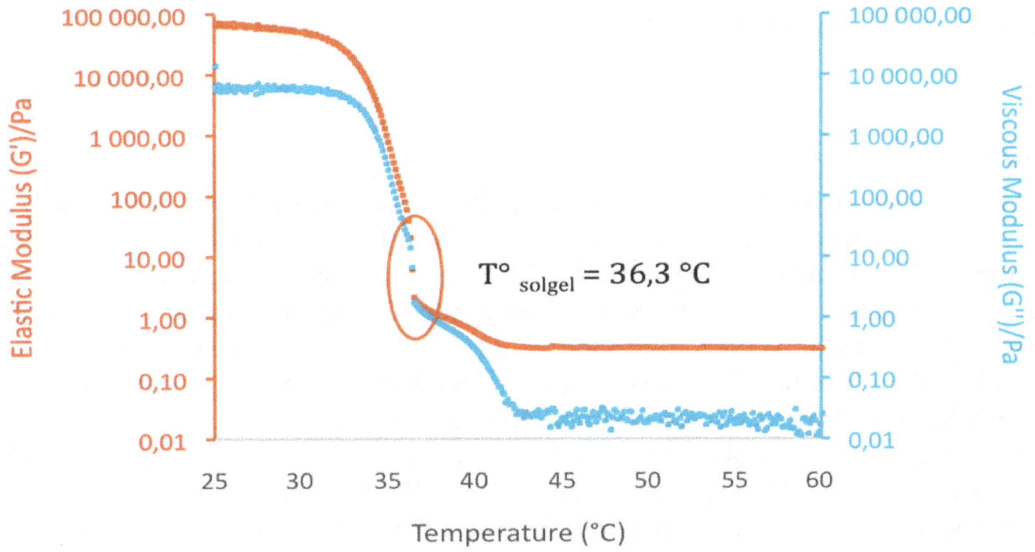

Figure 4. Gel–sol transition temperature determination for bolaamphiphile **7** hydrogel (4% (w/v)).

3. Conclusions

In order to expend the current family of glycosyl-nucleoside-lipids, we synthesized a new bolaamphiphile analog featuring carbamate moieties in six steps starting from thymidine. This bioinspired amphiphile, which possesses additional H-bonding capabilities, allows the formation of low-molecular-weight gels (LMWGs) in water at low concentrations (1% w/v). Importantly, the rheological investigations revealed that the new bolaamphiphile **7** stabilizes hydrogels with higher elastic moduli ($G' < 50$ kPa) than the diether analogs ($G' \approx 30$ kPa) reported previously [15], indicating that the carbamate function contributes favorably to the supramolecular network. As a matter of comparison, dipeptide-based LMWGs previously reported [17] show lower elastic moduli than the bolaamphiphile **7** (*aka* Glycosyl Nucleoside Bola Amphiphile carbamate or GNBA carbamate) hydrogels. The rheological studies also revealed thixotropic properties, emphasizing the possible injectability of the carbamate-based hydrogels.

The molecular approach developed in this report, namely the insertion of hydrogen bond functions in a bolaamphiphile architecture, will be used in the future for the design of new glycosyl-nucleoside lipids for biomedical applications.

4. Materials and Methods

4.1. General

All commercially reagents and solvents (Alfa-Aesar, Karlsruhe, Germany; Fluka and Sigma-Aldrich, Saint Quentin Fallavier, France) were used without further purification. For reactions requiring anhydrous conditions, dry solvents were used under an inert atmosphere (nitrogen or argon). Analytical thin layer chromatographies (TLC) were performed on pre-coated silica gel F254 plates with a fluorescent indicator (Merck, Fontenay sous Bois, France). The detection of compounds was accomplished with UV light (254 nm). All compounds were characterized using ^{1}H and ^{13}C nuclear magnetic resonance (NMR) spectroscopy (Bruker Avance DPX-300 spectrometer, Wissembourg, France; ^{1}H at 300.13 MHz and ^{13}C at 75.46 MHz). Assignments were made by 1H-1H COSY, DEPT, and HSQC experiments. Chemical shifts (δ) are given in parts per million (ppm) relatively to tetramethylsilane (TMS) or residual solvent peaks ($CHCl_3$: ^{1}H: 7.26 ppm, ^{13}C: 77.0 ppm). Coupling constants J are given in Hertz (Hz); peak multiplicity is reported as follows: s = singlet, bs = broad singlet, d = doublet, t = triplet, and m = multiplet. High-resolution mass spectra (HRMS) were performed by the CRMPO (Rennes, France) on a Thermo Fisher Q-Extractive mass spectrometer (Waltham, MA, USA) equipped with an ESI source and recorded in negative mode.

4.2. Synthesis

N-3-Propargylthymidine (**1**)

Propargyl bromide (80% soln. in toluene, 2.68 g, 18.5 mmol) was added to a mixture of thymidine (3 g, 12.3 mmol), potassium carbonate (2.5 g, 18.5 mmol) and tetra-n-butylammonium iodide (TBAI) (0.45 g, 1.2 mmol) in anhydrous DMF (60 mL). The reaction mixture was stirred for 12 h at room temperature. After concentration to dryness under reduced pressure, water was added to the solid residue, and the product was extracted with ethyl acetate. The organic layer was separated, washed (aqueous 10% KCl then brine), dried (Na_2SO_4), and evaporated to give a yellow oil used in the next step without further purification, with a crude yield of 2.82 g (81%).

^{1}H NMR ($CDCl_3$) δ 1.93 (s, 3H, CH_3), 2.17 (t, *J* = 2.4 Hz, 1H, CH propargyl), 2.28–2.37 (m, 2H, H-2′), 2.49 (t, *J* = 2.4 Hz, 2H, CH propargyl), 3.81–3.93 (m, 2H, H-5′), 3.98–4.01 (m, 1H, H-4′), 4.52–4.57 (m, 1H, H-3′), 4.68 (d, 2H, CH_2 propargyl), 6.29 (t, *J* = 6.6 Hz, 1H, H-1′), 7.58 (s, 1H, H-6 thymine).

1-Deoxy-1-azido-2,3,4,6-tetra-*O*-acetyl-β-D-glucopyranose (**2**)

Sodium azide (3.90 g, 60 mmol) was added to a mixture of 2,3,4,6-tetra-*O*-acetyl-β-D-glucopyranosyl bromide [18] (8.22 g, 20 mmol) in anhydrous DMF (100 mL). After stirring for 24 h at 65 °C, the mixture was cooled to room temperature and poured into water (200 mL). The resulting solid was filtered off, washed with water, and dried under vacuum, with a crude yield of 6.92 g (92%). NMR data were identical with those obtained from a commercial sample of the product.

N-3-[1-((β-D-Glucopyranosidetetraacetate)-1*H*-1,2,3-triazol-4-yl)methyl]thymidine (**3**)

Copper sulfate (0.260 g, 1.63 mmol) and sodium ascorbate (0.645 g, 3.26 mmol) were successively added to a degassed suspension of (**1**) (4.59 g, 16.3 mmol) and 1-deoxy-1-azido-2,3,4,6-tetra-*O*-acetyl glucopyranose (**2**) (6.085 g, 13.3 mmol) in 100 mL of ^{t}BuOH/H_2O (1:1). The mixture was stirred at 75 °C for 7 h. After cooling to room temperature, the solvents were removed under reduced pressure. The resulting solid was washed with water until the washings were colorless. After drying under high vacuum, the resulting white amorphous solid was used in the next step without further purification, with a crude yield of 8.65 g (81%).

^{1}H NMR (MeOD) δ 1.94 (s, 3H, CH_3 thymine), 1.81, 2.00, 2.06 (3s, 12H, OAc), 2.22–2.32 (m, 2H, H-2′), 3.72–3.85 (m, 2H, H-5′), 3.91–0.94 (m, 1H, H-4′), 4.15–4.26 (m, 2H, H-6), 4.30–4.36 (m, 1H, H-3′), 4.39–4.43 (m, 1H, H-5), 5.16–5.30 (m, 3H, H-4 + CH_2 triazole), 5.52 (dd, *J* = 9.3 Hz, 1H, H-3), 5.60 (dd, *J* = 9.3 Hz, 1H, H-2), 6.1 (d, *J* = 9.0 Hz, 1H, H-1), 6.33 (t, *J* = 6.7 Hz, 1H, H-1′), 7.89 (s, 1H, H-6 thymine), 8.17 (s, 1H, CH triazole).

5′-Deoxy-*N*-3-[1-((β-D-glucopyranosidetetraacetate)-1*H*-1,2,3-triazol-4-yl)methyl] azidothymidine (**4**)

Triphenylphosphine (825 mg, 3.15 mmol), sodium azide (950 mg, 13.1 mmol), and carbon tetrabromide (1.05 g, 3.15 mmol) were added to a solution of (3) (1.5 g, 2.63 mmol) in anhydrous DMF (25 mL). The reaction mixture was stirred at room temperature for 24 h and then treated with 5% aqueous $NaHCO_3$ (50 mL). After extracting with dichloromethane, the organic solution was washed with water, followed by drying over Na_2SO_4, and concentrated under a reduced pressure. The crude product was purified by column chromatography, eluting with a step gradient of MeOH in CH_2Cl_2 (0%–5%) to give the title compound as a white amorphous solid, with a yield of 652 mg (36%).

^{1}H NMR ($CDCl_3$) δ 1.94 (s, 3H, CH_3 thymine), 1.83, 2.01, 2.05, 2.08 (4s, 12H, OAc), 2.16–2.25 (m, 1H, H-2′a), 2.36–2.44 (m, 1H, H-2′b), 3.55–3.75 (m, 2H, H-5′), 4.01–4.05 (m, 1H, H-5), 4.06–4.10 (m, 1H, H-4′), 4.14–4.46 (m, 2H, H-6), 4.45 (m, 1H, H-3′), 5.14–5.25 (m, 3H, CH_2 triazole + H-4), 5.35–5.48 (m, 2H, H-2, 3), 5.84 (d, J = 9.0 Hz, 1H, H-1), 6.33 (dd, J = 6.7 Hz, 1H, H-1′), 7.37 (s, 1H, H-6 thymine), 7.87 (s, 1H, CH triazole); ^{13}C NMR ($CDCl_3$) δ 13.45 (CH_3), 20.31, 20.65, 20.85 (OAc), 36.02 (CH_2 triazole), 40.48 (C-2′), 52.33 (C-5′), 61.63 (C-6), 67.72 (C-4), 70.26, 72.82 (C-2, 3), 71.51 (C-3′), 75.15 (C-5), 84.49 (C-4′), 85.62, 85,70 (C-1, 1′), 110.57 (C-5 thymine), 122.60 (CH triazole), 134.04 (C-6 thymine), 143.82 (C-4 triazole), 150.68, 163.06 (C=O thymine), 168.97, 169.48, 170.13, 170.76 (C=O acetyl); HRMS m/z calcd. for $C_{27}H_{34}N_8O_{13}Na$ 701.21375 $[M + Na]^+$, found 701.2138.

5′-Deoxy-*N*-3-[1-((β-D-glucopyranoside)-1*H*-1,2,3-triazol-4-yl)methyl]azido thymidine (**5**)

A stirred suspension of (4) (1.1 g, 1.62 mmol) in anhydrous methanol (20 mL) was treated with a few drops of freshly prepared 1N sodium methoxide solution at room temperature. When TLC (CH_2Cl_2/MeOH 8:2) showed complete conversion (approx. 30 min), the reaction mixture was treated with Dowex® 50X2 (Sigma-Aldrich). After filtration, the solvent was removed under reduced pressure. The solid residue was applied to a column of silica gel, and the product eluted with 8:2 CH_2Cl_2-MeOH to afford the title compound as a white amorphous solid, with a yield of 0.853 g (77%).

^{1}H NMR (MeOD) δ 2.07 (s, 3H, CH_3), 2.41–2.43 (m, 2H, H-2′), 3.46–3.64 (m, 5H, H-3, 4, 5 + H-5′), 3.68–3.74 (m, 1H, H-6a), 3.85–3.92 (m, 2H, H-2 + H-6b), 4.06+4.13 (m, 1H, H-4′), 4.46–4.51 (m, 1H, H-3′), 5.36 (s, 2H, CH_2 triazole), 5.60 (d, J = 9.2 Hz, 1H, H-1), 6.45 (dd, J = 6.7 Hz, 1H, H-1′), 7.73 (s, 1H, H-6 thymine), 8.26 (s, 1H, CH triazole); ^{13}C NMR (MeOD) δ 14.12 (CH_3), 38.00 (CH_2 triazole), 41.22 (C-2′), 54.16, 71.69, 79.22, 81.92 (C-3, 4, 5, 5′), 63.21 (C-6), 73.21 (C-3′), 74.77 (C-2), 87.21 (C-4′), 88.02 (C-1′), 90.41 (C-1) 112.00 (C-5 thymine), 125.16 (CH triazole), 137.19 (C-6 thymine), 143.19 (C-4 triazole), 152.94, 165.64 (C=O); HRMS m/z calcd. for $C_{19}H_{26}N_8O_9Na$ 533.17149 $[M + Na]^+$, found 533.1716.

1,12-Dodecanediyl-bis-*N*-propargylcarbamate (**6**)

Triethylamine (1.06 g, 10.5 mmol) and propargyl chloroformate (1.18 g, 10 mmol) were added to a cooled (0 °C) suspension of 1,12-diaminododecane (1 g, 5 mmol) in anhydrous dichloromethane (40 mL) over 30 min, and stirring continued for a further 24 h at room temperature. The reaction mixture was poured into water 40 mL), and the organic phase separated. The aqueous phase was extracted with dichloromethane, the organic extracts combined, dried (Na_2SO_4), and the solvent removed under reduced pressure. The product was purified by column chromatography eluting with CH_2Cl_2/MeOH (99:1 then 98:2) and obtained as a white amorphous solid, with a yield of 0.728 g (40%).

^{1}H NMR ($CDCl_3$) δ 1.27 (s, 16H, CH_2), 1.49–1.53 (m, 4H, CH_2CH_2N), 2.49 (t, J = 2.4 Hz, 2H, CH propargyl), 3.20 (dt, J = 6.4, 13.3 Hz, 4H, CH_2CH_2N), 4.69 (d, 4H, CH_2 propargyl) ; ^{13}C NMR ($CDCl_3$) δ 26.66, 29.20, 29.46 , 29.83 (CH_2), 41.16 (CH_2N), 52.30 (CH_2 propargyl), 74.51 (CH propargyl), 78.38 (Cq propargyl), 155.41 (C=O).

1,12-Diaminododecanediyl-*N*,*N*′-bis-[5′-(4-methyloxycarbonyl-1*H*-1,2,3-triazole-1-yl)-*N*-3-((1-(β-D-glucopyranoside)-1*H*-1,2,3-triazole-4-yl)methyl)-5′-deoxy thymidine] (**7**) (GNBA-carbamate)

Copper sulfate (18.8 mg, 0.118 mmol) and sodium ascorbate (46.8 mg, 0.236 mmol), were successively added to a degassed suspension of (**6**) (0.215 g, 0.59 mmol) and (**5**) (0.546 g, 1.18 mmol) in 20 mL of H_2O/THF (1:1). The mixture was stirred at 65 °C for 4 h. After cooling to room temperature, the solvents were removed under reduced pressure. The resulting solid was applied to a column of silica gel eluting with CH_2Cl_2/MeOH (8:2 to 7:3) to give the title compound as a white amorphous solid, with a yield of 492 mg (65%).

^{1}H NMR (MeOD) δ 1.30 (s, 16H, CH_2), 1.42–1.51 (m, 4H, $CH_2CH_2NH(CO)$), 1.93 (d, J = 1.0 Hz, 6H, CH_3 thymine), 2.31 (dd, J = 6.2 Hz, 4H, H-2′), 3.08 (dd, J = 6.9 Hz, 4H, $CH_2NH(CO)$), 3.36–3.56 (m, 6H, H-3, H-4, H-5), 3.68–3.73 (m, 2H, H-6a), 3.85–3.92 (m, 4H, H-2, H-6b), 4.16–4.21 (m, 2H, H-4′),

4.40–4.46 (m, 2H, H-3′), 4.72–4.76 (m, 4H, H-5′), 5.13 (s, 4H, triazole CH_2O), 5.52 (s, 4H, triazole CH_2N), 5.56 (d, J = 9.2 Hz, 2H, H-1), 6.21 (dd, J = 6.7 Hz, 2H, H-1′), 7.24 (d, 2H, H-6 thymine), 7.97 (s, 2H, H triazole), 8.06 (s, 2H, H triazole); ^{13}C NMR (MeOD) δ 11.76 (CH_3 thymine), 26.39 (CH_2), 28.95, 29.19, 29.43 (CH_2 & CH_2CH_2O), 35.68 (CH_2N triazole), 38.13 (C-2′), 40.43 ($CH_2NH(CO)$), 51.17 (C-5′), 57.02 (OCH_2 triazole), 60.96 (C-6), 69.44, 77.00, 79.70 (C-3, C-4, C-5), 71.00 (C-3′), 72.52 (C-2), 84.23 (C-4′), 86.77 (C-1′), 88.18 (C-1), 109.80 (C-5 thymine), 122.84, 125.44 (CH triazole), 135.49 (C-6 thymine), 143.16, 143.37 (C-4 triazole), 150.54, 163.37 (C=O), 157.01 (OC=O); HRMS m/z calcd. for $C_{58}H_{84}N_{18}O_{22}Na$ 1407.58998 $[M + Na]^+$, found 1407.5897.

4.3. Rheology

All rheological measurements were made using a Kinexus® Pro+rheometer (Malvern Instruments Ltd., Orsay, France). The lower plate was equipped with a Peltier temperature control system and the upper plate was a steel plate–plate geometry (20-mm diameter, gap: 0.3 mm). All experiments were done at 25 °C ± 0.01 °C unless indicated otherwise. To prevent water evaporation and to control the temperature, a solvent trap was used. A sample of bolaamphiphile 7 was placed as a gel between the two plates before assays. All experiments were carried out within the linear viscoelastic regime (LVR). It was determined by an amplitude strain sweep experiment (0.01% to 10% at an angular frequency of 1 Hz or 6.283 rad·s^{-1}). Then, a frequency sweep assay (0.6283 to 62.83 rad·s^{-1}; applied strain of 0.04% within LVR) allowed the evaluation of elastic (G′) and viscous (G″) moduli. At least three replicates were analyzed for each sample.

Acknowledgments: The authors acknowledge financial supports from the "French National Research Program for Environmental and Occupational Health of ANSES" (2015/1/072). This work was realized within the framework of the Laboratory of Excellence AMADEus with the reference ANR-10-LABX-0042 AMADEUS.

Author Contributions: Laurent Latxague and Philippe Barthélémy conceived and designed the experiments; Alexandra Gaubert, David Maleville, Julie Baillet, and Michael A. Ramin performed the experiments, analyzed the data, or both; Philippe Barthélémy, Laurent Latxague, and Alexandra Gaubert wrote the paper.

References

1. Steed, J.W. Supramolecular gel chemistry: Developments over the last decade. *Chem. Commun.* **2011**, *47*, 1379–1383. [CrossRef] [PubMed]
2. Neelakandan, P.P.; Hariharan, M.; Ramaiah, D. A Supramolecular ON–OFF–ON Fluorescence Assay for Selective Recognition of GTP. *J. Am. Chem. Soc.* **2006**, *128*, 11334–11335. [CrossRef] [PubMed]
3. Griffin, D.R.; Weaver, W.M.; Scumpia, P.O.; di Carlo, D.; Segura, T. Accelerated wound healing by injectable microporous gel. *Nat. Mater.* **2015**, *14*, 737–744. [CrossRef] [PubMed]
4. Ziane, S.; Schlaubitz, S.; Miraux, S.; Patwa, A.; Lalande, C.; Bilem, I.; Lepreux, S.; Rousseau, B.; le Meins, J.-F.; Latxague, L.; et al. A theromosensitive hydrogel for tissue engineering. *eCM* **2012**, *23*, 147–160.
5. Tiller, J.C. Increasing the local concentration of drugs by hydrogel formation. *Angew. Chem. Int. Ed.* **2003**, *42*, 3072–3075. [CrossRef] [PubMed]
6. Godeau, G.; Barthélémy, P. Glycosyl-nucleoside lipids as low-molecular-weight gelators. *Langmuir* **2009**, *25*, 8447–8450. [CrossRef] [PubMed]
7. Godeau, G.; Bernard, J.; Staedel, C.; Barthélémy, P. Glycosyl-nucleoside-lipid based supramolecular assembly as a nanostructured material with nucleic acid delivery capabilities. *Chem. Commun.* **2009**, 5127–5129. [CrossRef] [PubMed]
8. Godeau, G.; Brun, C.; Arnion, H.; Staedel, C.; Barthélémy, P. Glycosyl-nucleoside fluorinated amphiphiles as components of nanostructured hydrogels. *Tetrahedron Lett.* **2010**, *51*, 1012–1015. [CrossRef]
9. Estroff, L.A.; Hamilton, A.D. Water gelation by small organic molecules. *Chem. Rev.* **2004**, *104*, 1201–1218. [CrossRef] [PubMed]
10. Brard, M.; Richter, W.; Benvegnu, T.; Plusquellec, D. Synthesis and supramolecular assemblies of bipolar archaeal glycolipid analogues containing a cis-1,3-disubstituted cyclopentane ring. *J. Am. Chem. Soc.* **2004**, *126*, 10003–11012. [CrossRef] [PubMed]

11. Meister, A.; Blume, A. Self-assembly of bipolar amphiphiles. *Curr. Opin. Colloid Interface Sci.* **2007**, *12*, 138–147. [CrossRef]
12. Yan, Y.; Lu, T.; Huang, J. Recent advances in the mixed systems of bolaamphiphiles and oppositely charged conventional surfactants. *J. Colloid Interface Sci.* **2009**, *337*, 1–10. [CrossRef] [PubMed]
13. Meister, A.; Blume, A. Single-Chain Bolaphospholipids: Temperature-Dependent Self-assembly and Mixing Behavior with Phospholipids. *Adv. Plan. Lipid Bilayers* **2012**, *16*, 93–128.
14. Blume, A.; Drescher, S.; Graf, G.; Kçhler, K.; Meister, A. Self-assembly of different single-chain bolaphospholipids and their miscibility with phospholipids or classical amphiphiles. *Adv. Colloid Interface Sci.* **2014**, *208*, 264–278. [CrossRef] [PubMed]
15. Latxague, L.; Ramin, M.A.; Appavoo, A.; Berto, P.; Maisani, M.; Ehret, C.; Chassande, O.; Barthélémy, P. Control of Stem-Cell Behavior by Fine Tuning the Supramolecular Assemblies of Low-Molecular-Weight Gelators. *Angew. Chem. Int. Ed. Engl.* **2015**, *54*, 4517–4521. [CrossRef] [PubMed]
16. Ghosh, A.K.; Brindisi, M. Organic Carbamates in Drug Design and Medicinal Chemistry. *J. Med. Chem.* **2015**, *58*, 2895–2940. [CrossRef] [PubMed]
17. Raeburn, J.; Cardoso, A.Z.; Adams, D.J. The importance of the self-assembly process to control mechanical properties of low molecular weight hydrogels. *Chem. Soc. Rev.* **2013**, *42*, 5143. [CrossRef] [PubMed]
18. Horton, D. *Methods in Carbohydrate Chemistry*; Lemieux, U., Whistler, R.L., Wolfrom, M.L., BeMiller, J.N., Eds.; Academic Press Inc.: New York, NY, USA, 1963; Volume II, pp. 433–437.

13

Droplets, Evaporation and a Superhydrophobic Surface: Simple Tools for Guiding Colloidal Particles into Complex Materials

Marcel Sperling and Michael Gradzielski *

Stranski Laboratorium für Physikalische & Theoretische Chemie, Institut für Chemie, Technische Universität Berlin, Straße des 17. Juni 124, Sekr. TC7, D-10623 Berlin, Germany
* Correspondence: michael.gradzielski@tu-berlin.de

Academic Editor: Clemens K. Weiss

Abstract: The formation of complexly structured and shaped supraparticles can be achieved by evaporation-induced self-assembly (EISA) starting from colloidal dispersions deposited on a solid surface; often a superhydrophobic one. This versatile and interesting approach allows for generating rather complex particles with corresponding functionality in a simple and scalable fashion. The versatility is based on the aspect that basically one can employ an endless number of combinations of components in the colloidal starting solution. In addition, the structure and properties of the prepared supraparticles may be modified by appropriately controlling the evaporation process, e.g., by external parameters. In this review, we focus on controlling the shape and internal structure of such supraparticles, as well as imparted functionalities, which for instance could be catalytic, optical or electronic properties. The catalytic properties can also result in self-propelling (supra-)particles. Quite a number of experimental investigations have been performed in this field, which are compared in this review and systematically explained.

Keywords: superhydrophobic surfaces; droplets; nanoparticles; evaporation; self-assembly; self-propelling; anisometric; colloids; supraparticles; functional materials; catalysis

1. Introduction

Self-assembly of colloids into bigger particles (from μm to mm) has been in the focus of colloid and material research for some time, as it allows for the fabrication of materials providing various functionalities [1–3]. Their production can be accomplished by the utilization of several different techniques that have been developed over the last few years, such as sedimentation, evaporation, adsorption, external force fields, bio-recognition or surface and droplet templating, which have been summarized elsewhere [4–6]. Especially, considering the ratio of estimated fabrication costs vs. the added value in terms of provided functionality is a crucial aspect, when it comes to potential applications or scale-up for industrial production. Hypothetically, this ratio has been estimated by Velev and Gupta as shown in Figure 1 [4]. They were also relating principally available techniques, also those mentioned above, to the accessible dimensionality (1D, 2D or 3D) and scalability of the procedure. One may state that usually, the lower cost processes are favored for scaling up [4]. Besides the applied technique, also the quality of materials plays an important role. When considering dispersed particles, especially on the micron- or even nano-meter scale, the efforts needed for their synthesis may increase dramatically when seeking a high degree of monodispersity, as for instance required in photonic applications [7]. In order to produce high quality photonic crystals [8], a defined control of inter-particle spacing is needed to achieve a high sensitivity for the desired diffracted wavelength. Thus, this sensitivity depends on the fluctuations in particle size as that determines the quality of the corresponding crystal lattice. This correlation of in-depth lattice spacing with the resulting diffracted wavelength is expressed by Bragg's law [9]. Moreover, besides pure size aspects of the assembling particles, also their shapes, either static [10–13] or dynamic [14,15] can be of major importance.

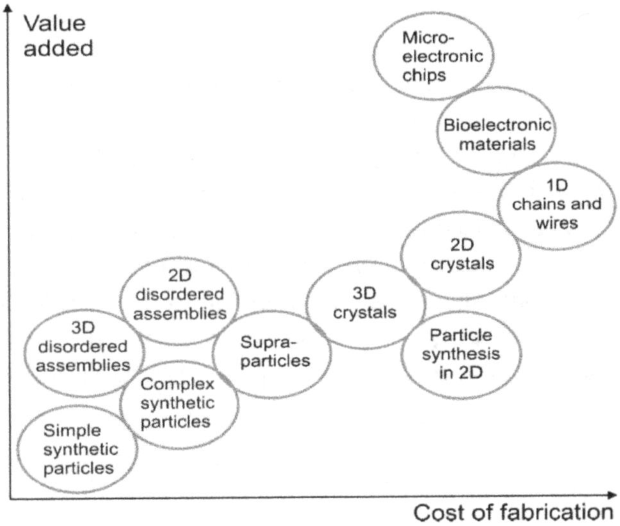

Figure 1. Hypothetical qualitative estimation of the value-to-price ratio for different products gathered by colloidal assembly.

There exist multiple ways of producing colloidal assemblies, such as supraparticles, that differ with respect to dimensionality, cost and scale. Many of these ways have been reviewed by Velev and Gupta [4]. In general, all of these methods start from confining the colloidal building blocks within a droplet (or generally a container). This could for example be done using emulsion droplets, where the solubilized liquid becomes removed by heating and other approaches. However, our main focus in this work will be set on evaporation-induced self-assembly (EISA) on super-repellant solid surfaces, typically superhydrophobic ones. In the following sections, we shall give a detailed overview of the basics of this technique, in terms of the materials applied and the condition parameters to be set, as well as the most recent materials fabricated by it.

In that context, EISA is a particularly simple and therefore appealing approach, in which one simply evaporates the solvent of a colloidal dispersion leading to a situation where the ingredients become more or less arranged during the drying process [13,14], thereby forming self-assembled nanostructures in a simple way. In this fashion, one may access structured thin films [16,17], but also supraparticles by the preparation from droplets deposited on a solvophobic surface. Probably the most common version of the latter are superhydrophobic surfaces applicable to water droplets [18–22].

In the scope of this review, we will present recent achievements in the field of supraparticle fabrication using aqueous suspension droplets dispensed and dried on superhydrophobic surfaces, i.e., using EISA and droplet templating. After some fundamental discussion of EISA and the droplet templating method, as well as superhydrophobic surfaces, we will show possible ways to create supraparticles of various shapes, hierarchical structures (like Janus-type) and functionalities. Furthermore, we will give some general discussion about applications for supraparticles that have recently been developed. It may be noted that in our review we focus on large supraparticles in the size range of hundreds of μm or even mm, neglecting the abundant work on smaller-sized assemblies.

2. Evaporation-Induced Self-Assembly and the Droplet Templating Method

The concept of evaporation-induced self-assembly (EISA) uses the controlled removal of a volatile component to trigger the defined aggregation of contained materials. This aggregation can either occur due to continuously limiting the available space for the dispersed components or simply due to their increasing concentration. One example for the latter can be found within mixed surfactant systems, in which the starting solution has a concentration below the cmc (critical micelle concentration), which is surpassed during subsequent evaporation of the solvent. This leads to spontaneous formation micellar aggregates [23,24]. In this article, we shall focus on systems of colloidal dispersions containing insoluble particles, which are forced to assemble thereby potentially promoting colloidal crystallization via EISA [5,8,25].

At first glance, a simple system to do so is represented by droplets with suspended colloidal components. These droplets can either float in an immiscible second liquid, e.g., water droplets in fluorinated oil, or be dispensed on a solid surface. Considering the latter, the resulting assemblies strongly depend on the type of surface used, as the interaction between the droplets' liquid phase and the solid substrate may vary substantially. Therefore, the wetting properties constitute the predominant influencing parameter, characterized by adhesion forces and contact angle (CA). The CA for a droplet deposited on a solid substrate depends on the interfacial tensions between the different phases (Figure 2) and is given by Young's equation (Equation (1)), with θ being the CA and γ the interfacial tension between the solid (s), gas (g) and liquid (l) phases, respectively.

$$\cos\theta = \frac{\gamma_{sg} - \gamma_{sl}}{\gamma_{lg}} \tag{1}$$

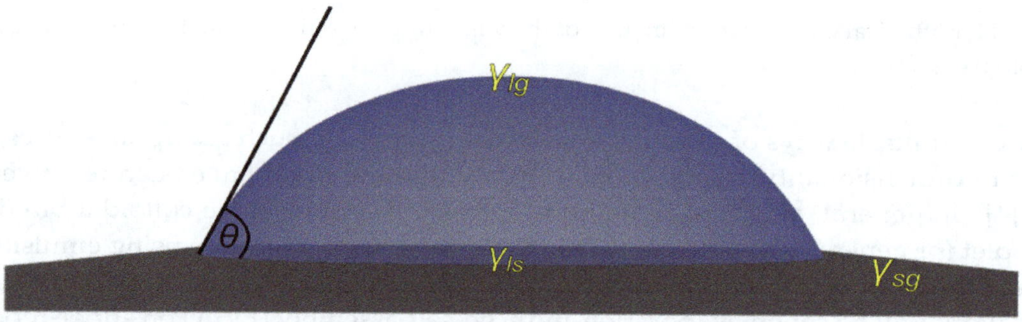

Figure 2. Scheme describing the contact angle (CA) θ of the liquid droplet on a solid substrate and its relation to the different interfacial tensions between the solid (s), liquid (l) and gas (g) phase, as related to each other by Equation (1).

A well-known phenomenon related to drying suspension droplets on solid surfaces at low CA is the so-called "coffee-ring effect" [26,27]. Closer investigation of the leading mechanism in these drying spherical cap droplets revealed that convective liquid transport occurs from the center apex down and outwards to the three-phase contact line (TPCL). This preserves the geometry upon strong evaporation at the droplet edge, which is driven by pinning of the TPCL due to the contained particles adhering to the surface, thus prohibiting transversal shrinkage during evaporation. Consequently, colloidal material accumulates and finally precipitates at the TPCL, leaving a solid ring-like pattern. Such a pattern can be quite characteristic for the dried sample, e.g., depending on its ionic strength [28,29]. This effect can for example be of interest for diagnostic purposes when investigating dried droplets of blood, as shown in Figure 3 [30,31].

The formation of "cracks" such as seen in Figure 3 is always an indication of rather small CAs and pronounced pinning of the droplets on the substrate. In that context, it might be mentioned that such cracks have also been observed very pronouncedly for drying droplets containing mixtures of DNA and small silica nanoparticles (Figure 4). The final solid films showed interesting surface patterns with unique shapes, which are induced by local segregation of the two colloidal components of DNA and silica [32]. For more complexly-composed colloidal mixtures, such local segregation is a further complication that may arise in the EISA process.

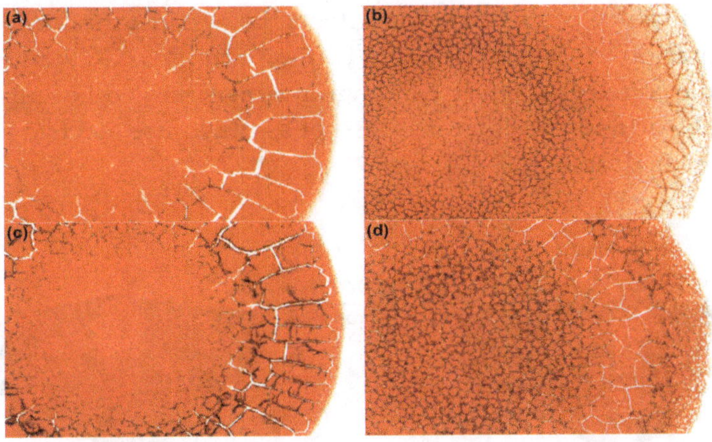

Figure 3. Blood samples dried on a glass surface from: (**a**) 27-year-old healthy woman; (**b**) a person with anemia; (**c**) a 31-year-old healthy man; (**d**) a person with hyperlipidemia.

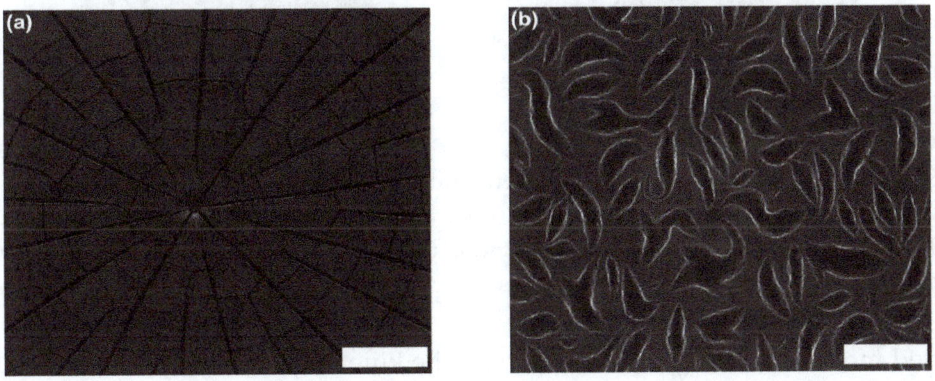

Figure 4. (**a**) SEM image of dried drops for a ratio of silica NPs (diameter: 10 nm)/DNA (20,000 bp) ratio 1:0.5. The scale bar is 200 μm; (**b**) HR-SEM images of dried drops for an NP/dsDNA ratio of 1:0.5. The scale bar is 2 μm.

The process of coffee-ring formation is counterbalanced by the Marangoni effect [33–35]. This effect describes convection processes arising from thermally-induced gradients in the droplet's interfacial tension (also called Bénard–Marangoni convection). In addition, convectional flow simply arising from a decrease in surface temperature due to evaporation is observed, which then in turn causes density differences (see also Figure 5). Usually, the occurrence of coffee-ring deposits indicates stronger evaporation closer to the TPCL and low or even absent Marangoni stresses [36]. It is also reported that experimentally, Marangoni convection in water droplets is often lower than expected from theoretical simulations, which can be related to the surface contaminants enriching at the water interface with ease and, hence, lowering the interfacial tension [35,37]. Generally, thermal conductivity and the latent heat of evaporation are important parameters that determine the extent of Marangoni flow in an opposing fashion due to the generation of a thermal gradient, but also concentration gradients of ingredients, e.g., surfactants, may induce such convective flow. However, for the case of EISA, the thermal gradient-induced Marangoni effect should play the predominant role. Furthermore, the Marangoni convective flow may also completely reverse the coffee-ring effect, as reported for octane droplets on glass [35].

The basic concepts of microfluidic flow inside sessile droplets are usually discussed for the case of a pinned droplet contact line, i.e., the "constant contact radius" (CCR) mode. Of course, one may also observe the complementary case, in which the droplet's contact line successively recedes at constant CA due to the absence of any pinning effects, i.e., "constant contact angle" (CCA) mode. The details of the evaporation process for the two different modes (CCR and CCA) are described in Figure 5. In contrast to the coffee-ring effect, Marangoni convection will occur for both CCR and CCA modes [33,34]. It might be added here that it is also possible to use electric fields to manipulate the wetting behavior of a sessile droplet [38–40].

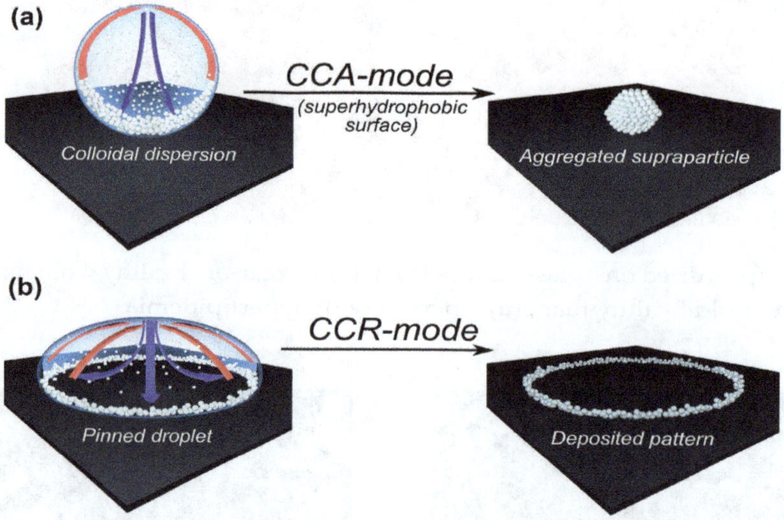

Figure 5. Schematic description of the different modes observed for a drying droplet on a solid substrate, with the mass flow of cooled water (blue arrows) and that due to the interfacial tension gradients (Marangoni flow; red arrows) being indicated for: (**a**) constant contact angle (CCA) mode, as observed on most superhydrophobic surfaces; (**b**) constant contact radius (CCR) mode.

The provided mode of wetting directly controls the kinetics of the evaporation process, which has been semi-empirically analyzed by Picknett and Bexon [41]. It was already postulated by Maxwell that the rate of mass loss $\frac{dm}{dt}$ of an evaporating sessile droplet is attributed to the product of the vapor diffusion coefficient D in air and the difference of the vapor concentration at the droplet surface c_s (which can be approximated by the saturation vapor concentration), its value in the surrounding c_∞, as well as the shape-dependent electrostatic capacitance C (Equation (2)) [42].

$$\frac{dm}{dt} = 4\pi DC(c_s - c_\infty) \tag{2}$$

Starting Snow's series [43], Picknett and Bexon used a polynomial evaluation for equiconvex lenses, being similar to those of sessile droplets at different CAs. As a result, an empiric expression was obtained giving C/r as a function of the CA, represented by $f(\theta)$, with r being the droplet radius [41]. In this way, one accounts for the interfacial blockage by the solid surface, which prohibits free vapor diffusion and thus reduces local evaporation. For $0.175 \leq \theta \leq \pi$, this translates into Equation (3).

$$f(\theta) = C/r = 0.00008957 + 0.6444\,\theta + 0.1160\,\theta^2 - 0.08878\,\theta^3 + 0.01033\,\theta^4 \tag{3}$$

At this point, it also becomes clear that the mode of evaporation (CCR or CCA) clearly affects the evaporation as for the case of CCA $f(\theta)$ remains constant, while for CCR, the changing CA will also have an effect. Now, for the case of a constantly-receding contact-line, i.e., in the absence of droplet pinning, Equation (2) translates to a simple expression (Equation (4)) for the $V(t)$-function, with V_0 being the initial volume and t_{total} the total drying time. However, it might be noted that in recent investigations on sessile droplets of aqueous saline solutions (CAs between 2° and 50°), high salt concentrations and small contact angles showed significantly lower evaporation rates than expected from simple diffusion-controlled evaporation. Particle tracking experiments proved that this lower evaporation has to be attributed to the Marangoni effect [44]. In contrast, it has been observed that the presence of nanoparticles increases the evaporation rate of aqueous droplets, as observed for anthraquinone nanoparticles (diameter: 285 nm) on a hydrophobized silica wafer (CA: 80°) [45]. Furthermore, for evaporating droplets with an initial CA larger than 90°, one may start with a CCA mode, then switch to a CCR mode and end the evaporation in a mixed mode [46]. In general, it may be concluded that the description of the evaporation processes of droplets can become quite complicated, especially for more complex geometries. Moreover, the change in composition of the drying droplet may also have a substantial impact on the interfacial properties and, hence, on the evaporation process.

$$V(t) = V_0\left(1 - \frac{t}{t_{\text{total}}}\right)^{\frac{3}{2}} \quad (4)$$

Obviously, in order to produce defined assemblies of micro- or nano-particles, which after production can be easily isolated from the surface, one will preferentially use surfaces with low adhesion forces, thus avoiding pinning of the liquid. Superhydrophobic surfaces represent a nice, biomimetic example for this purpose. Due to their large CA and, mostly, low adhesion forces, they represent an ideal substrate for the preparation of particle assemblies that allow for easy separation after drying. Hence, for the sake of completeness the basic concept of superhydrophobic surfaces will be discussed in the following section.

3. Superhydrophobic Surfaces

Superhydrophobic surfaces represent a class of biomimetic, nanostructured materials, where usually the term "superhydrophobicity" accounts for water contact angles (WCA) larger than 150° [18–22]. A common example from nature for water super-repellency is the "lotus-effect" [47] as seen for the leaf of sacred lotus [48], which also has been mimicked by artificial fabrication [49]. In general, it can be concluded that it is the combination of chemical hydrophobicity paired with surface texture on the micro- or even nano-scale, which is essential for achieving such high WCAs. Consequently, the main challenge for the artificial preparation of superhydrophobic surfaces is the introduction of surface roughness on the nm to µm scale in hydrophobic surfaces. A large variety of different technical approaches has been developed over the past years to address this challenge [50], and the bare number of publications available in the field of superhydrophobic surfaces is a statement of the success of this concept, as well as its relevance for applications. This is one reason why scientists' curiosity even pushed the development further towards the creation of superamphiphobic surfaces, capable of efficiently repelling both oil and water [51,52]. Starting Young's equation (Equation (1)), this is quite remarkable when considering the large range of interfacial tension values covered when comparing water to typical oils.

3.1. Wetting Modes

For superhydrophobic surfaces, a rough surface topography is essential to amplify hydrophobicity in order to reach WCAs of greater than 150°. Thus, the solid surface will consist of grooves on the micro- to nano-scale, where their ratio is proportional to surface roughness. Consequently, when depositing a water droplet on such a surface, one may imagine two limiting cases: either the liquid is sitting on top of the grooves, i.e., air stays entrapped within the grooves or the liquid is penetrating the same. The former state is known as Cassie's (Cassie–Baxter) [53] and the latter as Wenzel's [54] mode of wetting, as shown in Figure 6a,b, respectively. In the Cassie–Baxter state, the droplet is situated on top of the surfaces structure, while air stays entrapped within the interspacing provided by surface roughness (Figure 6a). Hence, at these regions, instead of wetting the surface, the droplet is in contact with air, which is the most hydrophobic (solvophobic) medium. Again, considering Young's equation (Equation (1)), this then favors a higher CA due to $\gamma_{sl} = \gamma_{lv}$. Note that for most polymeric or waxy substrates γ_{sl} is about equal to 30–50 mN/m, which is substantially lower than γ_{lv} (72.8 mN/m at

25 °C). Therefore, maximizing the fraction of interspatial air pockets entrapped within the solid surface leads to CAs approaching 180°. In contrast to that, in the Wenzel state, the droplet penetrates the interspatial volume and fully wets the surface (Figure 6b).

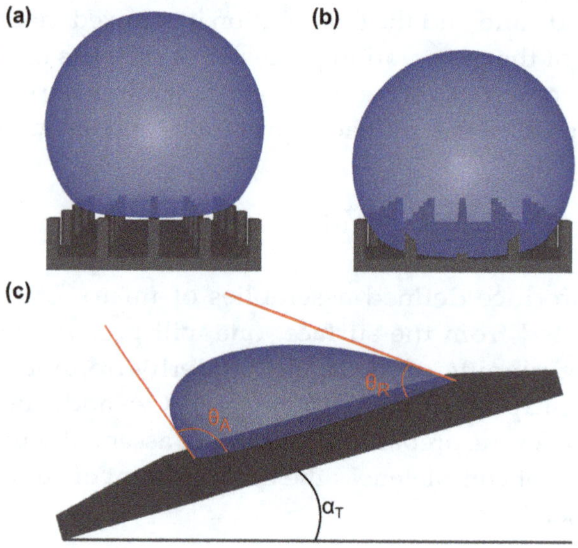

Figure 6. Wetting of superhydrophobic surfaces: (**a**) Cassie state with entrapped air within the surface grooves; (**b**) Wenzel state with liquid filling the grooves. Dynamic wetting strongly depends on the prevailing mode of (**a**) versus (**b**): (**c**) the difference of advancing (θ_A) and receding (θ_R) CA is the measure of the surface hysteresis and droplet adhesion.

Another important quantity is the CA hysteresis, especially when considering the self-cleaning effect of superhydrophobic surfaces. CA hysteresis is determined by the difference of the advancing (θ_A) and receding (θ_R) CA, as shown in Figure 6c. Low hysteresis equals a low tilting angle (α_T) being necessary to cause the droplet to roll off the surface. Accordingly, paired with low adhesion forces, these rolling droplets can collect dirt particles from the surface, thereby promoting self-cleaning, as known from many plant leaves. This effect has been the inspiration for numerous materials and applications [55], such as self-cleaning and dirt- and water-repellent coatings used on tiles, roofs or panels (for instance, in bathrooms or kitchens). Using such coatings on glass leads to the "anti-fog" effect, while also antimicrobial coatings based on superhydrophobic surfaces have been investigated [55].

Usually, CA hysteresis stays lower for droplets in the Cassie–Baxter as compared to the Wenzel state. However, this does not automatically imply low roll-off angles for the Cassie–Baxter state. Moreover, the roll-off angle depends on the pinning forces at the contact area between solid and liquid [56]. Generally, it may be stated that for supraparticle formation by EISA, the Cassie–Baxter state with a CA as high as possible is to be preferred.

3.2. *Production of Superhydrophobic Surfaces*

For the preparation of superhydrophobic surfaces, there exists a large range of different approaches, which also have been reviewed comprehensively in recent times [57,58]. Accordingly, we refrain here from describing these processes in further detail, but instead will briefly describe two typical approaches that also have been successfully employed in our work. One of them employs superhydrophobic surfaces produced by an electrochemical deposition (ECD) method [59]. The other is a soot-based method comprising a chemical vapor deposition (CVD) leading to surfaces that can either be rendered superhydrophobic or even superamphiphobic, the latter meaning super-repellant for water and oil [51].

ECD methods usually employ an electrochemical potential to deposit dissolved material from solution onto a surface substrate. This usually leads to a uniform and highly porous surface coverage. If the material deposited is not hydrophobic enough by itself, further chemical modification may serve to complete superhydrophobic surface formation. For example, a suitable approach was reported by Gu et al. utilizing activated copper surfaces. Here, activation usually just means surface polishing to remove the passivation layer of copper oxide or sulfide (typical reactions, when left in air) [59]. When immersed in a silver nitrate solution, a black precipitation layer of silver is formed on the copper surface. Thus, if choosing the appropriate time of immersion and silver concentration, highly porous, coral-like structures are achieved within the solid silver layer. Further functionalization with an aliphatic thiol, such as 1-dodecanethiol, then leads to a well-performing superhydrophobic surface with a CA above 150° and low CA hysteresis [59].

Another method well-known to most of us from childhood days is that of covering a heat-stable surface, such as a spoon, with soot by holding it into the flame of an ordinary candle. Unluckily, the adhesion of the as-deposited carbon layer is very poor, which results in its immediate rupture when exposed to a rolling water droplet (similar to the self-cleaning effect of lotus leaves [48]). Hence, Deng et al. searched a way to preserve the structure of the soot, while at the same time increasing its stability [51]. In order to do so, they covered the soot surface layer with a robust silica shell using CVD of a silica precursor (like tetraethyl-orthosilicate (TEOS)). After subsequent removal of the soot using calcination at high temperatures and functionalization with an appropriate aliphatic silane via CVD of a corresponding precursor, as a result, a superhydrophobic or, in the case of a fluorinated silane, a superamphiphobic surface could be achieved. Surfaces of this kind provide a CA above 150° and low CA hysteresis. Another aspect attributing quite some elegance to this method is the fact that the application of this layer can occur on any surface able to resist the temperatures used for the calcination step. Moreover, if carefully done, this surface also provides high transparency [51], which for example even allows experiments with an inverse optical microscope setup for observing the inside of the droplet from below the surface [60]. It might be noted here that the curvature of small-scale roughness was revealed to play a key role for achieving high resistance against wetting, thus being of major importance for acquiring superamphiphobicity [61].

For the sake of not losing our intended focus of this review, we will not further discuss the many other preparation methods available. Instead, the interested reader is referred to comprehensive reviews giving a nice overview of superhydrophobic surfaces and their different preparation methods that have been accomplished [19,22,50,52].

4. The Concept of Supraparticle Formation

Supraparticles can be created by self-assembly of colloidal components into larger, ordered arrays [62]. Such supraparticles can possess a large number of combined functionalities as they may contain many different colloidal materials of a specific nature, like proteins [63,64], photosensitive particles, such as semi-conductors [65,66], or magnetic materials [67,68].

With respect to their preparation, a controlled guiding of the assembly process is vital in order to create supraparticles in a defined way [69]. This means that typically, one employs droplets as the confining object within another, immiscible liquid or on a solid surface. Accordingly, besides assembling particles from freely-suspended solutions [65,70], several methods have been developed employing droplets as confining geometry. There are different methods suitable for generating droplets, such as emulsion techniques [71–74], microfluidics [75–78] or ink-jet printing [79–82], that besides precise control on the droplet size, also offer potential for scalability towards mass production. Apart from just serving as a container for the contained particles, the droplets may also actively promote the assembly process by drying induced shrinkage that constrains the colloidal components. One way to do so is dispersing aqueous colloidal suspension droplets in hydrocarbon or fluorinated oil with subsequent removal of the aqueous phase [25]. In such a setup, the position of the droplets can also be controlled using separately addressable electrodes constructed in an array allowing for defined droplet dielectrophoresis [83]. Some examples for resulting supraparticles achieved by such processes are illustrated in Figure 7 for the case of polystyrene (PS) latex particles dispersed in aqueous droplets floating in fluorinated oil.

The supraparticles shown in Figure 7a,b show a remarkably smooth and regular structure, which is also indicated by the colored appearance under top-light illumination. This color effect requires long-range ordering of the particles, which was corroborated by analyzing the colloidal structure with scanning electron microscopy (SEM). Here large arrays of regularly-ordered particles within hexagonal closely-packed planes and face-centered cubic lattices were observed, as illustrated in Figure 7c,d [25]. The driving force for this high degree of ordering was addressed by Denkov et al., who investigated 2D arrangements of monodisperse and micron-sized PS-latex particles on a solid surface [84,85]. They found that crystallization, i.e., the ordering process, always occurred when the height of the liquid's meniscus reached the particle diameter. Furthermore it was found that this fact neither depends on the ionic strength of the suspension, nor on the surface charge or initial concentration of the particles. Thus, electrostatic and van der Waals interactions could be excluded as governing factors [85]. Instead, capillary forces between the particles caused by the menisci formed and convective particle transport within the solution were found as the predominant control parameters. This extraordinarily high degree of long-range ordering also leads to a precise control of the porosity, e.g., by tuning the size of the particles or applying additives, such as glucose [86] or DNA to drying silica suspensions [32]. Using a microfluidic setup and analyzing the time-dependent solute release profile from similar sub-mm-sized supraparticles containing 320-nm diameter PS latex microspheres using a dye, Rastogi et al. showed that such assemblies provide a permeable matrix allowing for a more homogeneous and slower dye release compared to normal pellets [87]. The internal structuring of colloidal spheres within liquid droplets also was discussed based on geometrical constraints, surface tension and the interparticle potential. Depending on the detailed conditions, one may expect the formation of colloidal clusters, colloidosomes or Pickering emulsions [88]. Here, the latter two structures are most likely to be observed upon the conditions prevalent if the second phase is another liquid, i.e., in an emulsion.

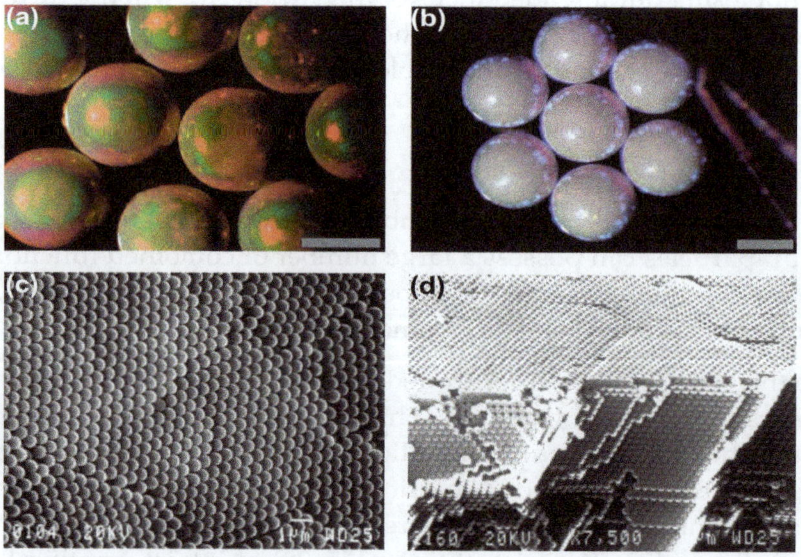

Figure 7. Typical examples for supraparticles prepared by drying of aqueous droplets containing polystyrene (PS) latex particles dispersed in fluorinated oil. Spherical supraparticles are formed showing different color patterns based on the size of the PS latex particles: (**a**) 270 nm and (**b**) 320 nm, scale bars = 500 μm. These patterns arise from light-diffraction due to long-range ordering of the particles as shown in (**c**) at the surface and (**d**) along the vertically-broken edge of a similar supraparticle; scale bars = 1 μm.

5. Supraparticles by EISA on Superhydrophobic Surfaces

In the following, we will focus on supraparticles obtained by drying aqueous suspension droplets on (solid) superhydrophobic surfaces. Recent work in that field has shown how simple procedures grant precise control over the shape and internal hierarchical structure in these supraparticles, thereby leading to new kinds of functional materials.

One major drawback of using a two-liquid system, as often employed, is the difficult isolation of the obtained products and their purification. This fact gave rise to the utilization of superhydrophobic surfaces. Due to their high WCA, these surfaces also provide spherical droplets and thus isometric surrounding conditions, but avoid the need of subsequent separation of a second (oil) phase from the dried supraparticles. Instead, those can readily be collected from the surface due to low adhesive forces (as evident from the high WCA and low roll-off angle). Note, that despite being superhydrophobic, these surfaces can be designed for both adhesive and non-adhesive properties by tuning pitch values and the density of micro- and nano-structures, respectively [89,90]. Using EISA on such superhydrophobic surfaces in combination with the droplet templating method, several new types of supraparticles of different structures and functionality have been produced. However, in order to be easily isolable, the droplets must not penetrate the surface texture. Consequently, the size range of supraparticles produced on such surfaces is limited to several μm [91] as a lower size limit and may reach about mm size [92–97], while for still larger droplets gravitational effects become dominant leading to their shape being no longer spherical.

In the following, we shall present recently developed methods for the creation of supraparticles of various kinds of shapes and functionalities based on their preparation on superhydrophobic surfaces.

5.1. Shaping of Supraparticles

In the simple case, the shape of the droplet will directly determine that of the finally obtained supraparticle, since the contained particles remain trapped within the provided geometry, i.e., one simply templates the initial shape. Therefore, the confining geometry is retained. As this geometry on superhydrophobic surfaces is to a first order spherical (except for gravitational effects), the resulting supraparticles for the ordinary case are also of a similar shape. Hence, in order to alter the final shape of the supraparticles, one has to change the droplet geometry. For the case of aqueous droplets dispensed and floating on fluorinated oil and containing PS-latex microspheres, this was done by using additional fluorinated surfactant. This led to the formation of dimpled, red blood cell-like or even doughnut-shaped supraparticles at higher concentrations [25], which is caused by the change of the interfacial tension between the liquids promoted by the surfactant. Lastly, a pronounced deviation from spherical geometry also occurred when lowering the particle concentrations below 20%, and a continuous transition from a spherical to disc-like shape was observed [25].

Superhydrophobic surfaces, for the case of up to μL-volumes, also provide spherical droplets, with some slight, but mostly not substantial bottom deformation due to gravitational effects. This can be deduced from the high CA and low Eötvös or Bond number B_0 (Equation (5)), which is a dimensionless number describing the relative influence of gravity and surface tension on the shape of a liquid drop. Here, $\Delta\rho$ represents the density difference between the droplet and its surroundings, g the gravitational constant, r the droplet radius and γ the interfacial tension [98].

$$B_0 = \frac{\Delta\rho g r^2}{\gamma} \quad (5)$$

Rastogi et al. used superhydrophobic surfaces to create supraparticles of a reduced symmetry, "doughnut"-like shape, as shown in Figure 8 [92]. The morphological loss of symmetry corresponds to the geometric alteration from spherical R_3- to cylindrical D_∞-symmetry. This interesting overall shape is accompanied by light diffraction effects that are similar as for the spherical supraparticles arising from the segregation of the different colloids contained in the supraparticle.

It has been reported for aqueous polymer solution droplets that a thin shell of concentrated polymer is formed at the water-air interface, due to internal transport flux. This flux is directed radially outward and significantly influences the droplet's shape upon evaporation due to increased surface elasticity promoted by the concentration gradient [29,98–101]. If this accumulation is stronger at the three-phase contact line (TPCL) than at the apex, self-pinning of the droplet occurs. This means that the transversal shrinkage is suppressed, which in turn results in exclusive shrinkage along the droplet height. Note that the origin of this self-pinning is notably different from the coffee-ring effect, where the droplets are pinned due to the surface wetting properties [26,27]. The importance of the CA for the shape of the formed dry supraparticles has been demonstrated by assembling silica microspheres of 300 nm via EISA on substrates with CAs ranging from 28°–105°. Depending on the CA, ring-like structures, doughnut-like structures or hemispheres were formed [102]. Such hemispherical assemblies could also be achieved by EISA on a substrate with a CA of 100° when employing monodisperse latex particles with diameters varying in the range of 300–1100 nm. Here, a very high degree of ordering of the latex particles was locally observed, which created pores of relatively high uniformity in theses supraparticles [103].

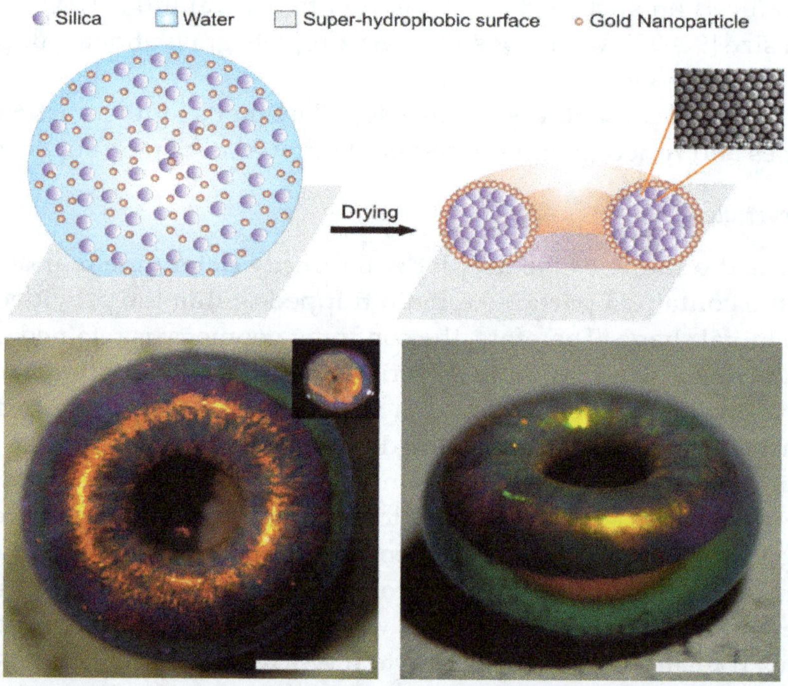

Figure 8. Top: Illustration of the evaporation-induced formation of "doughnut" supraparticles; the inset is showing an SEM of the particle lattice built by 330-nm diameter silica particles. Bottom: Optical micrographs of the final supraparticle from top- (left) and side-view; scale bars are 500 µm.

Large particles contained within the drying solution similarly undergo surface collection in analogy to polymers [28,60]. In the case of the doughnut silica supraparticles, the colloidal particles (330-nm diameter silica microspheres) get collected at the interface of the precursor droplets due to the shrinking surface, which is propagating faster towards the interior than can be counterbalanced by particle diffusion [92]. As the particles predominantly collect close to the TPCL, the droplets deform in a fashion that yields dimpled supraparticles for volume fractions of silica lower than 15%. However, when working at a silica volume fraction of 2.5%, the dimple evolved into a complete hole, hence yielding doughnuts. This effect vanishes when using 1000-nm instead of 330-nm silica particles, as in that case early sedimentation prohibited the doughnut-hole formation. Furthermore, the effect of surface concentration and the way the drying process affects the internal structure of the drying droplet has been studied by means of microbeam small-angel x-ray scattering (SAXS) on a hanging droplet of a colloidal suspension containing silica nanoparticles of ~10-nm diameter. These experiments showed that isotropic assembly is still possible for Péclet numbers significantly higher than one and an accumulation of silica at the droplet surface was only observed for high initial concentrations [104]. This observation is to be expected, as shell accumulation may only occur if the colloid movement by diffusion is slower than the rate by which the droplet surface is receding due to evaporation.

The symmetry of the supraparticles can substantially reduced be further towards anisometric, ellipsoidal shapes when fumed silica (FS) is used in aqueous colloidal suspension droplets instead of compact spherical silica nanoparticles. In contrast to the latter, FS has a very open and fluffy fractal structure of polydisperse particles with hydrodynamic radii of 100–200 nm and correspondingly large specific surface areas that typically are in the range of 50–400 m^2/g [105]. However, the formation of anisometric supraparticles is only observed once a certain concentration of salt is present and the extent of anisometry can be controlled by adjusting the ionic strength [93,96]. Typical examples for different FS particle concentrations are illustrated for the case of no deformation in Figure 9a at 0.001 mM and for the maximum anisometry in Figure 9b obtained at 25 mM ionic strength of the suspensions at the start of the drying process.

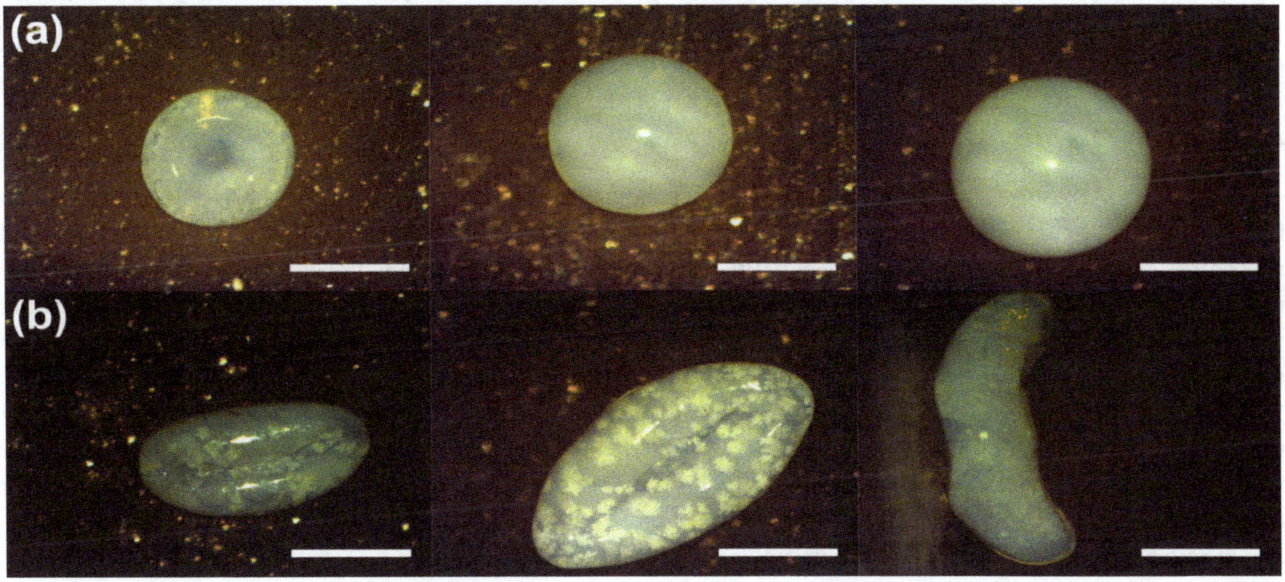

Figure 9. Examples for anisometric supraparticles obtained from drying fumed silica (FS) suspensions (from left to right 1.75%, 3.50%, 5.25% w/v) for an initial ionic strength of (**a**) 0.001 mM and (**b**) 25 mM using NaCl; the last image on the right side shows a side-view perspective. The scale bars are 500 μm.

Similar to previous findings for doughnut-shaped supraparticles, particle accumulation at the water-air interface occurs. This leads to a modification of the surface rigidity, which depends on the ionic strength, resulting in a higher rigidity for higher ionic strength. This could also be verified using time-resolved confocal laser scanning fluorescence microscopy. By labeling the FS and the water phase with different fluorescent dyes, the density profiles of the FS perpendicular to the droplet surface, i.e., in the radial direction, could be deduced throughout the drying process. Thereby, it was shown that the FS particles accumulate at the droplet surface and that the deformation of the droplet takes place once a certain effective thickness and density of this shell is reached [60]. In a somewhat related study, the evaporation process of an aqueous lysozyme solution on a superhydrophobic PMMA surface has been followed by microbeam SAXS. The evaporation process leads to hollow spherical residues, and the SAXS experiments show the increasing concentration at the droplet surface which leads to the precipitation of crystalline lysozyme nanoparticles towards the end of the drying process [106].

Coming back to the formation of the anisometric supraparticles, the main difference between the spherical silica microspheres used before and the amorphous ones of FS is the capability of intercalating and interconnecting during their aggregation, wherein this capability does not apply for spheres. In turn, FS can interact in a much more pronounced and cohesive way than silica spheres, where this attractive interaction will become dominant for sufficiently strong electrostatic screening. With increasing ionic strength and thereby lower Debye-screening length, the electrostatic repulsion between the FS particles becomes reduced. As an effect thereof, the shell rigidity arising from the intercalated FS particles at an ionic strength of above 0.5 mM increases such that the droplets can no longer shrink, while at the same time retaining their spherical shape. Accordingly, they become anisometrically deformed, just in the same fashion as a (spherical) football changes its shape when becoming deflated. The deformation leads to a droplet elongation, whose direction varies statistically, as it depends on the weakest spots of the as-formed shell. The resulting supraparticle anisometry, i.e., the ratio of principal axes lengths, could be precisely tuned within the range of 0.5–25 mM of ionic strength, and the observed anisometry is directly proportional to the logarithm of the ionic strength and reaches values of up to 1.6 [93,96]. A closer look into the formation mechanism revealed that the deformation occurs at a constant surface excess concentration of FS [95] and for an effective shell-thickness of about 22 μm, which corresponds to an interfacial FS volume fraction of 0.17 [60]. It might be noted that shell formation during evaporation can be related to buckling of the drying droplets. For instance, for μm-sized silica supraparticles obtained by spray-drying, the formation of doughnuts and dimpled/grainy spheres of buckyball appearance has been attributed to such surface instabilities. The buckling phenomenon could be arrested by the addition of PEO, which then allows for shape control [107].

The major drawback of producing anisometric supraparticles on flat surfaces is their statistical distribution of orientation after drying with respect to the surface placement. This renders it impossible to position a patch in a controlled fashion, as for instance is possible by using additional magneto-responsive ingredients within the precursor suspension droplets [92]. In a recent approach the process of anisometric supraparticle formation using FS [93,95,96] was modified by manipulation of the evaporation conditions using distinct surface geometries of the solid substrate [108]. More precisely, similar superhydrophobic substrates as utilized in the original experiments were bent at different angles into a V-shape. Monitoring the evolution of the droplet shape, a directed anisotropy perpendicular to the surface's bending axis was observed. This effect of orientation is caused by the evaporation rate being higher at the free sides of the droplet compared to the blocked ones. This anisotropic evaporation, indicated by the red arrows in Figure 10a, leads to more accumulation of FS particles, which at ionic strengths above the threshold of 0.5 mM promotes the controlled droplet elongation perpendicular to the bending axis. It might be noted that this directional orientation then is reproducible with high precision, i.e., within a few degrees with respect to the bending axis of the surface. With this well-predictable orientation of the supraparticles it is possible to position patches of colloidal assemblies at will at any point(s) of the as-produced supraparticle (e.g., for magnetic colloids by an external magnetic field, but also other external influences could be imagined). Thus, this achievement constitutes a substantial advance for accessing increasingly more complex and functional supraparticles.

Accordingly, besides a general improvement of the extent and reproducibility of the anisometry values, by taking advantage of the predictable orientation of the final anisometric supraparticles, it is possible to render them patchy in a sophisticated and defined manner. For instance, supraparticles with distinct patch locations can be obtained by employing additional magnetic colloidal components (Fe_3O_4) and positioning an external magnet parallel (Figure 10b) or perpendicular (Figure 10c) to the bent channel. Such anisometric patchy supraparticles can be of potential use for designing self-propelling systems or applications requiring orientation functionality.

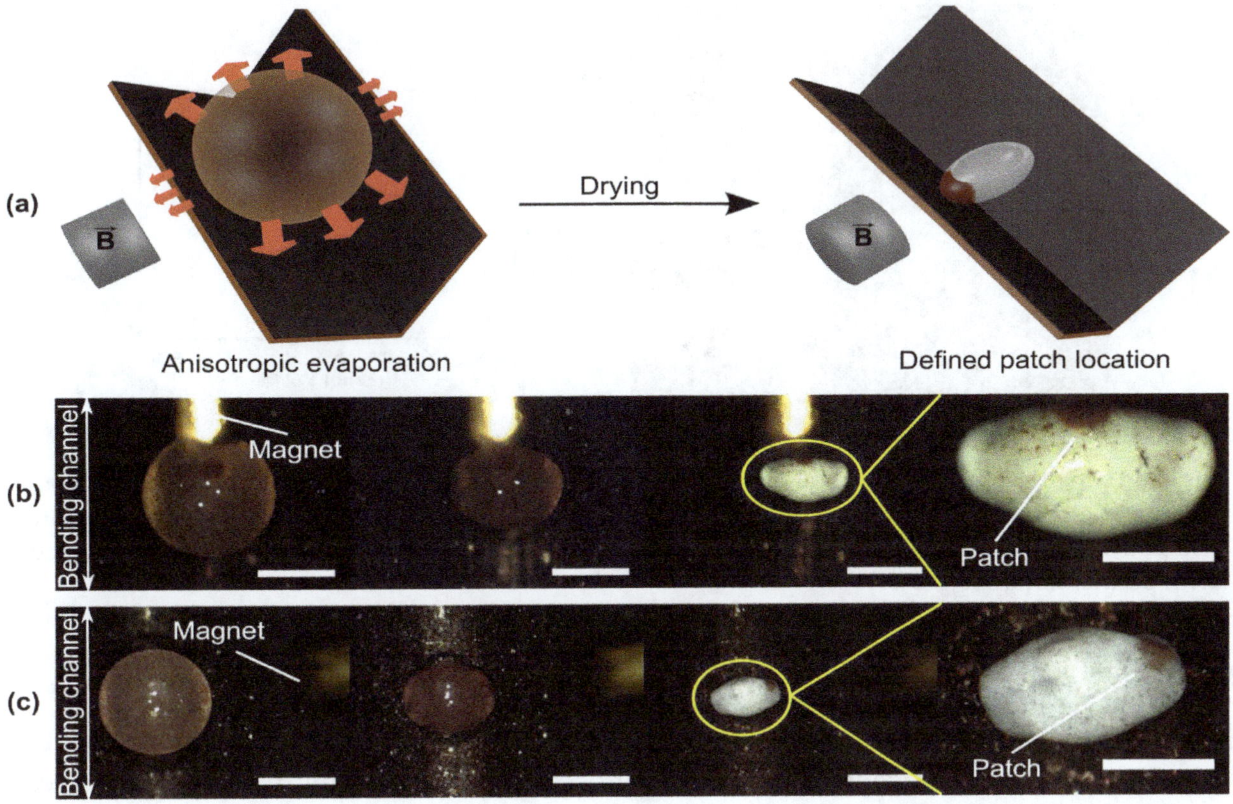

Figure 10. Controlled supraparticle orientation after drying: (**a**) anisotropic evaporation due to the surface bending. This allows for the creation of patchy anisometric particles with defined location of magnetic components (Fe_3O_4) by appropriately placing external magnets: (**b**) along the transversal or (**c**) longitudinal diameter of the ellipsoid. In (**b**) and (**c**), the images on the right side present zoomed-in excerpts; scale bars = 0.5 mm; other scale bars are 1 mm.

As already described in Section 2, Picknett and Bexon discussed the sessile droplet evaporation on solid surfaces using Maxell's approach [41,42]. This also explained the anisotropic evaporation of the droplets on the V-shaped surfaces by introduction of an apparent CA, created by the upwardly-directed surface legs. Comparing the values for both, the surface and apparent CA obtained by fitting the CCA-model to the experimentally-obtained evaporation rate ratios (perpendicular vs. parallel to the surface bending axis), yielded a good correlation with the measured CAs. This revealed that the additional blockage by the upwardly-bent surface significantly reduces local evaporative vapor flux. Hence, anisotropic particle accumulation is promoted, being more pronounced on the free droplet faces, which in turn causes the controlled supraparticle elongation.

It might be worth noting that the formation of the anisometric FS supraparticles occurs in a non-pinned state, i.e., CCA-mode. However, Zhou et al. discovered a method that applies controlled pinning of the droplet contact-line using an ethanol-water mixture containing PS nanoparticles. This led to anisometric receding of the TPCL, producing anisometric photonic crystals [109]. By successive increase of the ethanol content, it was shown possible to create anisometric supraparticles within an aspect ratio from 1.14 (2 vol % EtOH) to 1.28 (8 vol % EtOH), as shown in Figure 11 [109]. This is an alternative approach to the FS supraparticle production, where droplet elongation occurs due to shell deformation [93,96,108]. Moreover, in pure aqueous systems, the structure of the resulting supraparticles could be tuned from microbeads to dimpled or microwelled shapes by varying the colloid concentration.

Concluding, one may state that there exist different ways of symmetry breaking that lead to many quite different shapes, which are of interest for various potential applications.

Figure 11. Optical microscopy of anisometric photonic crystals (PC) obtained by using mixtures of water and ethanol with varying compositions: (**a**) 2; (**b**) 4; (**c**) 6 and (**d**) 8 vol % of EtOH; scale bars are 800 µm. The systematic dependency of aspect ratios, i.e., anisometry values, is given in (**e**). Adapted with permission from [109].

5.2. *Substructuring of Supraparticles*

When using the droplet templating method on superhydrophobic surfaces, though the overall geometry may be fixed, still internal structuring or porosity may be of potential interest. One well-known structural type within this context is that of Janus or, more generally, patchy, hierarchical- or surface-anisometric particles, respectively [110–113]. Accordingly, we want to address supraparticles having different types of colloidal constituents that are inhomogeneously distributed throughout the assembled supraparticles.

One straightforward way to achieve this kind of anisotropy in drying droplets is the utilization of magnetic components and the application of an external magnetic field [92,94,108]. The principle is shown in Figure 12, wherein besides the basic concept of forming single-patched supraparticles (Figure 12a), altering the magnetic field setup also allows for more complex patchy assemblies, such as presented by bi- and tri-patched supraparticles (Figure 12a,b).

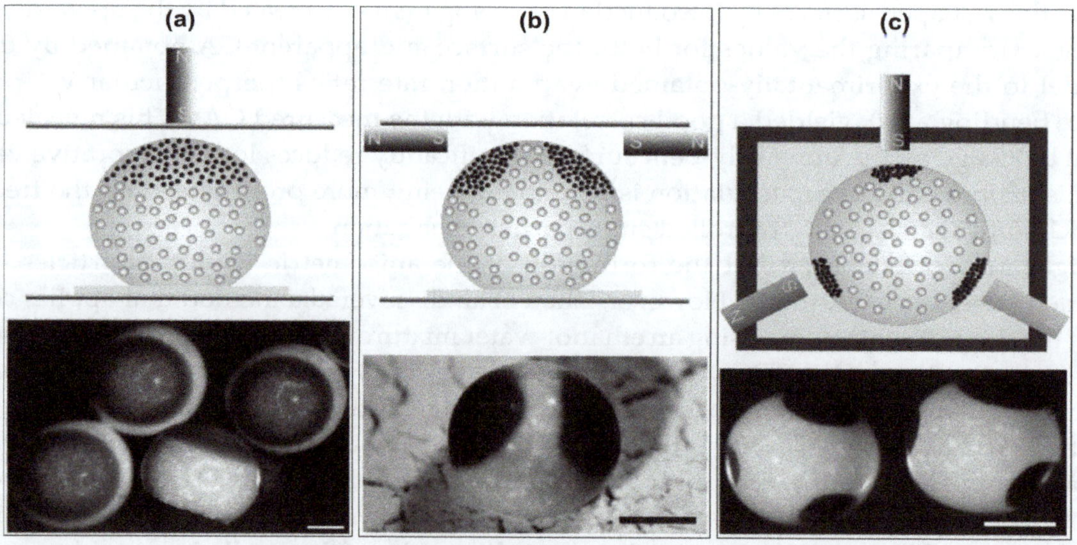

Figure 12. Assembly of (multi-)patched supraparticles in drying sessile droplets on a superhydrophobic surface: (**a**) single- (**b**) bi- and (**c**) tri-patched; scale bars are 500 µm.

Similarly, this patch formation can also be combined with anisometrically shaped supraparticles, as already described before and presented above in Figure 10b,c [108]. The combination of anisometric shape and controlled placing of patches on such particles in turn leads to much more complex supraparticles.

It is not necessarily required to use magnetic colloids with subsequent magnetic field guiding to generate hierarchically anisotropic, i.e., patchy particles. Rastogi et al. showed that by combining latex particles with diameters larger than ~300 nm (which collect on the droplet surface) with small gold nanoparticles (~22 nm), supraparticles, which exhibit remarkable optical features, as shown in Figure 13, can be obtained [97].

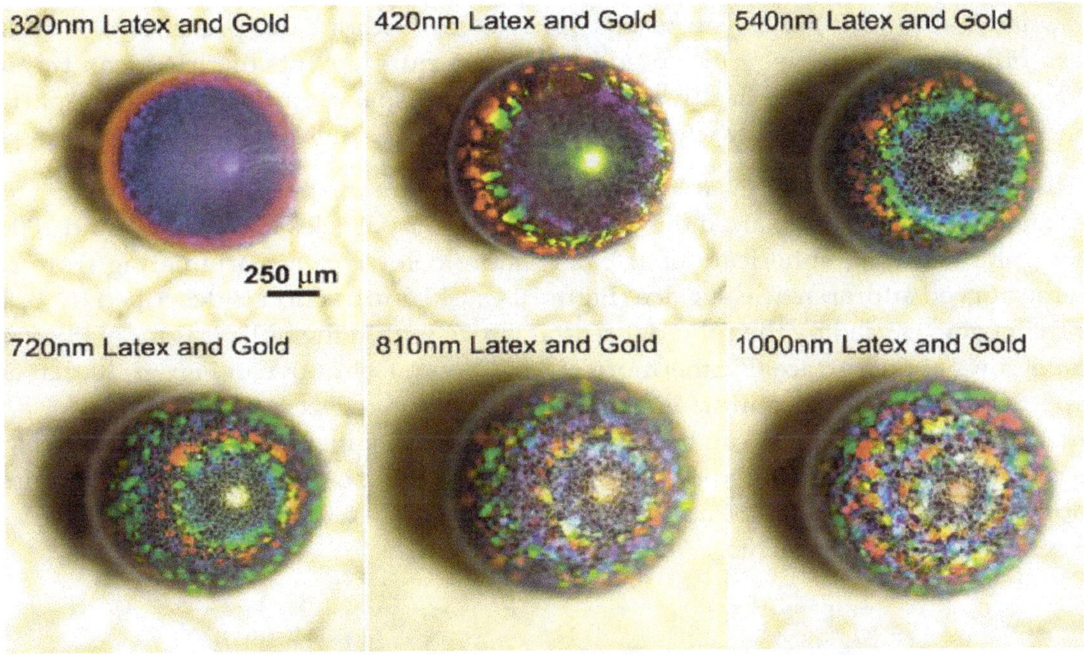

Figure 13. Optical microscopy images of patchy supraparticles assembled by EISA using variably-sized polystyrene (PS) latex nano-/micro-particles in suspension droplets generating highly light diffracting "opal balls". The gold nanoparticles are 22 nm in size.

The surface collection of the suspended particles was found to arise from a combination of Marangoni flow, already discussed in Section 2, and the comparably small size of the gold nanoparticles. Due to the droplet evaporation taking place mostly at the top (surface blockage at the bottom), the temperature gradient promotes Marangoni flux transporting particles to the top. Using very low concentrations of PS microspheres, Chang and Velev investigated this effect inside an aqueous droplet floating in a fluorinated oil [33]. In these droplets, even though the particles were allowed to sediment, they again collected at the droplet surface after restarting the evaporation (by removing the saturated atmosphere). Thereby, the consequence of the Marangoni effect/flow was visualized. Accordingly, if the concentrations are properly set and because the gold nanoparticles are small enough to travel in-between the inter-particle separations of the bigger microspheres, the resulting supraparticles are rendered patchy with the gold particles collected at the top, as shown in Figure 13 [97]. Of course, by that method, it is also possible to cover a well-defined surface area of the final supraparticles. This effect is independent of the supraparticle shape, as exemplified by the "glazed doughnuts", already presented in Figure 8 [92]. In summary, here, one faces a competition between sedimentation and induced Marangoni flow to which the dispersed colloids respond in a fashion depending on their size and density. This then leads to their segregation within the supraparticles and the observed internally inhomogeneous structuring, which allows for the creation of interesting and versatile supraparticles.

6. Applications of Supraparticles

As the preparation method leading to the final assembled supraparticles is not necessarily linked to their potential field of application, we will now extend our original focus and refer to supraparticles in a more general context. Nevertheless, for a start, we may first review some applications that emerged for supraparticles of the kinds described so far.

One example of potential applications is the field of photonics. Photonic applications require the precise control of optical properties of the materials, which can be achieved by proper nano-structuring. Defined spacing of monodisperse particles in highly ordered lattices allows for the diffraction of discrete wavelengths of the incident light leading to distinct coloring of the materials. Examples of this effect of defined coloring due to microstructuring can also be found in nature. Namely, the wings of *Morpho peleides* (butterfly) show an intense blue color without the presence of any dye [114]. Another example from nature is represented by the blue-purple fruits of an African tropical plant, named *Pollia condensata* [115]. Practically, assembling monodisperse particles into supraparticles with controlled inter-particle spacing allows for the fabrication of materials having tuned optical properties.

As a first example, Rastogi et al. used PS microsphere suspensions in drying droplets on a superhydrophobic surface and varied the diameter of the PS particles between 320 and 1000 nm [97]. After drying, the resulting spherical "opal balls" show colored rings originating from diffracted light from the curved supraparticle surface (Figure 14). This diffraction is described by Bragg's law, but rather than being caused by the colloidal crystal lattice within the bulk, it is more likely the result of the parallel rows at the surface. Furthermore, it was shown (see above in Figure 13) that additional gold nanoparticles of about 22 nm in size were not affecting the reflected colors, except for amplifying their intensity. Their color patterns can be precisely controlled by inter-particle spacing, which in turn is determined by the size of the microparticles. Of course, here, one may imagine even much more complex optical systems that could be achieved by appropriately designing hierarchical structures and using different colloidal components.

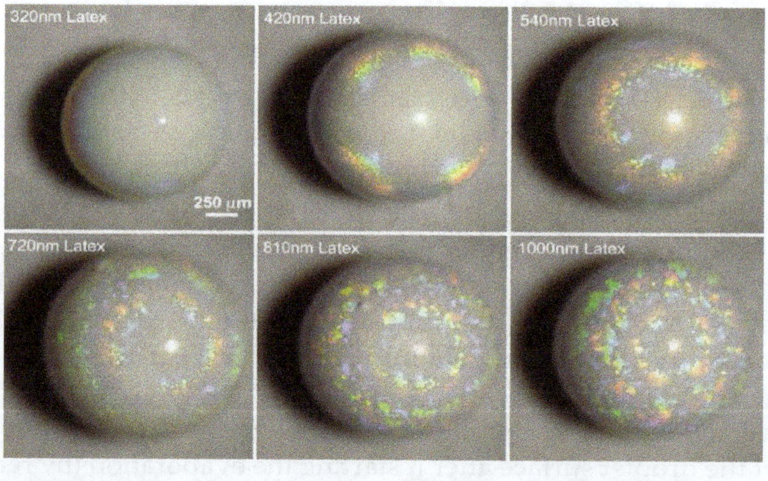

Figure 14. Optical microscopy images of patchy supraparticles assembled by EISA using differently-sized polystyrene (PS) latex microparticles in suspension droplets, thereby generating these highly light-diffracting "opal balls".

Supraparticles showing optical effects can be employed to produce colored films or layers. Such films containing buckled or spherical supraparticles show angle and strain independent reflections due to the colloidal matrices provided by the supraparticles. This leads to observable structural coloring [116]. Such films can be obtained from emulsion droplets using osmotic annealing and defined salt-concentration to produce buckled or spherical supraparticles and entrapping them in a silicon matrix. The resulting films are presented in Figure 15d–f.

Here, the angle and strain independence is simulated by free-standing (Figure 15a), contracted (Figure 15b) and stretched (Figure 15c) films. These films show similar structural coloring arising from the supraparticles embedded into the silicon matrix, where buckled supraparticles showed even better color quality than spherical ones.

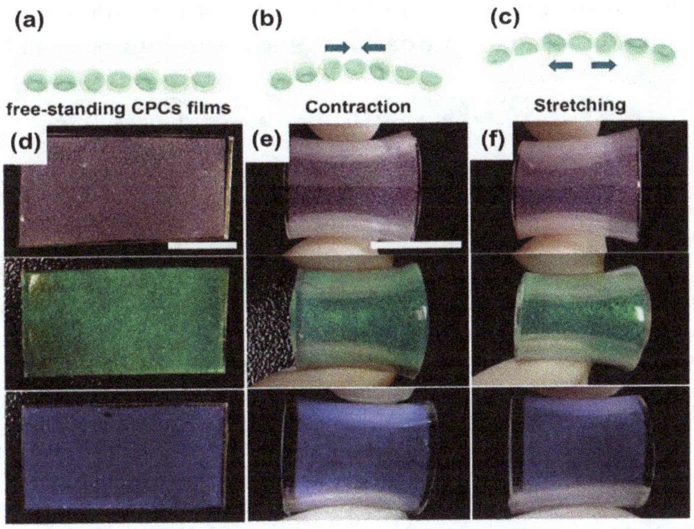

Figure 15. Colloidal photonic crystals (supraparticles) embedded in an elastomeric matrix in: (**a**) free; (**b**) contracted; (**c**) stretched state. Photographs of the resulting films are shown in (**d–f**) with supraparticles made of differently-sized PS particles. The reflected color is independent of the observing angle and strain on the films. The scale bars are 1 cm.

Switchable photonic materials may be of potential use for displays, indicators or similar devices. As an example for such a device, Janus supraparticles having hemispheres of different reflectance with one of them being magnetic were produced using microfluidics [117]. These supraparticles then can be oriented with an external magnetic field (Figure 16a) showing different reflected intensity for different light conditions (Figure 16b,c), i.e., forming a magnetically-switchable display.

Figure 16. Magnetic Janus supraparticles (**a**) switched using the different hemispheres at different light intensities: upwards directed (**b**) PS hemisphere at low light intensity or (**c**) Fe_3O_4-TMPTA hemisphere under strong light intensity; scale bars are 500 μm.

Similarly, also thermo-sensitive Janus supraparticles can be used in order to fabricate color changing displays [118].

Another interesting application of supraparticles is the development of bio-assays for antigen detection, which take advantage of the optical appearance of the supraparticles after drying depending on the exposure to the antigen and its concentration. Rastogi and Velev prepared an on-chip setup by drying droplets floating in fluorinated oil and containing PS microspheres and gold nanoparticles that were additionally functionalized with anti-rabbit IgG antibodies. Via the optical appearance after incubation, this system was able to serve as a sophisticated micro-bioassay for antigen detection [119]. This is a consequence of the antibody-antigen interaction, which is highly effective in terms of strength and specificity, thereby in turn controlling the particle interaction inside the dry droplets and consequently the resulting optical appearance. In other words, due to the antibody-functionalized gold and PS-latex particles, the patch formation is highly influenced by the presence of the corresponding antigen due to agglutination of the gold nanoparticles within the suspension. This effect also allowed for quantitative interpretation by optical comparison, shown in Figure 17 [119].

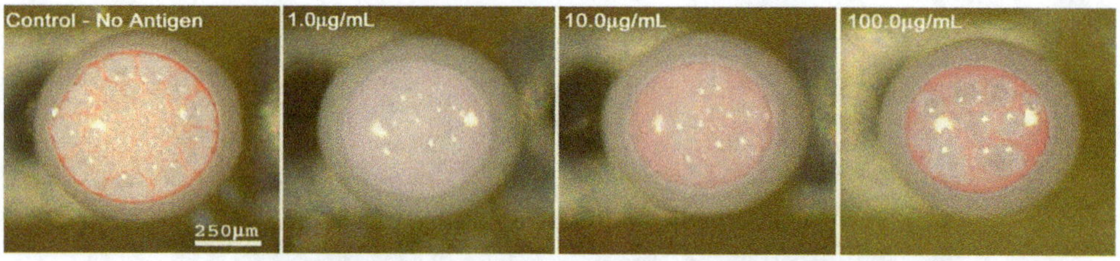

Figure 17. Resulting droplets for a 30-min incubation time and containing anti-rabbit IgG antibody functionalized gold nanoparticles at different concentrations of antigen (rabbit IgG); left to right: 0, 1.0, 10.0, 100.0 μg/mL.

Not only for diagnostic, but also for therapeutic purposes, supraparticles can serve as drug delivery systems [120–122]. This can be achieved by incorporating drugs meant to be released over an extended period of time. It has been shown by Rastogi et al. that supraparticles are permeable for solutes and able to continuously release contained solutes in a very controllable way [87]. Thus, taking advantage of analogous mesoporous structures made of silica and gelatin hybrids, it was shown that brain-derived neurotrophic factor (BDNF), a protein growth factor, and dexamethasone (DEX), a steroidal anti-inflammatory drug, could be easily loaded into the supraparticles and continuously be released over several days [120]. Using this concept, BDNF-loaded supraparticles were implanted into the inner ears of guinea pigs previously deafened by frusemide and kanamycin medication (damaging ion gradients between the auditory neuron network). The release of BDNF thereby showed significant rescuing of primary auditory neurons, potentially preserving residual hearing [121]. In another experiment, Park et al. immobilized a cysteine protease to porous calcium phosphate supraparticles [122]. This protease-supraparticle hybrid, in vitro allowed for systematic inactivation of the cytokine tumor necrosis factor-alpha (TNF-α), which is responsible for inflammatory effects potentially causing autoimmune diseases.

A quite different functionality of supraparticles is their potential ability to move. Self-propelling particles convert chemical energy into active, autonomous motion and have been developed and built in many different ways [123,124]. Able to perform complex tasks, such artificial vehicles have promising potential for applications in environmental treatment, like water remediation [125,126] or can serve as smart drug-delivery systems [127]. Yet, an interesting approach extending the field towards the millimeter scale has been taken using supraparticles prepared by EISA on a superhydrophobic surface [94]. These supraparticles were rendered patchy using Fe_3O_4 core nanoparticles decorated with Pt and a magnetic field during the synthesis. After additional surface hydrophobization and when placed in a H_2O_2 solution, these supraparticles generated adhering oxygen bubbles (for sufficient adhesion, the hydrophobization is essential), which increased the buoyancy. Once the buoyancy is high enough (typically one larger oxygen bubble is attached to the supraparticle), the whole supraparticle is lifted to the liquid's surface. There, it loses its oxygen bubble and drops down to the bottom of the liquid again, where the formation of a new bubble starts. The whole process then is repeated, and the supraparticle performs an oscillating, regular vertical motion, thus resembling "elevators", and this motion can continue for days. In addition, via the contained magnetic nanoparticles, the trajectory of the supraparticle can be steered by the application of an external magnetic field, which is illustrated in Figure 18.

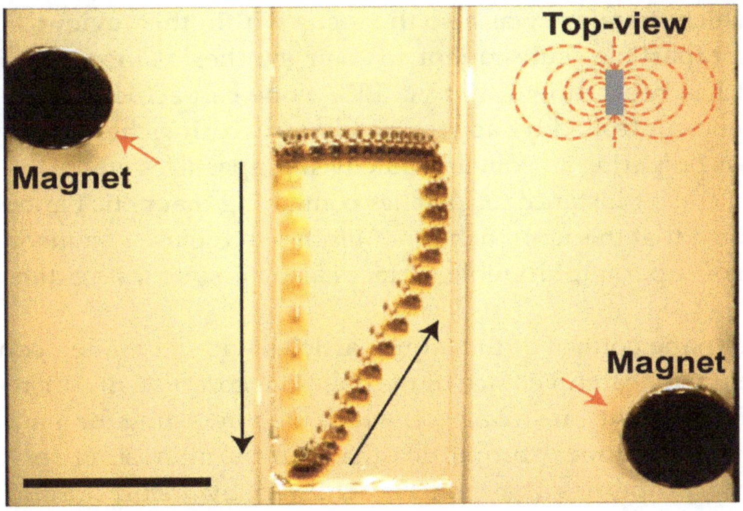

Figure 18. Oscillating elevator supraparticle in a wt % aqueous H_2O_2 solution. Starting at the top left, the elevator releases the oxygen bubble and falls down to the left bottom. After producing an oxygen bubble, it gets attracted by the bottom right magnet, following this attraction to the right during the way up. Having reached the surface meniscus, the left magnet pulls the elevator supraparticle back to its starting position, restarting the movement cycle; the scale bar is 1 cm.

Functionalizing the elevator supraparticle surface with the enzyme α-amylase, the catalytic decomposition of starch could be performed as a model reaction, showing potential applications of these elevator supraparticles in chemical catalysis [94]. Accordingly, this example proves the concept of self-propelled particles, whose movement can be steered by an external magnetic field and which are able to perform a chemical task (reaction) on their trajectory. Of course, the use of H_2O_2 limits the potential for applications substantially, but it might be noted that it is also possible to use more benign, non-toxic fuels, such as alcohols, for particle propulsion [128].

Finally, supraparticles are also interesting for the general field of catalysis, as they constitute rather versatile and small reaction containers. As an example in the field of catalysis, zinc sulfide supraparticles were developed for chemical dechlorination of 2,2′,4,4′,5,5′-hexachlorobiphenyl, an organic pollutant occurring in soils or groundwater [129]. Using UV irradiation, these supraparticles reductively degraded the model pollutant up to about 70% after 12 h in a solution of isooctane. Even more complicated, Xu et al. showed that Pt nanoparticles bound to a porous organo-polymer shell synthesized via soap-free emulsion polymerization onto Fe_3O_4 supraparticles can perform catalytic enantioselective hydrogenation of ethyl pyruvate [130].

Of course, this short section is far from complete with respect to the applications that have been explored for supraparticles, simply due to the fact that this is an ample field of research that allows for a vast number of options via appropriate structural and functional design by using various kinds of colloidal components that can be readily applied. Thus, many interesting developments have already been achieved and many more are to be expected in the near future to emerge from this field.

7. Conclusions

Supraparticles can be rich in terms of their size, shape and internal structure, and their properties can be varied largely via the choice of their colloidal constituents. In our review, we focused deliberately on supraparticles in the size range of hundreds of µm or even mm. Of course, colloidal assembly is not limited to this size range, and a much work has also been done on the assembly of nanoparticles, including their structured assemblies at surfaces [6,131–134]. However, in this review, we explicitly refrained from covering such investigations.

For the fabrication of supraparticles, a large number of methodologies has been developed, where in particular evaporation-induced self-assembly (EISA) on superhydrophobic surfaces has distinct advantages, which therefore was also the focus within this review. First, it is rather simple to collect the pure supraparticles subsequent to their synthesis since no other solvent has to be removed. Secondly, it has been shown that via EISA, one can achieve anisometric supraparticles whose orientation can be controlled by appropriately structuring the superhydrophobic substrate. This consequently allows preparing anisometric patchy supraparticles where the location of the patches can be precisely controlled (for instance for patches containing magnetic nanoparticles by an external magnet). It might be noted that this approach is not limited to the use of aqueous colloidal dispersions. Instead, the utilization of superamphiphobic surfaces also allows for its extension to a large range of organic solvents.

By combining the shape control of the supraparticles with a detailed control over the internal, e.g., hierarchical, structure by careful choice of the colloidal components, increasingly more complex systems with tailored functionality are available, which are interesting for a multitude of applications. These functionalities include optical properties that can be interesting for photonic applications and electric or catalytic properties that can be achieved by incorporating appropriately active (nano-)particles. Another functional feature is self-propulsion, which for instance can be achieved by incorporating nanoparticles and/or enzymes able to induce chemical reactions that lead to a momentum on the supraparticle. The movement then can be of a vertical and/or horizontal direction and also magnetically steered. One may also combine such a movement with other functional properties additionally incorporated in these supraparticles, such as catalytic activity for performing chemical reactions, thus providing "chemical microrobots". These "multi-tasking" supraparticles can be expected to become substantially further developed, and one may envision almost an unlimited potential via the combined functionalities that can be imparted.

In summary, there are rather simple ways to produce supraparticles, and by appropriate design, it is possible to integrate different functionalities into them, which can be independently combined and addressed. Accordingly, they represent smart systems, able to exhibit many interesting properties and being useful for several applications, like optical, electronic, chemical, mechanical ones, etc. However, the state of the art in this area certainly is still in its infancy, and one may expect many interesting developments leading to increased complexity in the structure and function of future materials.

Abbreviations

BNDF	brain-derived neurotrophic factor
CA	contact angle
CCA	constant contact angle
CCR	constant contact radius
cmc	critical micelle concentration
CVD	chemical vapor deposition
DEX	dexamethasone
ECD	electrochemical deposition
EISA	evaporation-induced self-assembly
FS	fumed silica
PEO	polyethylene oxide
PMMA	poly methyl methacrylate
PS	polystyrene
SEM	scanning electron microscopy
TMPTA	trimethylolpropane triacrylate
TNF-α	tumor necrosis factor-alpha
TPCL	three-phase contact line
UV	ultra-violet
WCA	water contact angle

References

1. Lu, Z.D.; Yin, Y.D. Colloidal nanoparticle clusters: Functional materials by design. *Chem. Soc. Rev.* **2012**, *41*, 6874–6887. [CrossRef] [PubMed]
2. Xia, Y.N.; Yin, Y.D.; Lu, Y.; McLellan, J. Template-assisted self-assembly of spherical colloids into complex and controllable structures. *Adv. Funct. Mater.* **2003**, *13*, 907–918. [CrossRef]
3. Stein, A.; Schroden, R.C. Colloidal crystal templating of three-dimensionally ordered macroporous solids: Materials for photonics and beyond. *Curr. Opin. Solid State Mater. Sci.* **2001**, *5*, 553–564. [CrossRef]
4. Velev, O.D.; Gupta, S. Materials Fabricated by Micro- and Nanoparticle Assembly—The Challenging Path from Science to Engineering. *Adv. Mater.* **2009**, *21*, 1897–1905. [CrossRef]
5. Li, F.; Josephson, D.P.; Stein, A. Colloidal Assembly: The Road from Particles to Colloidal Molecules and Crystals. *Angew. Chem. Int. Ed.* **2011**, *50*, 360–388. [CrossRef] [PubMed]
6. Grzelczak, M.; Vermant, J.; Furst, E.M.; Liz-Marzan, L.M. Directed Self-Assembly of Nanoparticles. *ACS Nano* **2010**, *4*, 3591–3605. [CrossRef] [PubMed]
7. Galisteo-Lopez, J.F.; Ibisate, M.; Sapienza, R.; Froufe-Perez, L.S.; Blanco, A.; Lopez, C. Self-Assembled Photonic Structures. *Adv. Mater.* **2011**, *23*, 30–69. [CrossRef] [PubMed]
8. Van Blaaderen, A.; Ruel, R.; Wiltzius, P. Template-directed colloidal crystallization. *Nature* **1997**, *385*, 321–324. [CrossRef]
9. Bragg, W.L. Diffraction of X-rays by Two-Dimensional Crystal Lattice. *Nature* **1929**, *124*, 125. [CrossRef]
10. Donaldson, J.G.; Kantorovich, S.S. Directional self-assembly of permanently magnetised nanocubes in quasi two dimensional layers. *Nanoscale* **2015**, *7*, 3217–3228. [CrossRef] [PubMed]
11. Li, R.; Bian, K.; Wang, Y.; Xu, H.; Hollingsworth, J.A.; Hanrath, T.; Fang, J.; Wang, Z. An Obtuse Rhombohedral Superlattice Assembled by Pt Nanocubes. *Nano Lett.* **2015**, *15*, 6254–6260. [CrossRef] [PubMed]
12. Van der Stam, W.; Gantapara, A.P.; Akkerman, Q.A.; Soligno, G.; Meeldijk, J.D.; van Roij, R.; Dijkstra, M.; de Mello Donega, C. Self-Assembly of Colloidal Hexagonal Bipyramid- and Bifrustum-Shaped ZnS Nanocrystals into Two-Dimensional Superstructures. *Nano Lett.* **2014**, *14*, 1032–1037. [CrossRef] [PubMed]
13. Choi, J.J.; Bian, K.; Baumgardner, W.J.; Smilgies, D.-M.; Hanrath, T. Interface-Induced Nucleation, Orientational Alignment and Symmetry Transformations in Nanocube Superlattices. *Nano Lett.* **2012**, *12*, 4791–4798. [CrossRef] [PubMed]
14. Nguyen, T.D.; Jankowski, E.; Glotzer, S.C. Self-Assembly and Reconfigurability of Shape-Shifting Particles. *ACS Nano* **2011**, *5*, 8892–8903. [CrossRef] [PubMed]
15. Epstein, E.; Yoon, J.; Madhukar, A.; Hsia, K.J.; Braun, P.V. Colloidal Particles that Rapidly Change Shape via Elastic Instabilities. *Small* **2015**, *11*, 6051–6057. [CrossRef] [PubMed]
16. Lu, Y.; Fan, H.; Stump, A.; Ward, T.L.; Rieker, T.; Brinker, C.J. Aerosol-assisted self-assembly of mesostructured spherical nanoparticles. *Nature* **1999**, *398*, 223–226.
17. Brezesinski, T.; Groenewolt, M.; Gibaud, A.; Pinna, N.; Antonietti, M.; Smarsly, B. Evaporation-Induced Self-Assembly (EISA) at Its Limit: Ultrathin, Crystalline Patterns by Templating of Micellar Monolayers. *Adv. Mater.* **2006**, *18*, 2260–2263. [CrossRef]
18. Feng, L.; Li, S.; Li, Y.; Li, H.; Zhang, L.; Zhai, J.; Song, Y.; Liu, B.; Jiang, L.; Zhu, D. Super-Hydrophobic Surfaces: From Natural to Artificial. *Adv. Mater.* **2002**, *14*, 1857–1860. [CrossRef]
19. Sun, T.; Feng, L.; Gao, X.; Jiang, L. Bioinspired Surfaces with Special Wettability. *Acc. Chem. Res.* **2005**, *38*, 644–652. [CrossRef] [PubMed]
20. Bhushan, B.; Jung, Y.C. Natural and biomimetic artificial surfaces for superhydrophobicity, self-cleaning, low adhesion, and drag reduction. *Prog. Mater. Sci.* **2011**, *56*, 1–108. [CrossRef]
21. Genzer, J.; Efimenko, K. Recent developments in superhydrophobic surfaces and their relevance to marine fouling: A review. *Biofouling* **2006**, *22*, 339–360. [CrossRef] [PubMed]
22. Li, X.-M.; Reinhoudt, D.; Crego-Calama, M. What do we need for a superhydrophobic surface? A review on the recent progress in the preparation of superhydrophobic surfaces. *Chem. Soc. Rev.* **2007**, *36*, 1350–1368. [CrossRef] [PubMed]
23. Brinker, C.J.; Lu, Y.; Sellinger, A.; Fan, H. Evaporation-Induced Self-Assembly: Nanostructures Made Easy. *Adv. Mater.* **1999**, *11*, 579–585. [CrossRef]

24. Grosso, D.; Cagnol, F.; Soler, G.J.D.A.; Crepaldi, E.L.; Amenitsch, H.; Brunet-Bruneau, A.; Bourgeois, A.; Sanchez, C. Fundamentals of Mesostructuring Through Evaporation-Induced Self-Assembly. *Adv. Funct. Mater.* **2004**, *14*, 309–322. [CrossRef]
25. Velev, O.D.; Lenhoff, A.M.; Kaler, E.W. A class of microstructured particles through colloidal crystallization. *Science* **2000**, *287*, 2240–2243. [CrossRef] [PubMed]
26. Deegan, R.D.; Bakajin, O.; Dupont, T.F.; Huber, G.; Nagel, S.R.; Witten, T.A. Capillary flow as the cause of ring stains from dried liquid drops. *Nature* **1997**, *389*, 827–829. [CrossRef]
27. Deegan, R.D.; Bakajin, O.; Dupont, T.F.; Huber, G.; Nagel, S.R.; Witten, T.A. Contact line deposits in an evaporating drop. *Phys. Rev. E* **2000**, *62*, 756–765. [CrossRef]
28. Kuncicky, D.M.; Velev, O.D. Surface-guided templating of particle assemblies inside drying sessile droplets. *Langmuir* **2008**, *24*, 1371–1380. [CrossRef] [PubMed]
29. Pauchard, L.; Parisse, F.; Allain, C. Influence of salt content on crack patterns formed through colloidal suspension desiccation. *Phys. Rev. E* **1999**, *59*, 3737–3740. [CrossRef]
30. Brutin, D.; Sobac, B.; Loquet, B.; Sampol, J. Pattern formation in drying drops of blood. *J. Fluid Mech.* **2011**, *667*, 85–95. [CrossRef]
31. Sobac, B.; Brutin, D. Structural and evaporative evolutions in desiccating sessile drops of blood. *Phys. Rev. E* **2011**, *84*, 011603. [CrossRef] [PubMed]
32. Joksimovic, R.; Watanabe, S.; Riemer, S.; Gradzielski, M.; Yoshikawa, K. Self-organized patterning through the dynamic segregation of DNA and silica nanoparticles. *Sci. Rep.* **2014**, *4*, 3660. [CrossRef] [PubMed]
33. Chang, S.T.; Velev, O.D. Evaporation-induced particle microseparations inside droplets floating on a chip. *Langmuir* **2006**, *22*, 1459–1468. [CrossRef] [PubMed]
34. Tam, D.; von Arnim, V.; McKinley, G.H.; Hosoi, A.E. Marangoni convection in droplets on superhydrophobic surfaces. *J. Fluid Mech.* **2009**, *624*, 101–123. [CrossRef]
35. Hu, H.; Larson, R.G. Marangoni Effect Reverses Coffee-Ring Depositions. *J. Phys. Chem. B* **2006**, *110*, 7090–7094. [CrossRef] [PubMed]
36. Hu, H.; Larson, R.G. Analysis of the Microfluid Flow in an Evaporating Sessile Droplet. *Langmuir* **2005**, *21*, 3963–3971. [CrossRef] [PubMed]
37. Hu, H.; Larson, R.G. Analysis of the effects of Marangoni stresses on the microflow in an evaporating sessile droplet. *Langmuir* **2005**, *21*, 3972–3980. [CrossRef] [PubMed]
38. Eral, H.B.; Augustine, D.M.; Duits, M.H.G.; Mugele, F. Suppressing the coffee stain effect: How to control colloidal self-assembly in evaporating drops using electrowetting. *Soft Matter* **2011**, *7*, 4954–4958. [CrossRef]
39. Krupenkin, T.N.; Taylor, J.A.; Schneider, T.M.; Yang, S. From rolling ball to complete wetting: The dynamic tuning of liquids on nanostructured surfaces. *Langmuir* **2004**, *20*, 3824–3827. [CrossRef] [PubMed]
40. McHale, G.; Brown, C.V.; Newton, M.I.; Wells, G.G.; Sampara, N. Dielectrowetting Driven Spreading of Droplets. *Phys. Rev. Lett.* **2011**, *107*, 186101. [CrossRef] [PubMed]
41. Picknett, R.G.; Bexon, R. Evaporation of Sessile or Pendant Drops in Still Air. *J. Colloid Interface Sci.* **1977**, *61*, 336–350. [CrossRef]
42. Maxwell, J.C. Diffusion. In *Collected Scientific Papers: Diffusion*; Encyclopedia Britannica: Cambridge, UK, 1877.
43. Snow, C. Potential Problems and Capacitance for a Conductor Bounded by Two Intersecting Spheres. *J. Res. Nat. Bur. Stand.* **1949**, *43*, 377–407. [CrossRef]
44. Soulie, V.; Karpitschka, S.; Lequien, F.; Prene, P.; Zemb, T.; Moehwald, H.; Riegler, H. The evaporation behavior of sessile droplets from aqueous saline solutions. *Phys. Chem. Chem. Phys.* **2015**, *17*, 22296–22303. [CrossRef] [PubMed]
45. Nguyen, T.A.H.; Nguyen, A.V. Increased Evaporation Kinetics of Sessile Droplets by Using Nanoparticles. *Langmuir* **2012**, *28*, 16725–16728. [CrossRef] [PubMed]
46. Yu, Y.S.; Wang, Z.Q.; Zhao, Y.P. Experimental and theoretical investigations of evaporation of sessile water droplet on hydrophobic surfaces. *J. Colloid Interface Sci.* **2012**, *365*, 254–259. [CrossRef] [PubMed]
47. Barthlott, W. Self-cleaning surfaces of objects and process for producing same. U.S. Patent 6,660,363, 9 December 2003.
48. Barthlott, W.; Neinhuis, C. Purity of the sacred lotus, or escape from contamination in biological surfaces. *Planta* **1997**, *202*, 1–8. [CrossRef]

49. Koch, K.; Bhushan, B.; Jung, Y.C.; Barthlott, W. Fabrication of artificial Lotus leaves and significance of hierarchical structure for superhydrophobicity and low adhesion. *Soft Matter* **2009**, *5*, 1386–1393. [CrossRef]
50. Feng, X.J.; Jiang, L. Design and creation of superwetting/antiwetting surfaces. *Adv. Mater.* **2006**, *18*, 3063–3078. [CrossRef]
51. Deng, X.; Mammen, L.; Butt, H.J.; Vollmer, D. Candle Soot as a Template for a Transparent Robust Superamphiphobic Coating. *Science* **2012**, *335*, 67–70. [CrossRef] [PubMed]
52. Chu, Z.; Seeger, S. Superamphiphobic surfaces. *Chem. Soc. Rev.* **2014**, *43*, 2784–2798. [CrossRef] [PubMed]
53. Cassie, A.B.D.; Baxter, S. Wettability of porous surfaces. *Trans. Faraday Soc.* **1944**, *40*, 546–551. [CrossRef]
54. Wenzel, R.N. Resistance of Solid Surfaces to Wetting by Water. *Ind. Eng. Chem.* **1936**, *28*, 988–994. [CrossRef]
55. Genzer, J.; Marmur, A. Biological and synthetic self-cleaning surfaces. *Mrs Bull.* **2008**, *33*, 742–746. [CrossRef]
56. Papadopoulos, P.; Mammen, L.; Deng, X.; Vollmer, D.; Butt, H.-J. How superhydrophobicity breaks down. *Proc. Natl. Acad. Sci. USA* **2013**, *110*, 3254–3258. [CrossRef] [PubMed]
57. Celia, E.; Darmanin, T.; de Givenchy, E.T.; Amigoni, S.; Guittard, F. Recent advances in designing superhydrophobic surfaces. *J. Colloid Interface Sci.* **2013**, *402*, 1–18. [CrossRef] [PubMed]
58. Yan, Y.Y.; Gao, N.; Barthlott, W. Mimicking natural superhydrophobic surfaces and grasping the wetting process: A review on recent progress in preparing superhydrophobic surfaces. *Adv. Colloid Interface Sci.* **2011**, *169*, 80–105. [CrossRef] [PubMed]
59. Gu, C.D.; Ren, H.; Tu, J.P.; Zhang, T.Y. Micro/Nanobinary Structure of Silver Films on Copper Alloys with Stable Water-Repellent Property under Dynamic Conditions. *Langmuir* **2009**, *25*, 12299–12307. [CrossRef] [PubMed]
60. Sperling, M.; Papadopoulos, P.; Gradzielski, M. Understanding the Formation of Anisometric Supraparticles: A Mechanistic Look Inside Droplets Drying on a Superhydrophobic Surface. *Langmuir* **2016**, *32*, 6902–6908. [CrossRef] [PubMed]
61. Tuteja, A.; Choi, W.; Ma, M.L.; Mabry, J.M.; Mazzella, S.A.; Rutledge, G.C.; McKinley, G.H.; Cohen, R.E. Designing superoleophobic surfaces. *Science* **2007**, *318*, 1618–1622. [CrossRef] [PubMed]
62. Phillips, K.R.; England, G.T.; Sunny, S.; Shirman, E.; Shirman, T.; Vogel, N.; Aizenberg, J. A colloidoscope of colloid-based porous materials and their uses. *Chem. Soc. Rev.* **2016**, *45*, 281–322. [CrossRef] [PubMed]
63. Il Park, J.; Nguyen, T.D.; Silveira, G.D.; Bahng, J.H.; Srivastava, S.; Zhao, G.P.; Sun, K.; Zhang, P.J.; Glotzer, S.C.; Kotov, N.A. Terminal supraparticle assemblies from similarly charged protein molecules and nanoparticles. *Nat. Commun.* **2014**, *5*, 3593.
64. Piccinini, E.; Pallarola, D.; Battaglini, F.; Azzaroni, O. Recognition-driven assembly of self-limiting supramolecular protein nanoparticles displaying enzymatic activity. *Chem. Commun.* **2015**, *51*, 14754–14757. [CrossRef] [PubMed]
65. Xia, Y.; Tang, Z. Monodisperse Hollow Supraparticles via Selective Oxidation. *Adv. Funct. Mater.* **2012**, *22*, 2585–2593. [CrossRef]
66. Yang, G.; Zhong, H.; Liu, R.; Li, Y.; Zou, B. In Situ Aggregation of ZnSe Nanoparticles into Supraparticles: Shape Control and Doping Effects. *Langmuir* **2013**, *29*, 1970–1976. [CrossRef] [PubMed]
67. Yu, S.; Wan, J.; Chen, K. A facile synthesis of superparamagnetic Fe_3O_4 supraparticles@MIL-100(Fe) core-shell nanostructures: Preparation, characterization and biocompatibility. *J. Colloid Interface Sci.* **2016**, *461*, 173–178. [CrossRef] [PubMed]
68. Guo, J.; Yang, W.; Wang, C. Magnetic Colloidal Supraparticles: Design, Fabrication and Biomedical Applications. *Adv. Mater.* **2013**, *25*, 5196–5214. [CrossRef] [PubMed]
69. Edwards, E.W.; Wang, D.; Möhwald, H. Hierarchical Organization of Colloidal Particles: From Colloidal Crystallization to Supraparticle Chemistry. *Macromol. Chem. Phys.* **2007**, *208*, 439–445. [CrossRef]
70. Xia, Y.S.; Nguyen, T.D.; Yang, M.; Lee, B.; Santos, A.; Podsiadlo, P.; Tang, Z.Y.; Glotzer, S.C.; Kotov, N.A. Self-assembly of self-limiting monodisperse supraparticles from polydisperse nanoparticles. *Nat. Nanotechnol.* **2011**, *6*, 580–587. [CrossRef] [PubMed]
71. Cho, Y.S.; Yi, G.R.; Kim, S.H.; Elsesser, M.T.; Breed, D.R.; Yang, S.M. Homogeneous and heterogeneous binary colloidal clusters formed by evaporation-induced self-assembly inside droplets. *J. Colloid Interface Sci.* **2008**, *318*, 124–133. [CrossRef] [PubMed]
72. Velev, O.D.; Furusawa, K.; Nagayama, K. Assembly of latex particles by using emulsion droplets as templates. 2. Ball-like and composite aggregates. *Langmuir* **1996**, *12*, 2385–2391. [CrossRef]

73. Velev, O.D.; Furusawa, K.; Nagayama, K. Assembly of latex particles by using emulsion droplets as templates. 1. Microstructured hollow spheres. *Langmuir* **1996**, *12*, 2374–2384. [CrossRef]
74. Cho, Y.-S.; Kim, S.-H.; Yi, G.-R.; Yang, S.-M. Self-organization of colloidal nanospheres inside emulsion droplets: Higher-order clusters, supraparticles, and supraballs. *Colloids Surf. A* **2009**, *345*, 237–245. [CrossRef]
75. Maeda, K.; Onoe, H.; Takinoue, M.; Takeuchi, S. Controlled Synthesis of 3D Multi-Compartmental Particles with Centrifuge-Based Microdroplet Formation from a Multi-Barrelled Capillary. *Adv. Mater.* **2012**, *24*, 1340–1346. [CrossRef] [PubMed]
76. Yu, Z.Y.; Wang, C.F.; Ling, L.T.; Chen, L.; Chen, S. Triphase Microfluidic-Directed Self-Assembly: Anisotropic Colloidal Photonic Crystal Supraparticles and Multicolor Patterns Made Easy. *Angew. Chem. Int. Ed.* **2012**, *51*, 2375–2378. [CrossRef] [PubMed]
77. Wang, J.T.; Wang, J.; Han, J.J. Fabrication of Advanced Particles and Particle-Based Materials Assisted by Droplet-Based Microfluidics. *Small* **2011**, *7*, 1728–1754. [CrossRef] [PubMed]
78. Brugarolas, T.; Tu, F.; Lee, D. Directed assembly of particles using microfluidic droplets and bubbles. *Soft Matter* **2013**, *9*, 9046–9058. [CrossRef]
79. Sowade, E.; Blaudeck, T.; Baumann, R.R. Self-Assembly of Spherical Colloidal Photonic Crystals inside Inkjet-Printed Droplets. *Cryst. Growth Des.* **2016**, *16*, 1017–1026. [CrossRef]
80. Sowade, E.; Hammerschmidt, J.; Blaudeck, T.; Baumann, R.R. In-Flight Inkjet Self-Assembly of Spherical Nanoparticle Aggregates. *Adv. Eng. Mater.* **2012**, *14*, 98–100. [CrossRef]
81. Chen, F.C.; Lu, J.P.; Huang, W.K. Using Ink-Jet Printing and Coffee Ring Effect to Fabricate Refractive Microlens Arrays. *IEEE Photonics Technol. Lett.* **2009**, *21*, 648–650. [CrossRef]
82. Wang, D.; Park, M.; Park, J.; Moon, J. Optical properties of single droplet of photonic crystal assembled by ink-jet printing. *Appl. Phys. Lett.* **2005**, *86*, 241114. [CrossRef]
83. Millman, J.R.; Bhatt, K.H.; Prevo, B.G.; Velev, O.D. Anisotropic particle synthesis in dielectrophoretically controlled microdroplet reactors. *Nat. Mater.* **2005**, *4*, 98–102. [CrossRef] [PubMed]
84. Denkov, N.D.; Velev, O.D.; Kralchevsky, P.A.; Ivanov, I.B.; Yoshimura, H.; Nagayama, K. 2-Dimensional Crystallization. *Nature* **1993**, *361*, 26. [CrossRef]
85. Denkov, N.D.; Velev, O.D.; Kralchevsky, P.A.; Ivanov, I.B.; Yoshimura, H.; Nagayama, K. Mechanism of Formation of 2-Dimensional Crystals from Latex-Particles on Substrates. *Langmuir* **1992**, *8*, 3183–3190. [CrossRef]
86. Lee, D.-W.; Jin, M.-H.; Lee, C.-B.; Oh, D.; Ryi, S.-K.; Park, J.-S.; Bae, J.-S.; Lee, Y.-J.; Park, S.-J.; Choi, Y.-C. Facile synthesis of mesoporous silica and titania supraparticles by a meniscus templating route on a superhydrophobic surface and their application to adsorbents. *Nanoscale* **2014**, *6*, 3483–3487. [CrossRef] [PubMed]
87. Rastogi, V.; Velikov, K.P.; Velev, O.D. Microfluidic characterization of sustained solute release from porous supraparticles. *Phys. Chem. Chem. Phys.* **2010**, *12*, 11975–11983. [CrossRef] [PubMed]
88. Manoharan, V.N. Colloidal spheres confined by liquid droplets: Geometry, physics, and physical chemistry. *Solid State Commun.* **2006**, *139*, 557–561. [CrossRef]
89. Liu, M.; Zheng, Y.; Zhai, J.; Jiang, L. Bioinspired Super-antiwetting Interfaces with Special Liquid-Solid Adhesion. *Acc. Chem. Res.* **2010**, *43*, 368–377. [CrossRef] [PubMed]
90. Bhushan, B.; Her, E.K. Fabrication of Superhydrophobic Surfaces with High and Low Adhesion Inspired from Rose Petal. *Langmuir* **2010**, *26*, 8207–8217. [CrossRef] [PubMed]
91. Wooh, S.; Huesmann, H.; Tahir, M.N.; Paven, M.; Wichmann, K.; Vollmer, D.; Tremel, W.; Papadopoulos, P.; Butt, H.-J. Synthesis of Mesoporous Supraparticles on Superamphiphobic Surfaces. *Adv. Mater.* **2015**, *27*, 7338–7343. [CrossRef] [PubMed]
92. Rastogi, V.; Garcia, A.A.; Marquez, M.; Velev, O.D. Anisotropic Particle Synthesis Inside Droplet Templates on Superhydrophobic Surfaces. *Macromol. Rapid Commun.* **2010**, *31*, 190–195. [CrossRef] [PubMed]
93. Sperling, M.; Velev, O.D.; Gradzielski, M. Controlling the Shape of Evaporating Droplets by Ionic Strength: Formation of Highly Anisometric Silica Supraparticles. *Angew. Chem. Int. Ed.* **2014**, *53*, 586–590. [CrossRef] [PubMed]
94. Sperling, M.; Kim, H.J.; Velev, O.D.; Gradzielski, M. Active Steerable Catalytic Supraparticles Shuttling on Preprogrammed Vertical Trajectories. *Adv. Mater. Interf.* **2016**, *3*, 160095. [CrossRef]
95. Sperling, M.; Velev, O.D.; Gradzielski, M. Formation of Anisometric Fumed Silica Supraparticles—Mechanism and Application Potential. *Z. Phys. Chem.* **2015**, *229*, 1055–1074. [CrossRef]

96. Sperling, M.; Velev, O.D.; Gradzielski, M. Kontrolle der Form verdunstender Tropfen über die Ionenstärke: Bildung anisometrischer SiO_2-Suprapartikel. *Angew. Chem.* **2014**, *126*, 597–601. [CrossRef]
97. Rastogi, V.; Melle, S.; Calderon, O.G.; Garcia, A.A.; Marquez, M.; Velev, O.D. Synthesis of Light-Diffracting Assemblies from Microspheres and Nanoparticles in Droplets on a Superhydrophobic Surface. *Adv. Mater.* **2008**, *20*, 4263–4268. [CrossRef]
98. Princen, H.M. *Surface and Colloid Science*; Matijevic, E., Ed.; Wiley: New York, NY, USA, 1969; p. 1.
99. Head, D.A. Modeling the elastic deformation of polymer crusts formed by sessile droplet evaporation. *Phys. Rev. E* **2006**, *74*, 021601. [CrossRef] [PubMed]
100. Tsapis, N.; Dufresne, E.R.; Sinha, S.S.; Riera, C.S.; Hutchinson, J.W.; Mahadevan, L.; Weitz, D.A. Onset of Buckling in Drying Droplets of Colloidal Suspensions. *Phys. Rev. Lett.* **2005**, *94*, 018302. [CrossRef] [PubMed]
101. Kajiya, T.; Nishitani, E.; Yamaue, T.; Doi, M. Piling-to-buckling transition in the drying process of polymer solution drop on substrate having a large contact angle. *Phys. Rev. E* **2006**, *73*, 011601. [CrossRef] [PubMed]
102. Gu, Z.-Z.; Yu, Y.-H.; Zhang, H.; Chen, H.; Lu, Z.; Fujishima, A.; Sato, O. Self-assembly of monodisperse spheres on substrates with different wettability. *Appl. Phys. A* **2005**, *81*, 47–49. [CrossRef]
103. Kuncicky, D.M.; Bose, K.; Costa, K.D.; Velev, O.D. Sessile Droplet Templating of Miniature Porous Hemispheres from Colloid Crystals. *Chem. Mater.* **2007**, *19*, 141–143. [CrossRef]
104. Sen, D.; Bahadur, J.; Mazumder, S.; Santoro, G.; Yu, S.; Roth, S.V. Probing evaporation induced assembly across a drying colloidal droplet using in situ small-angle X-ray scattering at the synchrotron source. *Soft Matter* **2014**, *10*, 1621–1627. [CrossRef] [PubMed]
105. Kätzel, U.; Vorbau, M.; Stintz, M.; Gottschalk-Gaudig, T.; Barthel, H. Dynamic Light Scattering for the Characterization of Polydisperse Fractal Systems: II. Relation between Structure and DLS Results. *Part. Part. Syst. Charact.* **2008**, *25*, 19–30. [CrossRef]
106. Accardo, A.; Gentile, F.; Mecarini, F.; Angelis, F.D.; Burghammer, M.; Fabrizio, E.D.; Riekel, C. In Situ X-ray Scattering Studies of Protein Solution Droplets Drying on Micro- and Nanopatterned Superhydrophobic PMMA Surfaces. *Langmuir* **2010**, *26*, 15057–15064. [CrossRef] [PubMed]
107. Bahadur, J.; Sen, D.; Mazumder, S.; Paul, B.; Bhatt, H.; Singh, S.G. Control of Buckling in Colloidal Droplets during Evaporation-Induced Assembly of Nanoparticles. *Langmuir* **2012**, *28*, 1914–1923. [CrossRef] [PubMed]
108. Sperling, M.; Spiering, V.J.; Velev, O.D.; Gradzielski, M. Controlled Formation of Patchy Anisometric Fumed Silica Supraparticles in Droplets on Bent Superhydrophobic Surfaces. *Part. Part. Syst. Charact.* **2017**, *34*, 1600176. [CrossRef]
109. Zhou, J.; Yang, J.; Gu, Z.; Zhang, G.; Wei, Y.; Yao, X.; Song, Y.; Jiang, L. Controllable Fabrication of Noniridescent Microshaped Photonic Crystal Assemblies by Dynamic Three-Phase Contact Line Behaviors on Superhydrophobic Substrates. *ACS Appl. Mater. Interfaces* **2015**, *7*, 22644–22651. [CrossRef] [PubMed]
110. Pawar, A.B.; Kretzschmar, I. Fabrication, Assembly, and Application of Patchy Particles. *Macromol. Rapid Commun.* **2010**, *31*, 150–168. [CrossRef] [PubMed]
111. Jiang, S.; Chen, Q.; Tripathy, M.; Luijten, E.; Schweizer, K.S.; Granick, S. Janus Particle Synthesis and Assembly. *Adv. Mater.* **2010**, *22*, 1060–1071. [CrossRef] [PubMed]
112. Walther, A.; Muller, A.H.E. Janus particles. *Soft Matter* **2008**, *4*, 663–668. [CrossRef]
113. Perro, A.; Reculusa, S.; Ravaine, S.; Bourgeat-Lami, E.; Duguet, E. Design and synthesis of Janus micro- and nanoparticles. *J. Mater. Chem.* **2005**, *15*, 3745–3760. [CrossRef]
114. Ding, Y.; Xu, S.; Wang, Z.L. Structural colors from Morpho peleides butterfly wing scales. *J. Appl. Phys.* **2009**, *106*, 074702. [CrossRef]
115. Vignolini, S.; Rudall, P.J.; Rowland, A.V.; Reed, A.; Moyroud, E.; Faden, R.B.; Baumberg, J.J.; Glover, B.J.; Steiner, U. Pointillist structural color in Pollia fruit. *Proc. Natl. Acad. Sci. USA* **2012**, *109*, 15712. [CrossRef] [PubMed]
116. Yeo, S.J.; Tu, F.; Kim, S.; Yi, G.-R.; Yoo, P.J.; Lee, D. Angle- and strain-independent coloured free-standing films incorporating non-spherical colloidal photonic crystals. *Soft Matter* **2015**, *11*, 1582–1588. [CrossRef] [PubMed]
117. Liu, S.-S.; Wang, C.-F.; Wang, X.-Q.; Zhang, J.; Tian, Y.; Yin, S.-N.; Chen, S. Tunable Janus colloidal photonic crystal supraballs with dual photonic band gaps. *J. Mater. Chem. C* **2014**, *2*, 9431–9438. [CrossRef]
118. Wang, H.; Yang, S.; Yin, S.-N.; Chen, L.; Chen, S. Janus Suprabead Displays Derived from the Modified Photonic Crystals toward Temperature Magnetism and Optics Multiple Responses. *ACS Appl. Mater. Interfaces* **2015**, *7*, 8827–8833. [CrossRef] [PubMed]

119. Rastogi, V.; Velev, O.D. Development and evaluation of realistic microbioassays in freely suspended droplets on a chip. *Biomicrofluidics* **2007**, *1*, 014107. [CrossRef] [PubMed]
120. Maina, J.W.; Cui, J.; Björnmalm, M.; Wise, A.K.; Shepherd, R.K.; Caruso, F. Mold-Templated Inorganic-Organic Hybrid Supraparticles for Codelivery of Drugs. *Biomacromolecules* **2014**, *15*, 4146–4151. [CrossRef] [PubMed]
121. Wang, Y.; Wise, A.K.; Tan, J.; Maina, J.W.; Shepherd, R.K.; Caruso, F. Mesoporous Silica Supraparticles for Sustained Inner-Ear Drug Delivery. *Small* **2014**, *10*, 4244–4248. [CrossRef] [PubMed]
122. Park, W.M.; Yee, C.M.; Champion, J.A. Self-assembled hybrid supraparticles that proteolytically degrade tumor necrosis factor-α. *J. Mater. Chem. B* **2016**, *4*, 1633–1639. [CrossRef]
123. Sanchez, S.; Soler, L.; Katuri, J. Chemically Powered Micro- and Nanomotors. *Angew. Chem. Int. Ed.* **2015**, *54*, 1414–1444. [CrossRef] [PubMed]
124. Ebbens, S.J.; Howse, J.R. In pursuit of propulsion at the nanoscale. *Soft Matter* **2010**, *6*, 726–738. [CrossRef]
125. Soler, L.; Sanchez, S. Catalytic nanomotors for environmental monitoring and water remediation. *Nanoscale* **2014**, *6*, 7175–7182. [CrossRef] [PubMed]
126. Gao, W.; Wang, J. The Environmental Impact of Micro/Nanomachines: A Review. *ACS Nano* **2014**, *8*, 3170–3180. [CrossRef] [PubMed]
127. Patra, D.; Sengupta, S.; Duan, W.; Zhang, H.; Pavlick, R.; Sen, A. Intelligent, self-powered, drug delivery systems. *Nanoscale* **2013**, *5*, 1273–1283. [CrossRef] [PubMed]
128. Yamamoto, D.; Takada, T.; Tachibana, M.; Iijima, Y.; Shioi, A.; Yoshikawa, K. Micromotors working in water through artificial aerobic metabolism. *Nanoscale* **2015**, *7*, 13186–13190. [CrossRef] [PubMed]
129. He, L.; Xiong, Y.; Zhao, M.; Mao, X.; Liu, Y.; Zhao, H.; Tang, Z. Bioinspired Synthesis of ZnS Supraparticles toward Photoinduced Dechlorination of 2,2′,4,4′,5,5′-Hexachlorobiphenyl. *Chem. Asian J.* **2013**, *8*, 1765–1767. [CrossRef] [PubMed]
130. Xu, S.; Weng, Z.; Tan, J.; Guo, J.; Wang, C. Hierarchically structured porous organic polymer microspheres with built-in Fe_3O_4 supraparticles: Construction of dual-level pores for Pt-catalyzed enantioselective hydrogenation. *Polym. Chem.* **2015**, *6*, 2892–2899. [CrossRef]
131. Wang, D.; Möhwald, H. Template-directed colloidal self-assembly—The route to 'top-down' nanochemical engineering. *J. Mater. Chem.* **2004**, *14*, 459–468. [CrossRef]
132. Renna, L.A.; Boyle, C.J.; Gehan, T.S.; Venkataraman, D. Polymer Nanoparticle Assemblies: A Versatile Route to Functional Mesostructures. *Macromolecules* **2015**, *48*, 6353–6368. [CrossRef]
133. Kotov, N.A. (Ed.) *Nanoparticle Assemblies and Superstructures*; Taylor & Francis: Abingdon, UK, 2016.
134. Kuemin, C.; Huckstadt, K.C.; Lörtscher, E.; Rey, A.; Decker, A.; Spencer, N.D.; Wolf, H. Selective Assembly of Sub-Micrometer Polymer Particles. *Adv. Mater.* **2010**, *22*, 2804–2808. [CrossRef] [PubMed]

Permissions

All chapters in this book were first published in MDPI; hereby published with permission under the Creative Commons Attribution License or equivalent. Every chapter published in this book has been scrutinized by our experts. Their significance has been extensively debated. The topics covered herein carry significant findings which will fuel the growth of the discipline. They may even be implemented as practical applications or may be referred to as a beginning point for another development.

The contributors of this book come from diverse backgrounds, making this book a truly international effort. This book will bring forth new frontiers with its revolutionizing research information and detailed analysis of the nascent developments around the world.

We would like to thank all the contributing authors for lending their expertise to make the book truly unique. They have played a crucial role in the development of this book. Without their invaluable contributions this book wouldn't have been possible. They have made vital efforts to compile up to date information on the varied aspects of this subject to make this book a valuable addition to the collection of many professionals and students.

This book was conceptualized with the vision of imparting up-to-date information and advanced data in this field. To ensure the same, a matchless editorial board was set up. Every individual on the board went through rigorous rounds of assessment to prove their worth. After which they invested a large part of their time researching and compiling the most relevant data for our readers.

The editorial board has been involved in producing this book since its inception. They have spent rigorous hours researching and exploring the diverse topics which have resulted in the successful publishing of this book. They have passed on their knowledge of decades through this book. To expedite this challenging task, the publisher supported the team at every step. A small team of assistant editors was also appointed to further simplify the editing procedure and attain best results for the readers.

Apart from the editorial board, the designing team has also invested a significant amount of their time in understanding the subject and creating the most relevant covers. They scrutinized every image to scout for the most suitable representation of the subject and create an appropriate cover for the book.

The publishing team has been an ardent support to the editorial, designing and production team. Their endless efforts to recruit the best for this project, has resulted in the accomplishment of this book. They are a veteran in the field of academics and their pool of knowledge is as vast as their experience in printing. Their expertise and guidance has proved useful at every step. Their uncompromising quality standards have made this book an exceptional effort. Their encouragement from time to time has been an inspiration for everyone.

The publisher and the editorial board hope that this book will prove to be a valuable piece of knowledge for researchers, students, practitioners and scholars across the globe.

List of Contributors

Mariana Carrancá Palomo, Victoria Martín Prieto and Plamen Kirilov
Université de Lyon (UCBL), Biologie Tissulaire et Ingénierie Thérapeutique UMR 5305–Vecteurs Colloïdaux et Transport Tissulaire, Institut des Sciences Pharmaceutiques et Biologiques, 8 Avenue Rockefeller, 69373 Lyon Cedex 08, France

Avinash J. Patil, Tomi Nissinen and Stephen Mann
Centre for Organized Matter for Chemistry and Centre for Protolife Research, School of Chemistry, University of Bristol, Bristol BS8 1TS, UK

Takumi Okamoto
Centre for Organized Matter for Chemistry and Centre for Protolife Research, School of Chemistry, University of Bristol, Bristol BS8 1TS, UK
Denso Corporation, 1-1, Showa-cho, Kariya, Aichi 448-8661, Japan

Luís J. del Valle, Angélica Díaz and Jordi Puiggalí
Barcelona Research Center for Multiscale Science and Engineering, Universitat Politècnica de Catalunya, Escola d'Enginyeria de Barcelona Est-EEBE, c/Eduard Maristany 10-14, Barcelona 08019, Spain

Peter T. Bähler, Michele Zanini and Lucio Isa
Laboratory for Interfaces, Soft matter and Assembly, Department of Materials, ETH Zurich, Vladimir-Prleog-Weg 5, 8093 Zurich, Switzerland

Giulia Morgese and Edmondo M. Benetti
Laboratory for Surface Science and Technology, Department of Materials, ETH Zurich, Vladimir-Prleog-Weg 5, 8093 Zurich, Switzerland

Satu Strandman and X.X. Zhu
Département de Chimie, Université de Montréal, C.P. 6128, Succursale Centre-ville, Montreal, QC H3C 3J7, Canada

Alexander Klaus Strzelczyk, Hanqing Wang and Stephan Schmidt
Institute for Organic and Macromolecular Chemistry, Heinrich-Heine-Universität, Universitätsstrasse 1, 40225 Düsseldorf, Germany

Andreas Lindhorst, Johannes Waschke and Tilo Pompe
Institute for Biochemistry, Universität Leipzig, Johannisallee 21-23, 04103 Leipzig, Germany

Christian Kropf and Benoit Luneau
Henkel AG & Co KGaA, Henkelstr 67, D-40589 Düsseldorf, Germany

Paula Malo de Molina
Centro de Física de Materiales (CSIC, UPV/EHU) and Materials Physics Center MPC, Paseo Manuel de Lardizabal 5, E-20018 San Sebastián, Spain

Niek Hijnen and Paul S. Clegg
School of Physics & Astronomy, University of Edinburgh, Peter Guthrie Tait Road, Edinburgh EH9 3FD, UK

Abdalla H. Karoyo and Lee D.Wilson
Department of Chemistry, University of Saskatchewan, 110 Science Place, Saskatoon, SK S7N 5C9, Canada

Jan Maarten van Doorn, Joris Sprakel and Thomas E. Kodger
Physical Chemistry and Soft Matter, Wageningen University & Research, Stippeneng 4, 6708 WE, Wageningen, The Netherlands

Masayuki Tokita
Department of Physics, Faculty of Science, Kyushu University, 744 Motooka, Fukuoka, Fukuoka 812-8581, Japan

Laurent Latxague, Alexandra Gaubert, David Maleville, Julie Baillet, Michael A. Ramin and Philippe Barthélémy
ARNA laboratory, Univ. Bordeaux, ChemBioPharm, INSERM, U1212, CNRS UMR 5320, F-33000 Bordeaux, France

Marcel Sperling and Michael Gradzielski
Stranski Laboratorium für Physikalische & Theoretische Chemie, Institut für Chemie, Technische Universität Berlin, Straße des 17. Juni 124, Sekr. TC7, D-10623 Berlin, Germany

Index